LANDSCAPING FOR WILDLIFE IN THE PACIFIC NORTHWEST

Landscaping for Wildlife

in the Pacific Northwest

Russell Link

University of Washington Press
Seattle & London

published in association with the

Washington Department of
Fish and Wildlife

Landscaping for Wildlife in the Pacific Northwest has been made possible with grants from the British Columbia Ministry of the Environment; the Environmental Protection Agency, Region 10 Office; the U.S. Fish and Wildlife Service; the Puget Sound Water Quality Action Team; and the Washington Wildlife and Recreation Foundation.

University of Washington Press
P.O. Box 50096
Seattle, WA 98145
www.washington.edu/uwpress

Library of Congress Cataloging-in-Publication Data
Link, Russell.
    Landscaping for wildlife in the Pacific Northwest / Russell Link.
        p.   cm.
    Includes bibliographical references.
    ISBN 978-0-295-97820-8 (alk. paper)
    1. Gardening to attract wildlife—Northwest, Pacific.   I. Title.
    QL59.L55   1999   99–14300
    639.9' 2—dc21   CIP

The paper used in this publication meets the minimum requirements of the American National Standard for Information Sciences—Permanence of Paper for Printed Library Materials, ANSI Z39.48-1984.

Design and production: Magrit Baurecht Design
Production assistance: Darrell Pruett
Cover design: Bob Hutchins and Magrit Baurecht Design
Pen-and-ink illustrations (except where noted): Jenifer Rees
Computer illustrations: Russell Link with assistance from Visual Dynamics, Inc.
Maps and color plates: Peggy Ushakoff
Color plates: Tom Boyden, David Hutchinson, William Leonard, Kelly McAllister, Dave Nunnallee, Dennis Paulson, Dave Pehling, Jim Pruske, Joy Spurr, and Idie Ulsh
Editorial review and index: Sigrid Asmus
Word Processing: Linh-Lan Nguyen, Kelly Smith

Front cover photo/Photographer: Keith Geller's Garden/Keith Geller
Back cover photos/Photographers (clockwise, starting from upper left):

Anna's Hummingbird/Jim Pruske
White-tailed Deer/Jim Pruske
Barn Owl/Idie Ulsh
Rough-skin Newt/Kelly McAllister
Western Tanager/Dennis Paulson
Red Fox/Joy Spurr
Western Tiger Swallowtail/Tom Boyden
Black-capped Chickadee/Jim Pruske
Eight-spotted Skimmer/Jim Pruske

Painted Turtle/William Leonard
Northern Flicker/Jim Pruske
Bumblebee/Dave Nunnallee
Great Blue Heron/Jim Pruske
Ceanothus Silk Moth/Tom Boyden
Pacific Treefrog/Jim Pruske
Green Lacewing/Dave Pehling
Western Bluebird/Jim Pruske
Painted Lady/Idie Ulsh

CONTENTS

INTRODUCTION

For many years the term landscaping brought to mind the image of a manicured lawn, a tidy shrub and flower border, a tightly sheared hedge, and a large shade tree—a civilized place distinctly different from the area of wild things.

Now many people create landscapes with the purpose of attracting and nurturing wildlife. This includes not only birds, bees, and butterflies, but also the multitude of less-often-seen inhabitants that make up the complex mix of animals that reside in, or are visitors to, landscapes for wildlife.

This landscaping style doesn't require "letting everything grow wild," giving up outdoor activities, or totally surrendering your property to nature. Rather it attempts to integrate human needs with those of wildlife. The techniques and approaches used to design and maintain a landscape for wildlife can be applied to any landscape style.

The number of ways that your outdoor areas can be made attractive for wildlife is almost endless, depending upon your imagination and resources. Many necessities for wildlife do not cost a lot of money—just a little time and a willingness to try something new.

You also don't have to see all the inhabitants to enjoy them. You can learn to sense the unseen presence of animals by the tracks and other signs they leave and by the sounds they make.

The location of your landscape will largely determine what wildlife species you attract. A landscape on property adjacent to a wild area will have the widest variety of visitors. And while a landscape in an urban area surrounded by development will not maintain the number or variety of animals that once lived there, it still has the potential to attract a wide array of species. It is important to remember that this type of landscaping will never replicate the naturally rich plant and animal communities that once existed in an area and it should not be considered a substitute.

Landscaping for wildlife offers many benefits, including:

- **Wildlife survival.** Not only are rare plants and animals becoming endangered, but many of the familiar species that we once took for granted are becoming uncommon. People with a landscape next to a wild area, such as a greenbelt, creek, or wetland, have an opportunity to help preserve and maintain areas needed by the wildlife species that use these places.

- **Educational opportunity.** Landscapes for wildlife can provide individuals or families with opportunities to explore, discover, and learn about wildlife during every season. Examples include noting the arrival of migratory birds in spring and beholding the appearance of a certain species of butterfly in summer. In fall, birds can be observed migrating through an area, and in winter animal tracks can be identified in the freshly fallen snow.

The knowledge and appreciation of nature are gifts you can give to children, gifts they may value all their lives and pass along to others. Any direct participation in wildlife-related activities around the home can also foster a larger ecological awareness.

- **Community involvement.** The value of wildlife habitat on your property can improve with the collaborative efforts of your surrounding neighbors. Public projects in larger areas can bring volunteers from the surrounding neighborhoods together to cooperate for the benefit of themselves and their community.

- **Opportunities for all.** Whatever your physical ability, design background, and knowledge of plants and animals may be, landscaping for wildlife offers possibilities for you. There are wildlife-watching opportunities directly outside your window and among the common landscape plants growing in your yard. For those who design and maintain landscapes professionally, the placement of wildlife features such as a pond, wildlife tree, brush pile, or nest box provides opportunities to think beyond ornamentation and to consider how wildlife will benefit from design and management decisions.

- **It's healthy and fun!** A wildlife landscape provides psychological value to those who take the time to enjoy it. In urban areas, peaceful, natural surroundings provide an escape from noisy crowded conditions. Emotional stresses are reduced in a natural environment, and, for many people, a visit to a natural area is a means of relaxation. By encouraging nature to exist around your home, you can have that environment outside your door. The wildlife landscape is especially satisfying to those who work or spend time at home.

Wildlife landscaping will reward you more and more each year as you discover new wild visitors that enrich your life.

The purpose of this book is to help you select, arrange, and maintain plants and other landscape components that fulfill wildlife needs in a landscape setting. Whether you are planting a yard from scratch or modifying some existing acreage, this book provides the information needed to design and maintain the area in such a way that it doesn't create a hazard to you or wildlife.

The book is arranged in five parts:

Wildlife habitat and landscaping basics are explained in the first part.

The second part provides information about the diversity of mammals, birds, reptiles and amphibians, and insects likely to be attracted to your property.

The third part provides details of special landscape features for wildlife.

The fourth part deals with potential wildlife problems and gives tips on how to watch wildlife.

The Appendices in Part 5 cover specific habitats and how to plant and maintain them, plants for particular settings, and figures of nest boxes and other habitat features. It also provides lists of references for further information.

ACKNOWLEDGMENTS

I offer my warmest thanks to the many people and organizations who have participated in the creation of this book.

For their continuous support throughout this project, I thank Dave Brittell, Tom Juelson, Lora Leschner, and Patricia Thompson from the Washington Department of Fish and Wildlife. In particular I am grateful to Steve Penland for his assistance early on and for his inspirational work on urban wildlife topics.

For their review of plant descriptions and lists, I thank Joe Arnett, Clayton Antieau, Susan Buis, Mary-Joe Buza, John Dixon, JoEllen Kassebaum,

Art Kruckeberg, Paulus Vrijmoed, and Fred Weinmann. I am also indebted to members of the Oregon Native Plant Society and the Washington Native Plant Society who have devoted time to this project. I take full responsibility for the inclusion of any questionable plant species and regret not being able to include many of your suggestions.

For sharing their expertise on specific habitat types, I thank Tim Brown, Binda Colebrook, Josh Kahan, Ivan Lines, Elliott Menashe, Richard Robohm, Jean Stam, Ron Vanbianchi, Bob Ziegler, and the World Forestry Center. For reviewing the infor-

mation on amphibians and reptiles, I thank Bill Leonard, Klaus Richter, and Kelly McAllister. For reviewing the information on birds, I thank Howard Ferguson and David Hutchinson. For reviewing the information on bees, butterflies, and other invertebrates, I thank Sharon Collman, Rod Crawford, Brian Griffin, Idie Ulsh, Dave Pehling, and Robert Pyle. For assistance with the section on ponds, I thank Sonneman Design.

Also, thanks go to Tom McCall, Jim Stephenson, Hal Michael, and Steve Kalinowski for their willingness to read large portions of the manuscript.

For his generous contribution of slides, I am grateful to Jim Pruske. For sharing his bird feeder designs, I am thankful to Ken Short.

For editing and technical writing assistance, I thank Dana Base, Susan Campbell, Chuck Gibilisco, Claire Hagen Dole, Donna Gleisner, Mark Goldsmith, and Hunter Thompson. Special thanks are due to Flora Johnson Skelly who donated so much of her time and talent to this undertaking.

I am deeply grateful to all the other people who have responded to my request for help throughout this project.

Finally, thanks to Kathy and Vanessa Link for their continuing support, patience, and humor.

—Russell Link

PART 1

WILDLIFE HABITAT DESIGN AND MAINTENANCE

CHAPTER 1

WILDLIFE HABITAT AND ITS STEWARDSHIP

Habitat is a wild animal's "support system." Within its habitat, an animal meets all its life requirements, which are the same as yours: food, water, shelter, and space.

As a wildlife steward, your goal is to meet the requirements of wildlife in a way that doesn't create problems for you or for the animals. This process begins around your home: if your property supplies habitat requirements in the needed quantity and quality, it can attract and sometimes maintain animals, and provide you with the opportunity to enjoy observing them.

To safely provide for wildlife, it's important to understand some basic concepts in wildlife and plant ecology, which are discussed in this chapter. Meanwhile, keep in mind that creating wildlife habitat is not a one-time project. It is a continuous exchange between your efforts and the responses of plants and wildlife. Your awareness and understanding of what will work in your outdoor space will grow in time and experience, as will your ability to enhance your landscape and make it even more attractive to wildlife.

Within a given area, whatever habitat ingredient is missing or in shortest supply is called the "limiting factor" because this factor sets a limit on what wildlife the area can support. For example, if you see adult swallowtail butterflies flying around your neighborhood—but not in your yard—your yard may not have the flowering plants that supply them with the nectar (food) they need. Even if all their other habitat needs are met on your property, you may not see swallowtails in your yard unless this limiting factor is addressed. In some areas, the constraints imposed by limiting factors may be severe. If you live in an area surrounded by houses, concrete, and little in the way of natural areas, for example, the numbers of many species may be severely limited. However, if you can identify the limiting factors for certain species and meet these needs, you may still be able to attract some wildlife. Even a small yard can meet the needs of at least some species.

Wildlife Habitat Requirements

All animals have these basic requirements:

Food

All wildlife obtain energy for survival by eating plants or other animals. Wildlife that eat plants are called "herbivores" (e.g., deer, chipmunks, and rabbits). Wildlife that primarily eat other animals are called "carnivores" (e.g., mountain lions and bobcats). Wildlife that eat both plants and other animals are called "omnivores." These include crows, jays, raccoons, and opossums. Some wildlife species, such as raccoons, opossums, and crows, also are termed "generalists" because they eat a wide assortment of foods.

As a general rule, animals with specific food requirements are more likely to have food as a limiting factor (see left) than are omnivores and generalists. The latter are often able to live successfully in a variety of habitats, including urban areas.

A property that contains plants that produce edible seeds, fruits, nuts, and flowering plants for nectar and pollen can help feed a variety of animals. Some of these animals could, in turn, become food for other animals. A property that attracts and maintains a large number of insects is especially likely to attract large numbers of birds because insects are an important food source for these species. Supplemental foods for birds may be provided in feeders.

Water

Water cleans and keeps cells in body tissue alive, and is vital for food production. Some songbirds and most mammals need to consume water daily.

Water is often a limiting factor in a backyard habitat. As a result, the provision of water can turn an average wildlife habitat into an extraordinary one.

While most species depend on water for drinking, some also require it for other reasons. Many birds need water to bathe in to keep their feathers in shape. The eggs and young of dragonflies, frogs, toads, and most salamanders need to be in water to develop. All amphibians need moisture in some form to keep their skins wet.

Although water can take the form of birdbaths, ponds, irrigation ditches, and natural landscape features, plants also provide water for wildlife. Rodents and rabbits usually obtain water from the leaves they eat. Many plant-eating mammals take advantage of the dew that is often present on plants at night or in early morning. Also, fleshy fruits and berries have a high water content.

Shelter

Shelter (also called cover) is a place to raise young, hide from predators, and avoid the heat, cold, and wind. Shelter also provides a place to feed, play, and rest safely.

The quality of shelter is particularly important for young animals in a nest. Unlike an animal that can flee when a predator approaches, young birds or small mammals must rely entirely upon the cover and the camouflage of the nest itself. The proximity of

cover while wildlife are feeding is also important, for this is a time when the animal's attention is partly occupied by matters other than safety.

One way to provide quality shelter is to preserve existing trees and shrubs and let them develop their natural form. Other examples of shelter include a rock or brush pile, any size dead or dying tree, leaf litter you let lie, or a section of grass you let grow. Further examples are nest boxes built specifically to shelter bats and songbirds.

Space

No matter how many ways you provide food, water, and shelter, space will always be a limiting factor for wildlife.

The space requirements for some wildlife species may be larger than you imagine. For example, a

pileated woodpecker covers about 300 acres in search of tree cavities for nesting and its main food source, carpenter ants. However, your property can be part of the home range of many wildlife species, and meet some, if not all, of their need for space. Furthermore, by providing natural food sources, water, and shelter, you may reduce the amount of living space required by an individual animal.

Sometimes the space requirements of a particular species can be met through joint effort, such as when neighbors get together to preserve a nearby hillside, ravine, greenbelt, or park. Neighbors can also work together to provide and protect corridors of space that allow wildlife to move between adjacent properties (Figure 1).

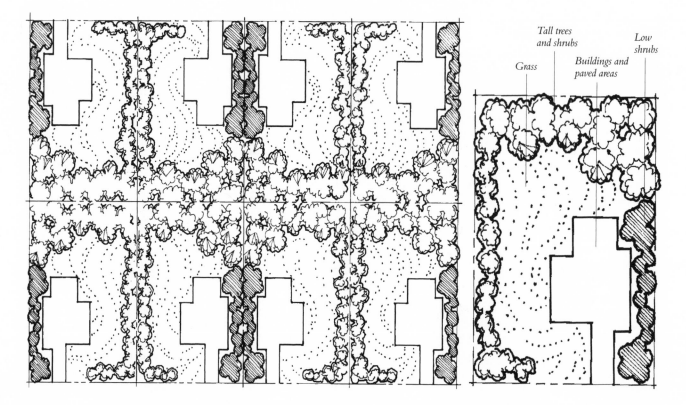

Figure 1. An example of how adjoining yards can be landscaped to connect and increase spaces for wildlife.

Wildlife Population Dynamics

All wildlife populations fluctuate; a given species may become abundant, then scarce, and then abundant again. However, despite ups and downs, wildlife populations tend to stay relatively stable in the long term if human intervention and natural catastrophes don't significantly change land use or vegetation where they live. This process is called population dynamics.

One way wildlife species ensure that their numbers will stay the same in the long run is by producing more individuals than can survive; some of these individuals then become food for other animals. Typically, the individuals taken by other animals are ones that are diseased or not as hardy as the others. In this way, natural predation may actually help prey species by removing less fit individuals, leaving stronger individuals to survive to reproduce. Although this may seem harsh from a human perspective, it is part of the way wildlife populations maintain a balance.

This delicately balanced system may be upset through human intervention. One example is when predatory species, such as foxes, owls, or hawks disappear because there is no longer adequate habitat for them; then populations of prey species such as mice and other rodents are likely to increase.

Another example is when exotic (non-native) animals are introduced into a habitat. Because they have not co-evolved with the other species in the habitat, introduced species may not be subject to the same checks and balances that keep native populations under control. Because of this, populations of some introduced species, such as house sparrows and European starlings, can increase rapidly following arrival in new habitat. Large numbers of introduced animals then compete with native animals for food, nest sites, and other resources; this can force native wildlife out of areas where they were once able to live.

Carrying Capacity

The size of your property, its proximity to a wild area, and the availability of quality food, shelter, water, and space determine the number of species that can survive there, as well as the population sizes of those species. The number of a given species that can survive on your property is called the property's carrying capacity. If the number of that species exceeds the property's carrying capacity, some will leave or die.

It is possible to increase the carrying capacity of your property by managing a limiting factor. For instance, you can quickly boost carrying capacity for certain songbirds by adding a bird feeder and a nest box. Long-term management includes increasing the number of plants that are known to provide food and shelter for songbirds and other species.

Habitat Quality

One way professional wildlife managers work with wildlife is by manipulating the habitat in which the animals are found; wildlife management is based on habitat management. You can increase the number and variety of wildlife species that are able to use your property by increasing the quality of the habitat. Ways to do this include:

Corridors for Space

Animals often need more space than a small backyard habitat can provide on its own. You may be able to give them that space, however, by linking areas of habitat on your property to one another and to habitat on surrounding properties. These travel lanes or "corridors" connect otherwise fragmented or isolated natural areas so that wildlife can travel among them. In effect, they make one large habitat area where previously there were many small ones.

There are many types of wildlife corridors. Examples include farm windbreaks and hedgerows, urban greenbelts, railroad rights-of-way, and ravines full of berry bushes and large trees. A smaller corridor can be formed by a group of connecting backyards planted with trees and shrubs (Figure 1). Wider corridors provide more wildlife benefits and, located along a waterway, protect water quality better than narrower ones do. Breaks in the corridor should be avoided.

With the large-scale fragmentation of habitat occurring due to development, preserving or creating corridors is especially important. Part of your role as a wildlife steward may be to educate your neighbors about the need for wildlife corridors and help them to create corridors on their own properties.

Low habitat diversity equals fewer wildlife species.

High habitat diversity equals more wildlife species.

Figure 2. *A landscape that includes a variety of plants will attract many more wildlife species than a landscape with only a few plant species. A property that has a variety of different soil types and sun exposures will provide opportunities to include a wider variety of plants.*

- Increasing structural diversity.
- Adding layers.
- Protecting and including native plants.

Structural Diversity

Different wildlife species will seek their life needs in different areas of your property. This is one way nature helps to reduce competition for the necessities of life. For example, you'll find that different wildlife species rarely eat the same food at the same time in the same place. Swallows catch insects in the air, chickadees on trees and shrubs, spiders on blades of grass, shrews on the ground, and moles in the soil.

If animals are associated with different parts of a habitat, it stands to reason that a habitat that can be divided into different parts will be able to support more different kinds of animals. A large grassy area can support a limited number of birds because there are limited ways that a grassland can be divided among them. However, when a tree is added to the grassland, as it matures wildlife have many more ways to use the habitat: lower trunk—upper trunk, large branches—small branches, upper foliage—middle foliage—lower foliage, inner foliage—outer foliage, cavities, bare ground under the tree, and so forth. Adding a tree to a grassland makes that habitat more structurally diverse.

Structural diversity can be provided in any landscape by adding a mixture of different types of plants and other special habitat components that provide food, water, shelter, and space for different species of wildlife. For instance, a monoculture of alder trees can be enhanced by replacing some of the alder with maple, red-cedar, or hemlock. Or a dense stand of Douglas spirea (hardhack) shrubs can be enhanced by creating openings or adding Sitka spruce, willow, cascara, alder, or other wetland plants. Even a suburban lot can be planted with a variety of different types of plants to provide food and shelter for wildlife at different times of the year (Figure 2). You can also add special habitat components, such as snags, rock piles, brush piles, or water sources. All of these add to the structural diversity of the habitat.

Layers in the Landscape

Wildlife not only select different kinds of habitats, but they also select specific parts of a given habitat. For example, some forest birds look for food or build their nests in the tops of tall trees (Figure 3). Others may do these things only in the mid-level portion of the canopy, while some birds are restricted to the low shrubs or the ground. Some birds may search for insects at the ends of the branches, while others poke amid the bark of the tree trunk. Some species spend most of their life in one layer, while others use several layers to supply their basic needs. Animals select these different levels, or layers of a given habitat, primarily to avoid competition with other species.

If you have a missing layer in an area, some wildlife species that are normally present may be missing. Not all layers may need to be present on your property, however, if they are present in the immediate area. If your neighbors' property contains a canopy of tall trees, birds and other wildlife can use their trees to perch and nest, and for shelter, and still use your property. Remember, however, that a neighbor's trees may be removed at any time.

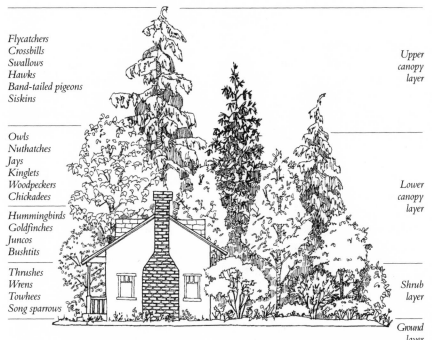

Flycatchers
Crossbills
Swallows
Hawks
Band-tailed pigeons
Siskins

Owls
Nuthatches
Jays
Kinglets
Woodpeckers
Chickadees

Hummingbirds
Goldfinches
Juncos
Bushtits

Thrushes
Wrens
Towhees
Song sparrows

Upper canopy layer

Lower canopy layer

Shrub layer

Ground layer

Figure 3. *Wildlife occur at different levels, or layers, in the landscape. By increasing the variety of plant species in your habitat and adding layers, you provide wildlife with the greatest opportunity to find the right combination of essential ingredients, and allow more wildlife to successfully coexist. The birds listed in this figure are the species typically seen using the corresponding layers.*

Native Plants

Native plants are those that have existed in an area since before the advent of Europeans. These plants have evolved side by side with insects, fungi, plant viruses, wildlife, and other native plants for thousands of years. Over time, native plants have improved their ability to attract helpful animals such as pollinating insects and seed-dispersing birds. They have also become adept at repelling or surviving attacks from destructive organisms such as plant viruses and munching insects.

When provided with the right conditions, native plants generally require minimum maintenance once established. They also have a subtle beauty: a variety of colors, forms, and textures that provide the characteristic look of the region where they grow.

It's crucial that you preserve what native plants you have on-

site. It's much harder to restore a wild area with native plants than to preserve the ones already present. Also, don't rely on nature to plant native plants for you. They may not be able to enter your landscape because the nearest mother plants may be too far away.

As a wildlife-friendly gardener, you will want to incorporate natives as much as possible in your planting plans, especially in Areas 2 and 3 of your landscape (see Chapter 2). Because the Pacific Northwest has many climatic and soil zones, try to select native plants from your specific part of the region rather than just any "Northwest natives." However, some natives such as the tall Oregon-grape and Douglas maple, are very adaptable and can be used in areas out of their natural range.

You don't have to plant natives exclusively. In most areas, growing native plants isn't an all-

or-nothing proposition. Many non-native plants, also called "exotics," boost the supplies of nectar, pollen, and fruit that attract wildlife close in for wildlife viewing.

In selecting non-native plants for the landscape, however, it is important to avoid certain plants that are known to cause problems if they spread into natural areas. This is especially true if you are landscaping at the edge of wild or relatively undeveloped areas. English ivy, for example, is a non-native plant that can grow rampantly in shady areas and will out-compete and kill stands of native plants. Purple loosestrife, formerly used as a garden plant in much of the country, is now considered a noxious weed in many places because it invades wetland areas and drives out native vegetation. Many non-native grasses are aggressive and can quickly displace native species. For a list of local plants *not* to include in landscapes, contact your local Native Plant Society, County Noxious Weed Board, or County Cooperative Extension Service (see "Public Agencies" in Appendix E for addresses and phone numbers).

Plant Succession

Without human or other interference, any patch of ground will tend to change over time as one type of plant cover is naturally replaced by another. Thus, for example, a patch of bare ground will be quickly colonized by pioneer plants including different annual herbs and grasses. Pioneer plants colonize newly disturbed areas, creating conditions that are more favorable to other plant species. Many of our common "weeds" are pioneer species—

Edges

Edges are transition areas where one type of vegetation meets and blends with another (Figure 4). Examples are where a shrub thicket along a fence line borders a field or where an area planted with landscape trees and shrubs meets a lawn. Bird-watchers and wildlife photographers often frequent edges because species from both areas may be present.

The best edge you can create is a layered one. This creates a gradual transition between the two different vegetation types and is better than one that creates a sudden change. When planting an edge, try to include plants that flower and fruit at different times of the year. In traditional landscapes, hedges often create edges where wildlife can exist. For information on hedges, see Chapter 19.

Although many species of wildlife benefit from edges, others do not. For information on the negative effects of edges, see "Woodlands and Woodland Landscapes" in Appendix A.

plants that are particularly good at colonizing soil that has been disturbed by human intervention. This is why, if you remove a well-established stand of vegetation and leave exposed soil, you are likely to find yourself with an invasion of "weeds."

These pioneer plants are then followed by perennial herbs and grasses and, over time, shrubs and brambles will take root in the area and eventually shade out the perennial plants. The different stages of this sequence are called successional stages (Figure 5).

In the forested areas of the Pacific Northwest, the stages of plant succession are generally as follows:

Stage 1. Bare ground.
Stage 2. Annual herbs and/or grasses.
Stage 3. Perennial herbs and grasses.
Stage 4. Shrubs and brambles.
Stage 5. Young woodland or forest.
Stage 6. Mature woodland or forest.

In some areas, factors such as soil or climate prevent succession from proceeding past a certain stage. For instance, in the dry areas east of the Cascade mountains, lack of water often prevents succession from proceeding past Stage 3. In such areas, Stage 3 would be considered the climax stage. However, where water is added in a landscape setting, trees can flourish and the area can proceed to a later stage of succession.

Human intervention can change or halt the succession of plant life in a given area. For example, when you remove existing vegetation and disturb the soil, you most likely will return your property to Stage 1 or 2.

Low-maintenance landscapes on your property can occur in more wild areas where plants are allowed to proceed naturally through their successional stages.

To encourage or maintain the presence of particular wildlife species, you can maintain a particu-

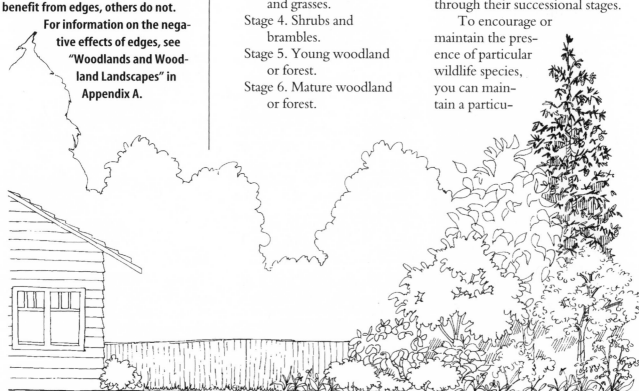

Figure 4. *Edges occur where different types of vegetation meet and are often places with a lot of wildlife activity. This illustration shows a woodland garden meeting a lawn area. A similar edge occurs where a forest meets a meadow.*

lar stage of succession to provide the best habitat for that species. For example, a field in Stage 2 will provide food and cover for small rodents eaten by hawks and owls. Deer may also find cover and bed in the grass if it is long enough, songbirds will feed on seeds and flying insects, and butterflies will perch on the long stalks. However, to keep the area in Stage 2 you will have to manage the area, perhaps by mowing once a year. If you were to stop mowing, succession will proceed; and in areas suitable for tree growth, the landscape will eventually be transformed into a forest.

You can speed up the process of succession by planting trees, shrubs, and other plants from later successional stages. You may be able to get clues to what plants to include and the conditions they require by observing nearby plant communities that have been relatively undisturbed. See Appendix A for additional information.

I Planted It and They Didn't Come

It is possible for you to do everything right for wildlife on your property—plant the right kinds of trees and shrubs, add a brush pile and rock pile, put in a small source of water, and put up some bird boxes and feeders—and still get

mostly house sparrows and English starlings as visitors. This depends, in part, on the location of your site and the explanation for this lies in the theory of island biogeography.

Your property may represent an "island" of natural habitat surrounded by a "sea" of residential and commercial development that separates it from the "mainland" of forest at the edge of the city. The number of animal species found on your property will then be a function of both its size and its distance from the "mainland," which is the source of immigrating animals. Urban parks, greenbelts, or backyards may be viewed as islands in which the wildlife species present are determined by these factors of size and distance.

| Stage 1 | Stage 2 | Stage 3 | Stage 4 | Stage 5 | Stage 6 |

Figure 5. *Vegetation in a particular area changes over time. Succession may take centuries, depending on the natural and human factors. When land is disturbed, the succession process may start over again.*

CHAPTER 2

LANDSCAPE DESIGN FOR YOU AND WILDLIFE

The best way to preserve your property's special qualities and to provide for wildlife is to make a plan. Getting ideas down on paper can help you establish priorities and properly guide your projects to completion. Any good introductory book on gardening or landscaping will give you information on how to design a layout plan. However, when landscaping for wildlife, there are a number of things to consider that may not be covered in a standard landscaping or gardening book.

Using the Area Approach

To establish a balance between the needs of wildlife and the needs of human beings who use your property, you may find it useful to take what is sometimes known as an "area approach" to your layout plan. This approach divides your property into three distinct areas based on the level of human activity each receives (Figure 1). Although the boundary between any two areas will be fuzzy, this approach assumes that the space around your home and other high-use zones will require the most mainte-

nance and will be best suited for certain wildlife-related landscape projects. As the distance from a high-use zone increases, human activity and maintenance requirements will lessen, and other landscape projects will be more appropriate.

In the area approach, the landscape can be divided into three areas:

Area 1 is where trees, shrubs, ground covers, and other plants create special landscape effects and attract wildlife within view. It has the most human activity and contains landscape and wildlife features that benefit from frequent visits and observation.

High-maintenance plantings should be placed here, where they will be easy to take care of. If you want a well-manicured lawn or other demanding landscape feature, this is the place to put it.

The majority of Area 1 should be within roughly 30 feet of the home; thus many urban yards will be mostly in Area 1. Area 1 may include:

- Trees and shrubs planted as food and shelter for wildlife, also to create beauty, shade, privacy, or screen an unattractive view.
- Specialty gardens for hummingbirds, butterflies, and other insects.
- An herb or vegetable garden, a portion of which can be used to grow sunflowers and other bird foods, and plants that attract beneficial insects.
- Nest boxes for mason bees and songbirds, including chickadees and violet-green swallows.
- A feeder for songbirds.
- A birdbath or a garden pond.
- A small brush shelter to attract songbirds within view.
- A rock retaining wall built to double as shelter for reptiles and amphibians.

Bird feeders
Birdbath
Nest boxes
Hummingbird garden
Butterfly garden

Area 1

Small snags
Brush and rock piles
Nest boxes

Bat boxes
Pond

Area 2

Large snags
Nest boxes

Area 3

Figure 1. *An example of how a property can be divided into different use areas.*

Area 2 has less human activity and contains landscape features that need less frequent visits than do most of those in Area 1. Part or all of Area 2 may contain a naturalistic landscape that blends smoothly into Area 3. Existing trees and shrubs should be preserved when possible. Area 2 may include:

- A layered landscape of trees, shrubs, and ground covers that are known to attract wildlife.
- A hedgerow located along a fence or property line planted with small trees and shrubs that flower and fruit at different times throughout the year.
- A snag, a rock shelter, a brush shelter, or other type of down wood.
- Nest boxes for birds, bats, and tree squirrels.
- A wild patch of grassland, shrubbery, or ground layer located where wildlife can use it.
- A wildlife-friendly ditch or drainage area, or a pond.

Area 3 has the least amount of human activity and contains landscape features that need little maintenance. Area 3 should be planted with native plants that historically have occurred in your local area. Area 3 may include:

- Undisturbed wild areas where the dominant plants are those native to your area.
- A buffer of native plants next to a wetland, creek, or other important wildlife area.
- A group of trees and shrubs to insure future privacy. (Never depend on your neighbors' woods or other vegetation— they may not be there tomorrow.)
- Snags, rock shelters, brush shelters, and other large down wood habitat.

- A corridor of vegetation that links a wild area on your property with a neighboring wild area.
- Fences, footpaths, and observation areas, such as a blind, added to control human access to sensitive wildlife areas.
- Nest boxes for flying squirrels and birds, such as woodpeckers and wood ducks, that need more secluded nest sites.

Creating a Layout Plan

Many people "plan" by trial and error. In our living rooms, we move couches and chairs about in an effort to find the right look. It isn't easy though, to move trees around to find the best landscape arrangement. That makes planning all the more important. It will help prevent costly mistakes and wasted time. Planning also helps you identify ecological relationships on your property that you may not have noticed before.

Don't be discouraged if you have never designed your own landscape, and don't worry if you don't have a degree in landscape architecture or wildlife science. Landscaping can be fun, especially with a bit of planning—and a layout plan is a good place to begin.

Before you begin your layout plan, it's important that you have a good understanding of your property and the area around it. Try to find a secret space on your property and visit it daily or weekly. By spending time in different areas at different times of the day and year, you will have the opportunity to observe the many things that will affect your planning decisions.

After you've become familiar with your property and its special

Be Up-front with Your Backyard Wildlife Sanctuary

Almost all outdoor activity in residential areas takes place in the backyard. Decks and patios are generally built on the rear of the house. Gardens, ponds, entertainment, and hobby activities all take place in the backyard. Drive around a suburb on a Saturday morning and often the only human activity you see is that of lawn mowing and car washing.

Front yards account for perhaps one-third of the land converted to housing. Consider using some of this precious area for wildlife. You can add plants known to attract butterflies and hummingbirds to existing flower beds, or include a feeder or bird-bath in view of a window. Take things a step further and transform your front lawn into a woodland garden to attract a wider array of birds and other wildlife. If a sprinkler system is already in place, so much the better for establishing new plants.

qualities, follow these next three steps to create your layout plan:
1. Create a Base Map
2. Create a Schematic Layout Plan
3. Create the Final Layout Plan

1. Create a Base Map

A base map identifies existing wildlife features and shows the general relationships of structures and other things on your property. A base map can be drawn on a large sheet of paper or a blown-up version of a survey map prepared for your property (Figure 2). For personal use, a sketch with estimated measurements is generally fine.

In addition to all structures, hard surfaces, and utility lines, you'll want to be sure to include features that are particularly valuable to wildlife. Examples, all of which are discussed in detail elsewhere in this book, include:

- Mature trees and shrubs; especially those favored by wildlife as a source of berries, nectar, shelter, etc.
- Water sources of any type.
- Major dead or partly dead trees (snags).
- Brush and rock shelters, large stumps, logs.
- Nest sites in trees and shrubs.
- Existing bird feeders, birdbaths, and nest boxes.
- Burrows and runways (mice, voles, and shrews make tiny paths through the grass).
- Wildlife travel corridors, including places where animals climb under or over a fence.

In addition, you'll want to take note of potential hazards to wildlife, such as areas where deer cross a busy road. You may also want to take special note of areas too steep, isolated, or otherwise difficult to maintain; these are ideal areas to

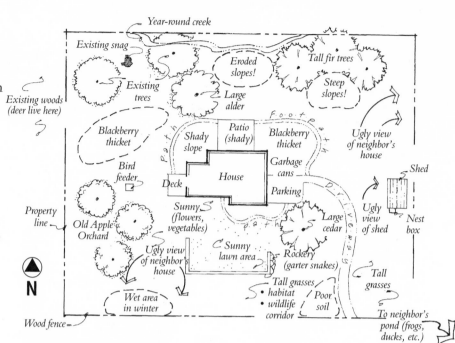

Figure 2. *A base map identifies existing wildlife features and shows the general relationship between things on your property.*

convert into low-maintenance wildlife habitat (Area 3).

2. Create a Schematic Layout Plan

To help design your property for compatibility with present and future wildlife, create several quick-and-simple layout plans. These schematic plans can be done on photocopies of your base map or on tracing paper laid over the map.

To make things easier you can use "bubble diagrams" (Figure 3). With bubble diagrams, you draw lines around areas as though they were "blobs" of space and then write your ideas for the use and contents of the space. Draw bubbles around areas you want to protect or create as wildlife habitat, as well as areas that you want to reserve for human activities. Also, draw circles, Xs, or other symbols where you want features such as a birdbath, bird feeder, or bench. Draw arrows where you want views, and dotted lines for poten-

tial pathways. Write in some of your ideas and objectives, such as building a deck and preserving the view of a feeder from a kitchen window.

Keep your ideas flexible and don't worry about wasting paper. The time and money you save at this stage will more than pay for your effort and the paper used.

Develop several different bubble diagram plans and don't spend too much time on any of them. You may end up with a "Plan A" that gives more space to wildlife and a "Plan B" that gives more space to human activities. Or you may develop three different plans that range from highest to lowest cost, or most change to least change. After you've developed several plans, choose one to develop into your final layout plan.

3. Create the Final Layout Plan

Once you have a general idea of how you want to lay out your property for you and wildlife, you can create a final layout plan (Figure 4). In the final layout plan, you can finalize the general location of new trees and shrub groupings, paths, special garden areas, and other features. Use colored pencils if it helps you distinguish different spaces or functions. If a more-detailed planting plan is desired or required, you can create one using the suggestions provided in the next section.

When it comes to landscaping for wildlife, no single "ideal" is applicable to all situations. This final layout plan is presented not because it might fit your location exactly, but to illustrate some important concepts about what wildlife find attractive. This landscape provides food, cover, and nesting sites in a variety of ways, and is richly stocked with the plants birds and other wildlife like the most.

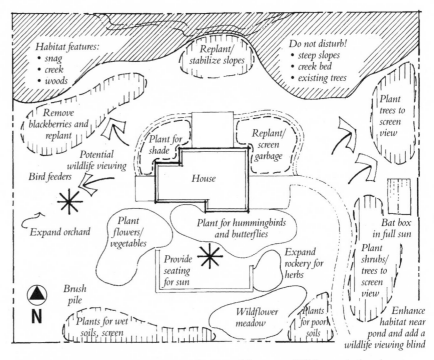

Figure 3. *An evaluation of existing conditions and future possibilities can be loosely sketched in with bubble diagrams. Using different shapes and signs will help you refine your ideas and prepare your final layout plan. Testing ideas on paper is easier than rearranging features in the landscape.*

Creating a Planting Plan

To help you plan for the selection and placement of new plants, create a planting plan. As with a layout plan, a planting plan allows you to experiment with the locations of plants on paper before any work begins.

A planting plan can be drawn to include as much detail as you choose. A detailed plan can be drawn to scale (not recommended for the beginner) and may include plant species, quantities, spacing, and locations. While the process described here won't provide you with such detail, it can give you a general idea of how plants might be combined. Final adjustments to any planting plan will always occur at planting time.

To create a planting plan, you can draw on tracing paper laid over your final layout plan. How-

ever, it will be easier to draw at a larger scale and create a separate planting plan for the area(s) you are designing.

When you start to draw your planting plan, don't be concerned with specific plants—only types, such as conifers for a screen, tall deciduous trees for summer shade, evergreen shrubs next to the house, and perennials for the butterfly garden (Figure 5). You can draw freehand or use a circle template to make circles of sizes that roughly represent the sizes of plants. As with the layout plan, keep your ideas flexible and don't worry about wasting paper.

Choosing a Style

As you begin deciding which plants to place where, consider choosing a planting style. There is no one planting style to use when landscaping for wildlife, and most

Design Principles to Keep in Mind

Chapter 1 contains important ecological principles you'll want to consider as you develop your layout plan. Also, keep the following principles in mind:

Maximize undisturbed areas: The larger the area of undeveloped land, the greater the chance that wildlife will use it as a feeding site, as a nesting area, or for travel needs. As much as possible, provide large areas without buildings, paving, or paths. Pay particular attention to providing undisturbed buffer areas and safe travel corridors for wildlife along a back fence, around a pond or wetland, and along creeks and streams.

Concentrate and contain human activity: Disturbance to wildlife can

continued on next page

be reduced if areas with a great deal of human activity are put close together and kept as small as possible. Avoid putting human activities other than those that are wildlife-related, such as watching wildlife from a bench or a blind, in good wildlife habitat. Ensure that travel corridors for wildlife bypass, rather than lead to, human areas.

Preserve existing vegetation: Think twice before you remove existing trees and shrubs no matter what they look like. These may be food and cover plants for wildlife and give you a start on landscaping for wildlife. Trees, including fruit trees that are old and tangled, are especially valuable. Avoid putting new structures where they will damage tree roots. Remember that a tree's roots grow far out from its trunk, and construction too close to the roots may affect the tree. Minimize removal of understory shrubs; they serve as important food and shelter plants for wildlife.

Work with nature and not against it: Slopes, shady areas, exposed sunny areas, rocky areas, and wet areas all offer different opportunities for wildlife habitat. Such features are only problems when they interfere with other plans you have for the site. But the problem may be an asset if you can accept it for the many values it possesses. Work with the land: "Don't fight the site" is a good adage. The less you alter the existing landscape, the less chance that you will inadvertently create problems.

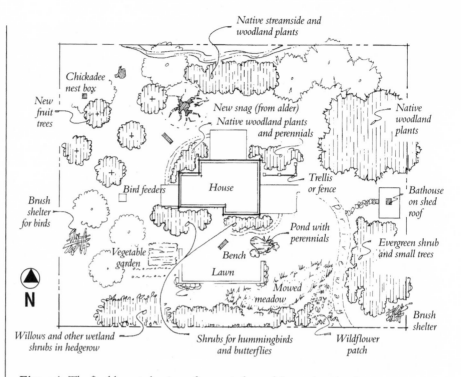

Figure 4. *The final layout plan is a refinement of one of the previous drawings.*

landscapes end up using a variety of styles. Your choice will depend on your taste, the location on the property, the conditions of the site you are planting, and your expectations for it.

If you are using the "Area Approach" described in this chapter, each area can have a different planting style. Styles requiring heavy maintenance (English-style perennial borders or cottage gardens) should be placed in Area 1. In contrast, the style chosen for Area 3 plantings should require little or no maintenance once any new plants are established.

In nature, plants are typically found in "communities" or "associations"—for example, the community of plants characteristically associated with a woodland, grassland, or wetland. Plants found together in communities have similar growing requirements and may actually help one another flourish. In landscape design, we often imitate naturally occurring

plant communities, either deliberately or simply because it seems "right" to us. Not only does this help to ensure that plants "look right" together but it also makes maintenance much easier, because all the plants in the same area have similar needs.

The Naturalistic Style

A popular design style used in many landscapes for wildlife is the naturalistic style. Naturalistic planting designs seek to emulate the appearance of one or more wild plant communities. Your property was probably once part of a larger forest, wetland, or grassland, and may still include remnants of such a plant community. If you are lucky enough to own property that retains some natural area, you should plan around it; and you may wish to expand this area.

However, sometimes it's desirable to use the naturalistic design approach to create a wild-

Table 1. The general spacing between landscape plants

Small-scale ground covers (wild strawberry)	12 to 24 inches
Larger-scale ground covers (kinnikinnik, cotoneaster)	2 to 3 feet
Small shrubs (salal, evergreen huckleberry)	3 to 6 feet
Large shrubs (tall Oregon-grape, red-flowering currant)	4 to 8 feet
Small conifers (mountain hemlock, Rocky Mountain juniper)	8 to 10 feet
Large conifers (red-cedar, Douglas-fir, ponderosa pine)	10 to 20 feet
Small deciduous trees (Douglas maple, vine maple)	10 to 15 feet
Large deciduous trees (paper birch, red alder)	15 to 20 feet

life habitat that is different from the one that's already present or occurred there historically. For example, in hot, dry areas of the Pacific Northwest where trees usually don't occur naturally, many people choose to plant a woodland landscape around their home. This provides shade, protection from cold winds, visual interest, and increased availability of food and shelter for wildlife that might not otherwise be found there. In creating such a woodland, you'd use the most characteristic tree, shrub, and ground-layer species of the woodland habitat you were using as a model. Even a small woodland landscape can suggest the patterns of light, colors, shapes, spacing, and textures of its natural counterpart. For information on woodland, wetland, grassland, and streamside landscapes, see Appendix A.

Don't assume that it will be easy to replace a plant community that once existed on your property if it has been substantially modified by development, management for timber or grazing animals, or by the introduction of non-native plants such as reed-canary grass, Himalayan blackberry, or Russian olive.

A Layout Plan Without the Paper

It's possible to create a layout plan directly on the ground. All you need is a bag of white flour or agricultural lime, and a windless day. It can also be done with some rope or stakes.

Take the bag or the container of the marking material and slowly spill your design on the ground. Use this approach to shape your flower beds, paths, etc. You can also spot future trees and shrubs by creating large circles. To make changes, simply brush the lines away and redo them. Have friends stand and pretend they are trees and then draw around them. The lime or flour will not stay around after water hits the site so take some notes or photographs to refer to for future reference.

A rope or hose laid on the ground, or stakes pounded in the ground can be used to make more long-lasting shapes. Two-foot-long stakes are very useful when experimenting with the layout of planting beds, paths, and other landscape design. When placed approximately 4 feet apart, they can provide a good indication of what the outline of an area might look like. Larger stakes can be used to signify trees or large shrubs.

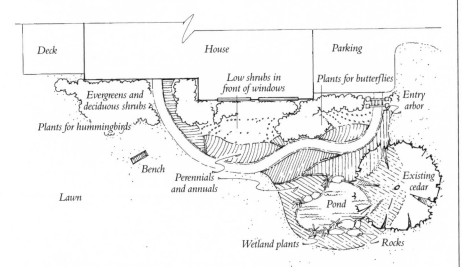

Figure 5. *An example of how to create a planting plan for the area in front of the house shown in Figure 4. Experiment with the locations of plants without being concerned about the specific species. Think of the plants in terms of their eventual shape and size.*

Identifying Existing Plants

By identifying existing plants on your property and getting to know something about them, you can learn important information about your landscape. Plants can tell you what the soil is like in that area and, when you are designing a naturalistic landscape, can give you a starting point for a landscape style.

There are a variety of good field guides that can help you to identify the existing plants on your property. Trees often can be identified by their leaves, needles, bark, and size. Most other plants are best identified by their flowers, fruit, or seed.

If you want to have a plant identified for you, snip off 6 inches or so and take it to a Native Plant Society meeting, a local nursery, a nearby arboretum or nature center, or a Cooperative Extension Master Gardener for identification. To keep a sample fresh, place it in a plastic bag and store it in the refrigerator until it's ready to be identified.

A fun way to get to know the wild plants growing on your property (and get ideas for what to add) is to go on one of the many guided nature walks offered by the above-mentioned or by other conservation groups. Look for offerings in your newspaper or mail, on the Internet, or call a local conservation group.

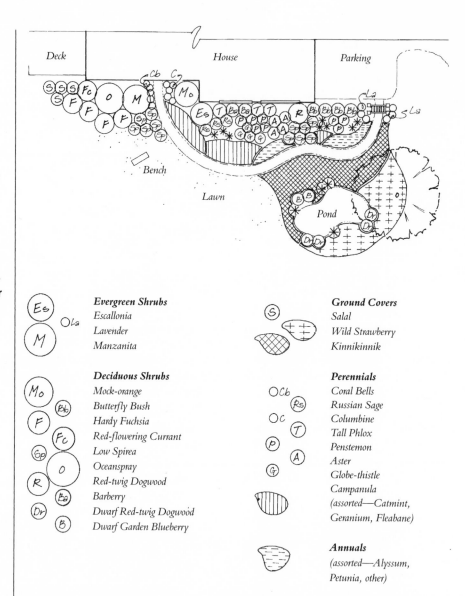

Evergreen Shrubs
Escallonia
Lavender
Manzanita

Deciduous Shrubs
Mock-orange
Butterfly Bush
Hardy Fuchsia
Red-flowering Currant
Low Spirea
Oceanspray
Red-twig Dogwood
Barberry
Dwarf Red-twig Dogwood
Dwarf Garden Blueberry

Ground Covers
Salal
Wild Strawberry
Kinnikinnik

Perennials
Coral Bells
Russian Sage
Columbine
Tall Phlox
Penstemon
Aster
Globe-thistle
Campanula
(assorted—Catmint, Geranium, Fleabane)

Annuals
(assorted—Alyssum, Petunia, other)

Figure 6. *The final planting plan is a refinement of one of the previous planting plans. This plan uses symbols to show where each plant is located.*

The Final Planting Plan

After creating several schematic planting plans, designating only the types or shapes of plants you want for specific areas, select a plan to turn into a final planting plan (Figure 6).

Next, list the specific plants you want to include in the plan. (Chapter 3 describes the different types of plants to use and in Appendix B you'll find several help-ful lists.) It's easiest to list the trees first, then shrubs, perennials, vines, ground covers, and other plants. You may want to list more than one species of plant that fits a situation because factors such as cost and availability will influence your final selection.

In nature, plant spacing often appears to be random. One way to create the same effect in a flower border or a naturalistic woodland is by using the clump-gap-mosaic planting pattern. Place several plants from each species together in a "clump," varying the spaces between plants slightly. Next, place some of the same plants at a slight distance from the group. Overlap similar clumpings of numerous species to create a "mosaic." This will avoid a haphazard, spotted, or over-regimented appearance.

Designing from a Photo Can Pay for Your Film and Maybe Your Camera

Color copy machines offer a quick, fun way to experiment with different landscape designs. First, using color print film, take several photos of the area you're designing. Take shots from angles from which you're likely to view the area. After the film is developed, have a print or prints enlarged on a color copier. A 11 x 17-inch sheet is a useful size. You may want several copies.

Next, use colored pencils to sketch in the general shapes and colors of plants (and other objects) you want to include in your landscape. You can draw directly on the color copy or on tracing paper placed over the copy. You can use the pencils to cover up features on the color duplication and add new ones. For instance, you can "erase" existing shrubs against the house by penciling the house color over them, or modify the plants by making them larger or adding other plants in front of them. You needn't be too careful with these drawings—bulk and texture can be created with crude strokes and scribbles.

CHAPTER 3

PLANTS IN THE WILDLIFE LANDSCAPE

All wildlife owe their existence to trees, shrubs, grasses, weeds, crops, aquatic plants, and other forms of vegetation. Animal-eating wildlife are no exception—their dependence on plants is, at most, a step or two removed. The long-term association of wildlife with plants is reflected in some of their common names: cedar waxwing, sage grouse, rabbitbrush, duckweed.

Not only are all plants eaten by something but all parts of plants are eaten at some time or another. Fruits, nuts, seeds, nectar, leaves, branches, twigs, and roots are consumed by some animal.

Plants as well as animals benefit from this arrangement. In the course of collecting food from flowering plants, for example, butterflies, hummingbirds, bees and other pollinators pick up pollen (genetic material), and carry it from one plant to another, insuring cross-fertilization. The nutritious flesh of fruits—often covered with an eye-catching covering—attracts birds, squirrels, foxes, coyotes, and other wildlife that eat the fruit, digest the flesh, and either regurgitate or pass the seeds at some distance from the parent plant. This ensures that a new generation of this plant will grow where it will not compete with the parent plant.

The availability of plant-based wildlife foods is greatly affected by the seasons. In the spring, swollen buds, flowers, tender vegetation, and a few rapidly maturing fruits become available. By summer many more kinds of fleshy fruits ripen and seeds mature. These supplement the abundant insects that many birds and other wildlife consume at this time of year. Fall is the season of plenty, when seeds, nuts, and fleshy fruit can be obtained without much trouble. Winter, especially late winter, can be a time of hardship for many wildlife species, particularly birds. The fleshy fruits of many plants are usually gone. The supply of insect and other plant food decreases markedly as the weather becomes colder and the first frosts arrive. Both plant and insect-eating birds must search more intensely over wider areas to find food, unless they fly south.

The availability of plant and animal food varies not only from season to season but also from year to year. Conifers and other trees produce bumper crops of seeds some years, and in others the yield is comparatively small. Because life is a complex web, these fluctuations affect all living things.

Noticing the interrelationships between plants and wildlife in landscapes adds richness to your role as a wildlife steward.

Plants for Wildlife Landscapes

Here are brief descriptions of the key types of plants and the role each plays in the overall plan of a wildlife landscape:

Trees are the most important plant choice for your landscape. The tallest element in the landscape, they generally live the longest, and their wildlife-attracting abilities improve with age. Even as they die, trees continue to provide wildlife with places to find food, shelter, and nesting sites. Deciduous trees grow quickly and can provide rapid visual screening and summer shade, and many have beautiful fall color. Therefore, carefully consider tree selection, planting, and preservation.

Shrubs, like trees, come in many shapes and sizes. In small landscapes many more types of shrubs than trees can be included. Shrubs provide colorful fruit, flowers, and leaves at different times of the year. Tall shrubs can screen unwanted views, and medium-sized shrubs can create peninsulas in the landscape around which paths can wander. When planted in close groups, low-growing shrubs serve well as a ground cover because few weeds will grow between them after they get established.

Ground covers are low-growing evergreen and deciduous plants grown to cover the ground. They can be planted on hot, dry slopes, over a wall, or in amongst trees and shrubs. Ground covers can keep the soil cool and prevent weed growth by blocking out light. Many also produce fruits which are eaten by ground-feeding birds, create good cover for insects and other invertebrates, and produce flowers that attract a variety of flying insects. Slow-growing ground covers don't belong in a planting where competing weeds (particularly grasses) will quickly overcome the new plantings. Fast-growing ground covers shouldn't be planted where they will require frequent maintenance to keep them from encroaching on other areas.

Vines are trailing or climbing plants that can grow on the ground or over trellises, fences, and buildings. In nature they often grow up and into tall shrubs and trees. Some species can also be grown as ground covers. There are evergreen and deciduous vines and some annual vines that can be grown each year from seed. Quick-growing vines are useful for hiding unsightly walls and softening the look of fences, snags, stumps, and rock or brush shelters. Vines provide nesting habitat for many birds, including hummingbirds and wrens. Some birds, such as song sparrows and robins, will hatch a second clutch in a sturdy vine after it has put on its spring growth, which creates the shelter these birds need. Many vines produce berries or seed tufts that are eaten by songbirds. Flowering vines may be frequented by hummingbirds and by other pollinators seeking nectar.

Annual plants such as zinnias and cosmos complete their life cycle in a year or less. *Perennial plants* such as asters and lupine are non-woody and live for more than one year. Both annuals and perennials can be used in flower borders and wild areas such as a wildflower meadow. Many annuals and perennials are particularly attractive to flying pollinators, including bees, butterflies, moths, and hummingbirds. Many attract predatory insects, such as parasitic wasps and flies, that keep problem insects in check. Annual and perennial flowers offer you the opportunity to be creative with color and texture combinations. They also allow you to concentrate flowers in a small area. Even an apartment dweller can have an abundance of color in pots, hanging baskets, and planter boxes. These plants dramatically extend the blooming season and fill gaps when other plants, such as bulbs and shrubs, have finished blooming.

Grasses are the main component of grasslands. Grasses help support tall wildflowers, stabilize the soil, add seasonal color and texture, and provide valuable food and cover for wildlife. Native bunch grasses leave spaces between them in which wildflowers can become established.

Ferns are at home in a wooded landscape. The taller species make an excellent backdrop for the colorful spring display of woodland wildflowers. Ferns' many shades of green provide a cool respite from the intense light of summer. Ferns are food for most woodland animals that feed on green vegetation, including deer, elk, rabbit, mountain beaver, and ruffed grouse. Some ferns remain green after most vegetation has been frost-killed, making them especially valuable during fall and winter.

Plants to Look For

Because a lot of time and effort will go into planting and maintaining the plants in your landscape, it's important to select the right ones for you and for wildlife. A key consideration is to select a wide enough variety of plants so that different wildlife species will have food and shelter available throughout the year. Here are some specifics to keep in mind:

Include native plants whenever you can. Make an effort to use plants that are native to your particular area in the Pacific Northwest. These plants have adapted to local climate, elevation, and soil conditions, making their survival more likely in the area where they have evolved. See "Native Plants" in Chapter 1 for more information on native plants; see the "Native Plant Occurrence Chart" in Appendix B when choosing natives for a specific area of the Pacific Northwest

Plan to include both deciduous and evergreen trees and shrubs. Deciduous trees and shrubs come in many shapes and sizes and provide food in the form of flowers, pollen, nectar, nuts, and fruits. The leaves are also more palatable than are tough and bitter conifer needles, so more animals, including many insects, eat them. With insects come their predators, including insect-eating birds. However, conifers and other evergreen plants can be lifesavers to wildlife during winter. Because most conifers keep their needles on all year long, their dense branches provide crucial shelter in winter for birds and other wildlife after deciduous trees have dropped their leaves. In addition, they are an important source of seeds for squirrels and many birds, including grosbeaks and chickadees. Many species of birds glean overwintering insects from needles, twigs, and the crevices of their bark. Conifers also serve as excellent nest and roost sites for many birds, including hummingbirds.

Thorny and thicket-making plants, such as wild rose, also are valuable in the wildlife landscape. A tangle of thorny stems can provide invaluable cover for ground-nesting birds and other wildlife.

Select shrubs and ground-covers that produce fruits at different times of the year. Although fruits such as huckleberries, wild cherries, serviceberries, and elderberries may not be as nutritious as animal flesh, insects, or seeds, they are easy to gather and generally available in quantity. Fleshy fruits can also be an important source of water for wildlife during the summer.

Include some shrubs with fruits that remain on the plants into early winter or even longer, such as snowberry and wild rose. Because fruits and seeds that fall to the ground are often washed away, buried in leaves or soil, or covered with a blanket of snow, plants that hold onto seeds, fruits, or other edible parts into winter are especially useful to wildlife. Seeds or fruit that are borne high on plants are unusually valuable sources of food in areas where there is abundant snow.

Nut-producing plants are also an important addition to the wildlife garden. Nuts are fruits with a dry, hard exterior. Like seeds, nuts are high in fat and rich in protein and minerals. In addition, because they do not spoil readily, they're available to wildlife over long periods. You won't see results right away when you plant a nut tree, but you can be sure your good deed will last for generations.

Many seed-producing plants are also valuable additions. Seeds are high in calories and a good source of energy and stored fat for wildlife. They are the main

dietary ingredient for many small mammals and common songbirds. Seeds from pines, firs, and other conifers rank high in food use. Weed seeds are often unwanted by humans but are important food sources. The seeds of agricultural crops, such as wheat, rye, and corn, are especially attractive to wildlife, and sunflower and millet seed are often used in bird feeders. Trees such as birch and alder, which drop their tiny seeds onto the snow or open ground in winter, provide winter food for ground-feeding wildlife such as juncos, finches, and sparrows. Annual and perennial plants and grasses often produce seeds that are eaten by birds and other animals if seed heads are left on the plants in the fall.

Plants with blooms that are rich in nectar attract bees, butterflies, hummingbirds, and other nectar-seeking wildlife. Many trees, shrubs, and annual and perennial plants are good nectar-producers.

Other points to consider when making plant choices include the following:

- Be sure the plant is appropriate for your hardiness zone. (See the "Plant Hardiness Zone Map" in Appendix B.)
- Be sure the soil, light, and water conditions match the cultural requirements of the plant. (See Appendices B and C.) Knowing what plants are already growing on your site can give you an idea of the growing conditions that exist there.
- Choose plants that can be allowed to attain their natural form without shearing or severe pruning. Native plants often grow faster and larger in a maintained landscape than they do in the wild, so be careful when using them in small areas, next to a

curb or walkway, and in soil that is highly amended. A native wild rose, for example, can quickly reach 8 feet in both height and width in soil that is prepared for perennials.

- Don't plant a tree with a network of aggressive surface roots next to a flower or vegetable garden, paved patio, entryway, or parking strip.
- Don't plant a tree where leaves, flowers, and fruit dropping under and around it may be a problem, for example next to a patio. Similarly, in regions that receive regular high winds or consistent annual snowfall, avoid planting trees described as having weak or brittle wood or weak crotches.
- Don't plant trees and shrubs that may be plagued by particular insects or diseases if this will compel you to use pesticides.
- Don't plant or spread non-native species that are known to invade natural areas and displace native plants.

Plants to Avoid

Some plants do not make especially good wildlife habitat. These include many hybrid plants which have been bred for floral appearance rather than for qualities that appeal to wildlife. Many hybrid roses, for example, produce neither abundant nectar nor the large "rose hips" (actually fruits) that help make wild roses good wildlife plants. Hybrid annuals and perennials may have fancy blooms that are constructed in such a way that, even if the bloom contains nectar, hummingbirds and pollinating insects are unable to reach it. In choosing plants for wildlife, it is usually preferable to select older, less hybridized varieties—better yet, try to find the wild plants from which later plants were developed. These "species" varieties are most likely to have qualities that appeal to wildlife.

Other plants to beware of in designing a wildlife landscape are those that tend to spread over large areas, displacing other species to form one-species plant communities ("monocultures"). Because of their lack of variety, monocultures typically don't support more than a few wildlife species, and the wildlife they do support may not be wildlife you want to encourage. Many vigorous, sod-forming grasses, including Kentucky bluegrass, are used in lawns precisely because they form monocultures, with aggressive, dense root mats that prevent other plants from growing through. This is one reason why lawn is not generally considered to be good wildlife habitat; it tends to attract only a few species, most of which (e.g., crane flies and starlings) are considered undesirable by most people.

Obtaining Plants

You can buy plants from nurseries, contractors, and yearly plant sales held by community groups. You can also grow your own plants from seeds or cuttings, or transplant them from a nearby area. Each source has advantages and drawbacks.

Many retail nurseries are becoming more familiar with native plants and wildlife plants, and can help you select the right plants for your specific project. Some mail-order nurseries have catalogs that can be good sources of information. If you want plants for a large project you may be able to save money by buying from a plant broker or landscape contractor. Either of these should be able

to provide you with plants for much less than the retail price. Plant brokers and landscape contractors normally are used for large projects and have a minimum charge for plants and delivery. Get bids from two or three different brokers or contractors; they're usually listed in the telephone directory.

Know Where Your Native Plants Come From

Collecting native plants from their natural settings is usually strongly discouraged because this practice has brought some rare plants and local plant populations to the brink of extinction. In fact, collecting many plant species in the wild is illegal. Also, this effort is frequently unsuccessful because transplants require much more care than nursery stock with healthy, compact root balls that have been prepared to withstand replanting.

Most reputable nurseries grow their native plants from cuttings, divisions, or seed. However, because some nurseries still carelessly collect live plants from the wild, try to determine where your nursery gets their plants and buy only from those that use environmentally sound means. Insist that plants be nursery-grown or obtained from authorized, supervised salvage activities. Do not buy field-collected plants.

When plants are threatened with destruction, however, you may want to consider salvaging some. If this is done with care, you may be able to get mature-looking, inexpensive plants for your landscape project. For information on how to salvage plants, see "Woodlands and Woodland Landscapes" in Appendix A.

How to Transform Your Lawn into a Wildlife-Friendly Landscape

Lawns are appropriate for areas that receive heavy foot traffic or aggressive use by pets and children. They are also nice for lying down on and they have a tidy appearance. However, lawns can be water hogs if they're kept green in summer, traditional maintenance requires heavy use of herbicides and fertilizers, lawn mowers are a major source of air and noise pollution, and a lawn is not particularly good wildlife habitat. Furthermore, grass is not the most appropriate or easiest-to-maintain ground cover in areas that are narrow, steep, or shady.

Many people include large lawn areas in landscape designs only because they aren't aware of the alternatives, or because they assume that a large lawn is necessary for a landscape with high visual impact. In fact, small lawns can have a stronger total visual impact than large ones.

Try reducing the size of your lawn and plant the area with ground covers and other plants that are wildlife-friendly, low maintenance, and appealing to you. For information on how to remove grass from an area, see "Fields, Meadows, and Grassland Landscapes" in Appendix A.

Instead of fighting moss in shady areas, let the leaves from nearby trees and shrubs fall where they will, eventually smothering the grass. Remove any grass that remains and create a woodland

continued on next page

glade of ferns, shade-tolerant flowers, and mosses.

Consider leaving parts of your lawn unmown. It's best to start this no-mow regime in your backyard or in a small area first. Edges can be mowed or weed-whacked to keep the area looking tidy. If the area is larger than 500 square feet, you may want to cut a pathway into it. Dig out small planting areas to introduce meadow plants such as daisies, black-eyed Susans, asters, golden-rod, or native grasses. Instead of using tiny seedlings, start with established plants strong enough to fight their way through the grass roots. (See "Fields, Meadows, and Grassland Landscapes" in Appendix A for more information.)

Preparing the Site for Planting

The way you prepare a site for planting will depend on the size of the area, the conditions of the site, and the types of plants you want to add. When planting trees and shrubs here and there in a large area, you may simply need to add "planting pockets." Create soil pockets by scalping out patches of grass or other material with a shovel or mattock. The size will depend on the size of plants you use. You can mulch around the new plants to help keep grasses from intruding.

Smaller areas for garden annuals and perennials, and areas where construction has taken place will require more intensive preparation. After you've roughly removed unwanted plants in the area, you'll want to eliminate remaining vegetation through techniques such as smothering (with black plastic, cardboard, or other materials), stripping (removing the top 2–3 inches of vegetation, roots, and soil), cultivating (digging or rototilling the earth at least three times to a depth of 4–5 inches), or using a glyphosate herbicide. See "Fields, Meadows, and Other Grassland Landscapes" in Appendix A for a description of all these techniques.

Depending on what you want to plant and on your soil conditions, you may have to add soil amendments or even bring in topsoil. Details on these methods are provided in many standard works on gardening.

In a large area with poor soil, you'll be much better off choosing plants that are adapted to the site conditions, rather than spending time and money to create the right conditions for a particular plant or group of plants. However, do not

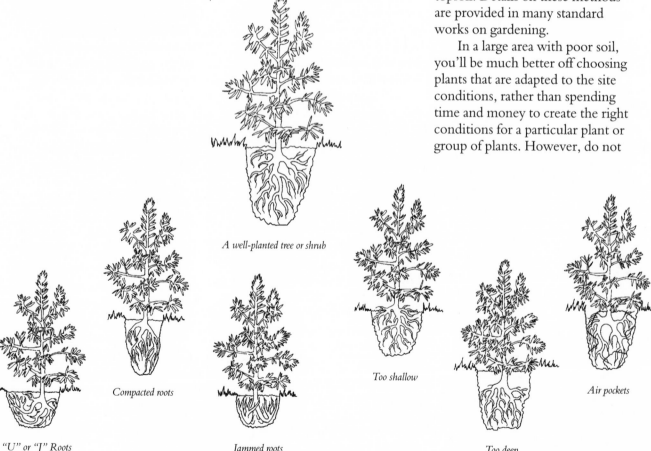

A well-planted tree or shrub

Compacted roots

Too shallow

Air pockets

"U" or "J" Roots

Jammed roots

Too deep

Figure 1. *When planting a bare-root tree or shrub, pay particular attention to how the roots get situated in the planting hole.*

Figure 2. An example of a properly planted and staked balled and burlapped tree. Note that the plant is slightly higher than the surrounding soil and the mulch is kept away from the crown. Never stake a tree except where the wind is likely to be very strong, the soil is loose, or the plant clearly cannot stand on its own. If you do stake, make sure the ties are loose and low on the trunk. Remove all stakes within 6 to 12 months to create a sturdy tree and eliminate problems such as girdling and rubbing, which occur when wire and stakes are left on a plant.

assume that native plants are the best choice for poor soils. No plant will thrive if its growing conditions aren't met, and many sites have been so altered by human activity that they no longer will support native plants that once grew there.

Removing Unwanted Vegetation in a Wildlife-Friendly Way

One way to remove weeds in a large area that is dominated by aggressive weeds such as Himalayan blackberry, Russian olive, or scots broom is to spread the effort over several years. Birds and other wildlife will then have time to move to another area before their food and cover are eliminated. By the time all the original cover is

eliminated, new cover plants will be well-established and will have started to fill in.

For example, in the first growing season, thoroughly remove all targeted plants in the section. That fall or the following spring, make sure the targeted plants are gone or under control and then plant the area with preferred plants. Repeat this process for three years or until all three areas have been replanted.

Getting Help with Your Project

Chances are you'll have questions you can't answer during the process of creating a plan and later doing landscape work. Ask for help

when you need it. Don't let your frustration force you to make an uneducated choice that may later destroy valuable wildlife habitat. Short-term solutions can create long-term problems and destroy what you wish to save. A ten-minute talk with an expert may help you avoid a major problem.

Before you begin any landscape projects from your layout plan or planting plan, you may want to have someone review it. An "outside" evaluation from a Cooperative Extension Master Gardener, a landscape designer, or a knowledgeable friend can be helpful in spotting oversights or potential problems.

If you can afford it, you can hire a professional to help with all

Mulch

Figure 3. When planting a bare-root tree, make sure the roots are spread out.

or part of the design and installation of your landscape project. A professional can complete certain parts, and you can spend time on the parts of the project you enjoy and can do well. Here are some tips for working with a landscape design professional:

- Make sure your basic landscaping philosophy is understood and respected.
- Be specific and talk freely about the type of landscape you have in mind.
- Look at examples of previous work, ask for photographs, get addresses and references of past clients.
- Clearly communicate what your budget is.
- Provide a written list of the plants and other items you want to include if you have definite ideas on these.
- Be clear about what plants you do not want to include in the plan. Ask for a second choice if the plants your designer suggests don't meet your exact requirements.

Sample Fee Structure for Professional Landscape Services

There are a number of different ways contractors can structure their fee to meet your needs and budget.

Consultant with a Straight Hourly Fee

This is the best approach for people who don't need complete or detailed landscape drawings, are looking for conceptual advice and information, or who have a project that is open-ended without clear tasks or limitations.

Estimate with a Straight Hourly Fee

If you have a good idea of the scope of work and the level of detail, contractors can estimate their fees based on anticipated design time or square footage of the area to be designed. Estimates are only ballpark figures, and the actual fees may be higher or lower than the estimate.

Not-to-Exceed Figure with a Straight Hourly Fee

When you have a limited budget, you may want to consider this approach. The not-to-exceed figure is similar to an estimate, but design fees will not exceed the agreed-upon figure.

Lump Sum Fee

There are no surprises with a lump sum fee. This approach is best when the scope of work and the individual tasks are clearly defined. A dollar amount is agreed upon for each task. The totals from the tasks add up to equal a total lump sum fee.

CHAPTER 4

MAINTAINING HABITAT

Your yard doesn't have to look wild to attract wildlife. But how you choose to maintain your property can have a tremendous effect on the species that live there, or that may be unable to live there, if your maintenance activities destroy crucial habitat.

Many routine garden chores actually consist of removing landscape elements that are vitally important to wildlife. For instance, leaves, twigs, and other plant debris provide a loose, moist, organic cover for diverse forms of life including spiders, earthworms, salamanders, and birds of the underbrush, such as towhees and winter wrens. A dead branch left on the sunny side of a shrub may be used as a perch by songbirds. A patch of uncut vegetation along a path can serve as shelter for garter snakes, a nectar source for butterflies, and a seed source for birds and other animals.

A thoughtfully maintained wild area can be as much a beauty spot as any other part of the landscape. But the key word here is "thoughtfully." Think before you act and ask yourself: Will this activity improve or destroy wildlife habitat? If portions of the landscape must be maintained to a certain standard to satisfy neighbors, is it necessary to maintain the entire property in the same way? Decisions about maintenance are best made only when you are intimately familiar with your property and the wildlife it attracts. Prior to any large-scale maintenance task, slowly walk through the area to survey it for nests and other signs of wildlife.

a more natural look. A common argument against so-called "naturalistic" landscapes is that they can be perceived as weedy and unmanaged. However, there is a big difference between a managed landscape that appears natural and a traditional landscape that has been neglected. A naturalistic, as opposed to neglected, landscape offers many advantages. Not only can it be highly aesthetically desirable, but it is also likely to require much less maintenance than would a traditional "manicured" landscape. For wildlife, the naturalistic landscape typically provides more shelter and more varied food sources.

Another way to make maintenance easier in your landscape is to pick the right plant for the right location. Once established, plants that are well-sited, with sun, soil, and moisture conditions that are right for them, are much more pest- and disease-resistant. Well-sited, established trees, shrubs, and ground covers are able to survive with only minimal maintenance—at most, occasional weeding, mulching, protection from animal or other damage, and perhaps watering during prolonged dry spells.

Flower beds containing annuals and perennials require more maintenance, including replanting and dividing. To minimize time spent on landscape maintenance, plant more trees, shrubs, and ground covers, reserving flower beds for small areas close to windows, decks, and other locations where you can enjoy them easily.

Starting Off Right

The best way to "do" maintenance on many properties is to arrange things so that you don't have to do much maintenance at all.

Often, when people move to an area where there is still natural growth, they "park it out," removing all understory plants, and possibly even many trees. After they've cleared the plants that might have been growing in harmony with local wildlife for years, the now "landscaped" area no longer supports the wildlife that had lived there. What's more, the area now needs planting, watering, fertilizing,

mulching, and "pest" control. These homeowners have traded a beautiful, maintenance-free, natural landscape full of birds and other interesting animals for one that will be a drain on time, money, and resources such as water.

For a landscape that you can maintain with ease, consider whether some parts of your property can be left in a natural state. This is true low-maintenance landscaping. (See the discussion on "Using the Area Approach" in Chapter 2.)

Over time, you may be able to reduce the maintenance requirements of landscaped parts of your property by giving them

Minimize Potential Conflicts with Neighbors

(Adapted with permission from Wild Ones, P.O. Box 23576, Milwaukee, WI 53223-0576)

In general, people prefer a sense of order and purpose. A wild yard can conflict with that preference and can cause discord among neighbors. However, there are ways to help prevent this:

Create a border. Place a border between your yard and the sidewalk or a neighbor's property to make it clear that your yard is a product of intent and effort, not neglect. The border can be a hedge, a series of low native plants, or a path.

Recognize the rights of others to be different. Remember that although you have a right to your naturalistic plantings, your neighbor has the right to clipped lawns and plastic geraniums. Don't be a self-righteous natural landscaper.

Advertise. You have good reasons to naturally landscape your yard—let others know what they are. Educating your neighbors and local officials before you begin your project is essential. Signs can tell passersby that your yard is intended to be the way it is, that your yard is a special place.

Start small. You will reduce expense, enjoy your efforts more, and engender less hostility from neighbors if you move in small steps. Also, allow yourself time to complete a project in stages—with time to recover between stages. Think in terms of years when it comes to attaining a landscape that meets your needs and the needs of wildlife.

The Wildlife-Friendly Approach to General Maintenance

Although the maintenance requirements of a naturalistic landscape are likely to be relatively low, maintenance will be required, especially in the first few growing seasons. In hot, dry regions, this establishment period may be even longer.

Weeding

There are as many definitions of what a "weed" is as there are gardeners—indeed, one definition of a weed is "any plant that is where the gardener doesn't want it to be." In this context, however, a weed will be defined as any plant that has low utility to wildlife and that, because of its ability to outcompete, threatens the survival of plants that are beneficial to wildlife. Also, a weed is any plant that, if it were to escape from your landscape, might threaten the survival of populations of native plants in wild or natural areas.

Many plants gardeners are accustomed to thinking of as "weeds" don't fit either of these definitions. Many less aggressive plants that may occur naturally in your garden are actually beneficial to wildlife. The seeds of lamb's-quarters and chickweed, for example, are eaten by house finches. Goldfinches and juncos eat the seeds of goldenrod, and grouse eat the leaves. Goldfinches use the cottony parachutes of dandelion seeds—after they have eaten the seeds—to line their nests. The multiple hairs that make mullein leaves so soft are gathered by hummingbirds to line their nests. Insect larvae feed on mullein seeds and overwinter by the hundreds within a single flower stalk. Woodpeckers and chickadees seek out these dried stalks in winter and early spring in search of these insects.

Plants commonly thought of as "weeds" are often those that have adapted to sterile soil, compacted ground, scorching sun and drought, and that survive where other plants won't. These naturally occurring plants may be the first to colonize inhospitable soil, changing it so that it will later support other plant species. A "weed" such as dandelion, for example, is capable of driving its taproot through extremely compacted soil, breaking it up and bringing nutrients to the surface for other plants to use. This can be advantageous if you have an area where "nothing will grow."

Before you take measures to remove an area of naturally occurring plants, consider that they may be providing wildlife with food and shelter. Also, accept that it is impossible to completely eradicate "weeds." A more realistic and sustainable policy toward naturally occurring plants involves deciding which ones need to be controlled and focusing on managing—not eradicating—those.

For long-term weed control, the best approach is to plant your landscape with desirable plants that will form a living mulch. Until you establish such a landscape, you can effectively control weeds by mulching, removing them by hand, or mowing.

Although nonchemical controls are preferred, you may decide you need an herbicide for aggressive weeds, particularly if labor is not available for other control methods. Chemical controls may be effective in ini-

Table 1. Common weeds, their problems, and how to control them

Plant Species	Symptom/Problem	Control Methods
Quackgrass, *Agropyron repens*	A perennial grass with white fleshy roots that form dense, wide-spreading mats. When pulled by hand, broken pieces of root easily resprout. Mulch is rarely effective.	Tightly cover area with black plastic for one or two growing seasons. When preparing soil for planting, remove all root pieces. If chemical control is required, paint or spray the leaves with a systemic herbicide when the plant is actively growing.
Morning glory, *Convolvulus* spp.	A perennial vine that can climb up, cover, and smother plants. Some species even cover small trees. Plants are spread by seed or roots. Fleshy roots can travel long distances just below mulch or the soil surface.	Carefully hand-weed to remove all roots. Any broken piece of root will sprout new growth. Repeated digging from midsummer to late fall as the new growth sprouts will starve the roots. Don't put plants in compost piles. For chemical treatment, wait until the plants are large enough to bundle up in a group. Wrap a string around them as you would sheaves of corn or wheat. Cut just above the string and immediately paint the cuts with systemic herbicide.
Scots broom, *Cytisus scoparius* and Gorse, *Ulex europaeus*	Scots broom is a fast-growing shrub that can quickly form a dense stand and shade out other plants. It spreads by seeds, which develop in pods. These burst open, scattering up to 15 feet. Seeds can lay dormant for 50 years. Gorse is a similar, equally aggressive plant in some areas.	Remove the entire plant and as much of the root as possible. A "weed wrench," which makes removal easier, is available from forestry supply and some rental stores. Repeatedly pull, or cut any seedlings with a weed-eater in August, when the plants are experiencing the greatest water stress, and/or during spring bloom, when the resources in the roots are at their lowest levels. If chemical control is required, spray or paint the leaves in the fall with a systemic herbicide, or paint all cut-off stumps and branches.
Russian olive, *Elaeagnus angustifolia*	A fast-growing small tree that out-competes native plants along streams and creeks east of the Cascade mountains.	Girdle to leave a snag, or cut at the base. Leave roots alone where they may be helping to stabilize soil.
English ivy, *Hedera helix*	A fast-growing common ornamental vine that forms a dense layer even in shady areas. The thick growth can prevent all other plants from getting established. Vines also grow up trees, stunting their growth or killing them. Containment is difficult because the plants are partially spread by birds, which eat the berries, and by people who dump garbage that includes ivy clippings.	Pull vines off trees in winter or late summer. Dig out roots whenever possible. For chemical treatment, see quackgrass. Residents surrounding wild areas should be requested not to dump ivy cuttings in public areas.

Plant Species	Symptom/Problem	Control Methods
Purple loosestrife, *Lythrum salicaria*	A flowering perennial that seeds profusely and roots aggressively. It displaces native plants and causes the decline of wildlife species by decreasing plant diversity along the edge of water.	Mowing, flooding, or hand-pulling may have success in small, young stands if repeated for three years. Particular care must be taken when plants are seeding. Clip seed heads off and seal in a plastic bag before digging out the plant. For chemical treatment, see quackgrass.
Reed-canary grass, *Phalaris arundinacea*	A tall perennial grass that spreads rapidly in wet areas that aren't permanently flooded. Plants grow fast and easily shade out and crowd out other plants. Attempting to plant within an area dominated by reed-canary grass may be futile.	Dig out roots and then continue to eliminate any new growth to starve the roots. Continual, persistent removal of the new growth will be necessary for at least a year following your initial efforts. (For chemical treatment, see quackgrass.) For long-term control, shade out plants. Plant 6-foot or taller trees after a 3-foot weed-free area is created for each tree. Slow-release fertilizer tablets in the hole and a water-permeable, woven weed mat around the plants will help trees get established.
Giant knotweed, *Polygonum cuspidatum*	A large perennial that can grow 4–8 feet tall in one year. Spreads vigorously by roots and by seed. Plants form dense stands and shade out native plants. Roots are very difficult to remove.	Hand removal is the best control. Dig out the roots and cut the stems to ground level for two or three seasons. For chemical control, see quackgrass.
Himalayan blackberry, *Rubus discolor* and Evergreen blackberry, *Rubus laciniatus*	Aggressive trailing brambles that spread with the help of birds and wherever a vine touches the ground. They crowd out other plants, dominating ravines and other waterways, wet meadows, and other open areas. Attempting to plant within an area dominated by blackberries may be futile.	In summer, cut down canes with loppers and dig out as many roots as possible. Wear leather gloves and protective clothing to shield yourself from the thorns. The following spring remove young plants as they appear. Continual, persistent removal of the new growth will be necessary for several years following your initial efforts. After plants are cut down in summer, a fall application of a systemic herbicide to the tips of the cut stems or the leaves gives good control.

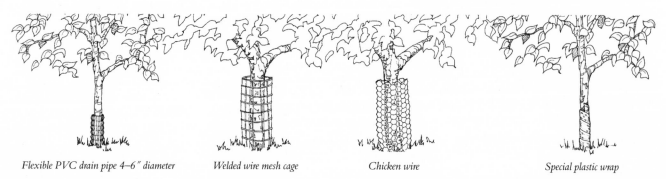

Flexible PVC drain pipe 4–6″ diameter   *Welded wire mesh cage*   *Chicken wire*   *Special plastic wrap*

Figure 1. *Place protective guards around the trunks of trees and shrubs to prevent damage by weed-eaters, mowers, and rodents.*

tially clearing out an infested area, but they should not be relied upon for long-term maintenance. Chemical controls should be used as little as possible, and targeted as closely as possible to the particular plant you wish to control. Avoid the use of weed-and-feed mixtures or any other type of herbicide that is spread over a wide area.

Mulching

In a woodland area, leaves and needles cover the ground; in a grassland or shrubland, plant litter from the previous season's growth does the same thing. Worms, beetles, and other invertebrates slowly move this vegetative matter into the soil, aerating it and replenishing it with nutrients. This is nature's own mulch.

But mulch offers many benefits to the landscaper, even when you have to apply it yourself. Mulch applied around new plants where there is intense competition from weeds—particularly grasses—will increase survival rate by at least 50 percent. Mulch also:

- Covers weed seeds.
- Conserves soil moisture in the summer by preventing evaporation.
- Cools soil in the summer, allowing shallow-rooted plants to survive.
- Insulates the soil in the winter, preventing freezing roots and minimizing frost heaving, which can expose and kill roots of small plants.
- Prevents crusting of the soil, thus improving percolation of water through to the roots and providing ideal conditions for earthworms.
- Helps prevent soil erosion and compaction.
- Adds nutrients to plants as the mulch slowly decays.

- Enhances the appearance of the landscape and sets plants off visually so they can be seen and not stepped on or mowed.

Mulch does have its disadvantages. Although straw and hay are inexpensive, lightweight, and easy to apply, they will always contain weed seeds. Often these are aggressive grasses. Also, mulch piled up around a plant's crown can eventually kill the plant by slowly rotting the crown or creating a pathway for insects to bore into a plant's trunk. It can also encourage rodents, especially meadow mice (voles), which may damage young trees. For these reasons, always pull back any mulch that comes within 2–4 inches of a plant's root crown.

Mulch can be added any time of the year if there isn't snow on the ground. There are many varieties of mulch. You can have commercial products (bark, compost, cardboard) delivered in bulk to your home. To find a product, look in the telephone directory under "Landscape Supplies." A local tree service company may be able to provide you with free wood chips. Look in the Telephone Directory under "Tree Service."

Watering

Watering new plants is important and will at least double their survival rate in areas where they have to compete with weeds. However, a well-designed wildlife landscape, once established, should not require regular watering outside of Area 1 (as defined in Chapter 2), and perhaps not even there. Consult a reference on gardening for information on how to water those areas that do require supplemental water.

Grooming the Flower Garden

Regular grooming of flowering annuals and perennials increases flower quantities and extends the flowering period. This not only makes the garden attractive to you for a longer period but also extends food supplies for nectar-loving butterflies, hummingbirds, and other pollinators. To groom your flower garden:

Remove heads from faded flowers using your fingers or by snipping them with hand pruners. Known as "deadheading," this procedure increases the number of flowers produced by a plant and extends bloom time. However, do not deadhead all plants that produce seeds eaten by birds and other wildlife (cosmos, goldenrod, and sunflowers), particularly in fall. These plants can provide visual interest and valuable seed heads for wildlife through the winter; spent seed heads can be clipped off in early spring.

Pinch back new, succulent stem growth by 2–6 inches. This can make plants bushier and encourage the growth of more flowers.

Pruning Large Conifers

(Adapted with permission from Elliott Menashe, Greenbelt Consulting, PO Box 601, Clinton WA 98236)

Because trees are often a critical part of the wildlife landscape, consider the following alternatives to removing a conifer that is obscuring an attractive or needed view, is at risk from wind or excess weight, is creating too much shade, or is considered an unattractive shape:

Skirting up involves removing tree limbs to a certain level above the ground. This technique is useful for creating a view when the tree is located high on a bluff face or hillside. Relatively more branches can be removed with this technique because the lower branches contribute fewer nutrients to the tree than do the higher branches.

Interlimbing involves the removal of an entire limb or individual branches at one level throughout the canopy to allow a view or to increase available light. Interlimbing also reduces wind resistance. This can be done in combination with windowing.

Windowing allows views through the existing foliage of the tree's canopy.

Topping any tree is poor practice and is not recommended. Retaining a topped tree's reduced size is an expensive, recurring maintenance requirement which is harmful to the tree. An exception is to create a wildlife tree. See Chapter 16 for information on how to do this.

Drip irrigation is an excellent watering technique, especially for trees and shrubs planted over a large area. It allows you to water only the plants you want to water, thereby reducing weed growth while conserving water. If you have ever put something together with Tinkertoys, you have the basic ability to put together a drip system. Small, backyard systems can be purchased in most hardware stores. For a drip system that waters over 50 plants at one time, you should get the components you need from an irrigation supply center. See "Drip Irrigation Basics" in Chapter 19 for additional information.

Fertilizing

If you select plants appropriate for site conditions, you generally won't need to fertilize. However, a nitrogen deficiency is common in soil that is heavily mulched with a wood product such as bark. The result is slow plant growth and possible yellowing of the foliage. In cases where you suspect a nutrient deficiency in the soil is inhibiting plant growth, use an all-purpose fertilizer and follow the recommendations on its label. A liquid, organic fertilizer tends to work best because there is little risk of burning the plant, and nutrients are immediately available. Where liquid fertilizers are not practical because of the scale of the project, consider the use of slow-release fertilizer tabs which are inserted into the soil. Granular fertilizer that is broadcast on top of the soil is not recommended because ground-feeding birds mistake it for food.

Pruning

Landscape plants for wildlife require minimal pruning. The graceful, spreading, natural shape of a plant maximizes the food and shelter it can provide for wildlife. The low skirt of branches around a conifer, for instance, provides important cover for many wildlife species. A loosely growing hedge welcomes songbirds, and provides

Figure 2. *Alternative pruning practices for large trees include skirting up, interlimbing, and windowing.*

areas for nesting, perching, and protection from predators.

You may want to prune plants in front of your home to meet your personal or neighborhood landscape standards. Continue pruning plants, such as fruit trees, that have been pruned routinely for years, since a sudden change in maintenance may weaken them or cause them to look out of place. (An old, unpruned fruit tree is like an elevated brush shelter and will attract a wide variety of birds. A sign attached to the trunk that designates it as a wildlife tree will help observers to appreciate it.)

When you do prune, do so in the winter when wildlife activity is low. If you must prune between March 15th and July 1st, the period during which birds are nesting, first examine the plants for nests and other signs of wildlife. Although your pruning activities may not physically disturb a nest, a drastic change in the appearance of the surroundings may lead the parents to desert the site and their young.

Avoid tree maintenance that involves cutting out rotting wood, filling cavities, and using wound paint. These techniques are unnecessary for plant health and eliminate potential habitat for cavity-nesting wildlife. Look upon rotting wood as a plus: It attracts insects that, in turn, provide food for woodpeckers and songbirds.

Occasional pruning of trees and shrubs may create a large volume of twigs and branches. These can be used to benefit wildlife if they are made into a brush shelter. A brush shelter also will save you the trouble of hauling or burning prunings, and add to the diversity of habitat on your property. A brush shelter will attract songbirds and other wildlife,

which use these areas as perching sites and escape cover from cats and other predators. For information on brush shelters, see Chapter 15.

Mowing Options to Improve Wildlife Habitat

Lawns

In the days before herbicides, sprinklers, and commercial fertilizers, lawns were what are now sometimes termed "good-enough" lawns. Good-enough lawns are still common where there isn't pressure to keep a lawn weed-free and green all year. The grass gets an occasional mowing and no water except for what nature provides (and that's good enough). Less-frequent mowing allows some common lawn weeds to flower, and this adds diversity, visual interest, and nectar sources for a variety of insects, including the common skipper butterfly. Unwanted lawn weeds can be removed by hand before they seed.

With a good-enough lawn, you can mow the grass to about 3 inches and leave clippings to decompose on site. The 3-inch height provides shade, helps preserve moisture, and contributes to the habitat for invertebrates living in the mulch. If chemicals are excluded from the lawn, the life in the soil will decompose the clippings and recycle their nutrients back to the grass.

Other Grasslands

A field or other grassland can be improved for wildlife by appropriate mowing. The goal is to reduce the harmful effects that mowing can have on wildlife and to create different habitats within the area.

For example, because repeated low mowing tends to keep all plants at one height and has the potential to harm wildlife, low-mow areas should be limited to high-use spots, firebreaks, and paths. Paths in public areas can be made wide enough for two or three people to walk side by side, but need not be wider. Often, a narrow path is preferred, especially if it includes occasional wide areas for groups or a bench. Garter snakes bask in these sunny mowed areas and rabbits and deer use them to find new shoots to eat.

Here are some other things to consider about mowing:

- Some grasslands that have been routinely cut over many years may have developed a distinctive group of plants and associated wildlife. Chances are these plants wouldn't be able to survive a different mowing regime. If possible, these areas or a portion of them should be managed in the same way to preserve the associated species.

- A wide range of small wildlife species use grassland areas. A mowing height of 4–6 inches will help prevent harm to newts, salamanders, snakes, or small mammals, which can find shelter in the stubble. This also reduces injury to clumping grasses and perennial flowers. A line-trimmer (weed-eater) is best for areas next to permanent water sources, where these animals are often found, and is often more practical than a lawn mower in other areas.

- Time your mowing to allow desirable plants to flower and to prevent unwanted plants from flowering or seeding. This may involve frequent spot mowing of unwanted plants.

- A single cut once annually can

Figure 3. A landscape designed for wildlife needs evidence of care. If there are no obvious signs of management, people may think the area has been abandoned. Litter and vandalism may increase, creating a downward spiral of damage and neglect. One way to show signs of maintenance in a large grassy area is to mow a strip between a path and meadow indicating that the grass is long by design rather than as a result of neglect. In a wooded area, make sure low branches are trimmed so people won't have to duck, which can make them feel annoyed or unsafe.

control invading woody plants, such as scots broom, even though it may not eradicate them.

- To reduce disturbance to most wildlife, including insects, make an annual cut as late in the year as possible, preferably after the first frost.
- Forgo a spring mowing to prevent problems with wet soil and so as not to disturb over-wintering insects (including butterflies).
- Forgo a winter mowing to attract seed-eating birds and to create visual interests for viewers.
- Forgo a late-summer mowing so flowers such as asters can bloom and attract their associated insects.

Although infrequent mowing with several months between cuttings may save you time and maintenance costs, the extra height of the plants that results may require use of a special mower, scythe, or line trimmer. If you attempt the job with your usual lawn mower, first remove large unwanted seedling trees and shrubs by hand.

Rotational Mowing

Rotational mowing is a wildlife-friendly approach to mowing because it involves cutting only a portion of a grassland area at one time. This minimizes disturbance to the wildlife using the uncut area, adds a variety of cut and uncut habitats, and helps prevent any aggressive trees, shrubs, or brambles from gaining overall dominance. An example of this management technique is to cut one-third or one-quarter of the total area at one time and continue this on a three- or four-year cycle until all the areas have been cut once.

With rotational mowing, fields and large lawns that were once routinely and completely mowed may eventually have hawks and owls hunting in them, and swallows, nighthawks, and bats will swoop above them catching flying insects. Unmowed sections preserve nesting areas for ground-nesting birds and over-wintering butterfly, moth, and other insect pupae and eggs, which would be destroyed in a complete mowing. Grassland plants that are allowed to flower can provide nectar for humming-birds, bees, butterflies, hover flies, and other insects. Their tall stems can provide structure for insects that require these parts of the plants. Birds use tall stems for perches and for gathering seeds and insects.

Try to develop a flexible, imaginative rotational mowing plan. For example, try leaving islands of uncut vegetation in certain areas or allow a patch of special plants to bloom and seed.

A Wildlife-Friendly Approach to Control of Harmful Insects

A typical yard may contain as many as 1,000 insect species. Fewer than one percent of these insects are considered pests. Rather than reaching for the spray bottle to indiscriminately destroy insects, why not try to maintain a balance in which, although some harmful insects are present, they are kept in check by a healthy population of predators? In other words, let the predacious insects, songbirds, and other insect-eating wildlife take care of unwanted insects for you.

More often than not, you will have to do little more than carefully watch certain plants to ensure that certain insects, such as aphids, remain within acceptable numbers.

Decide How Much Damage Is Tolerable

In order to maintain healthy wildlife habitat, you may have to tolerate some harmful insects and the damage they cause. The level of acceptable damage will depend on your standards, the importance of the plant to your landscape, and the destructive ability of the insect. Only a small percentage of plants will show significant insect damage in a wildlife landscape where a variety of plants attracts a variety of insects.

Consider the severity of the damage caused by insects. If you inspect plants around your property, you are likely to find that, although signs of insect damage are present, the plants' general condition is still good. Even fruit trees can have 25 percent or more of their leaves chewed off without suffering harm. Healthy trees and shrubs can also survive enormous

Figure 4. A 2-liter pop bottle with the top cut off above the label makes a good slug trap. Put dry cat food, beer, or a sugar, water, and yeast mixture in the bottle.

infestations of sucking insects, such as aphids, with little long-term damage. Birches and maples, for example, often are covered in the spring by aphids and by the sugary honeydew they excrete. These aphids, however, usually are eaten by other insects, songbirds, and other natural predators. The damage that occurs is little more than leaves covered with a harmless, sooty mold that soon may weather off.

Identify Problem Insects and Their Predators and Parasites

Monitoring your plants for insect damage will be more efficient and interesting if you identify the common, harmful insects likely to be present and the damage they cause. Also, you will want to learn what insect predators and parasites look like in both larval and adult stages and watch for them as well. The State University Cooperative Extension program provides bulletins with good pictures to help you identify these species (see "Public Agencies" in Appendix E). You also can take insect specimens to a local nursery or Master Gardener's clinic for help in identifying species.

Predators prey upon others. They are potentially a great help

Table 2. Plants that attract predator and other beneficial insects

Evergreen trees: bay, Douglas-fir, incense-cedar, madrona

Deciduous trees: alder, birch, cherry, cascara, dogwood, garden fruit trees, hawthorn, maple, oak , sumac, willow

Deciduous shrubs: deer brush, elderberry, oceanspray, potentilla, red-flowering currant, red huckleberry, serviceberry, snowberry, wild-buckwheat, spirea, wild rose

Evergreen shrubs: coffeeberry, coyote brush, evergreen huckleberry, hopsage, manzanita, mountain balm, Oregon-grape, rabbitbrush, sagebrush, salal

Garden flowers: alyssum, candytuft, coreopsis, cosmos, daisy, evening-primrose, feverfew

Wildflowers: angelica, baby-blue-eyes, goldenrod, fireweed, pearly everlasting, yarrow

Vegetables and herbs: carrot flowers, mustard family flowers, catnip, catmint, coriander, dill, fennel, hyssop, lemon balm, mint, parsley, rosemary, rue

to gardeners attempting to combat problem insects. In the insect world, ladybugs, lacewings, syrphid fly larvae, and ground beetles (see Figure 1 in Chapter 9) all prey upon harmful insects. Spiders feed almost entirely on insects. Songbirds, amphibians, and bats also are natural insect predators.

At times, particularly in the spring, it may seem that harmful insects are out of control. This

Pesticide-caused Problems

A National Academy of Sciences study revealed that home landscapes receive more pesticides per year than almost any other type of land in the United States.

The enormous amounts of urban- and suburban-applied chemicals may harm or kill wildlife in two ways: directly, if wildlife are exposed to the poisons during or immediately after application, and indirectly, if wildlife drink tainted water or eat exposed food, such as earthworms, flying insects, ants, seeds, buds, blossoms, and nectar.

Wildlife can come in contact with pesticides in a variety of ways. For instance, the bottoms of most birds' feet and the skin of frogs and salamanders are permeable to pesticides encountered on surfaces where these animals walk or crawl. Birds also may fly through pesticide spray and be poisoned later as they preen their feathers. They may also ingest pesticide when they eat insects crawling on sprayed plants, or when they sip the water droplets from a just-sprayed lawn.

A bird that can't use its legs, seems weak and slow-moving, and doesn't eat may be a victim of poisoning. You'll rarely observe a poisoned bird because it will die quietly, often in hiding. After death, it may be quickly eaten by another animal, which, in turn, may accumulate the toxins and die, a process known as bio-accumulation.

Pesticides also can harm wildlife by reducing prey. Bats and most birds, for instance, depend on a healthy population of flying insects. Elimination of these insects may lead to the direct loss of their predators.

is because predators are slow to reproduce. They've evolved that way in order to make sure there will be plenty of insects to eat when their young are born. If you are patient and leave their food supply alone, their populations will soon catch up with the populations of pest species and a balance will be restored.

To encourage the natural population of predator insects already on your property:

1. Don't use pesticides that will kill beneficial insects along with the pests.
2. Plant flowers that are known to attract predator insects for their nectar and pollen.

Using Pesticides

The term pesticide refers to any material used to kill pests. Such materials include herbicides for weeds, insecticides for insects, and fungicides for fungi.

Even "biologically safe insecticides," such as insecticidal soap and *Bacillus thuringiensis* (Bt), should be used with caution, as they can harm wildlife they were not intended to target.

One form of Bt is used to control problem insects in their caterpillar stage. It is applied when the caterpillars are actively feeding on leaves in warm weather. The caterpillars eat the foliage, sicken, stop feeding immediately, and die in a few days. However, this kills all caterpillars, including those that would have transformed into desirable butterflies and moths. In addition to enhancing the beauty of a landscape, these butterflies and moths provide food for bats and swallows.

Since no pesticide spray can be applied without some harmful consequences, limit all pesticide

use. Before deciding to spray, ask yourself:

- What other species may be harmed by contact with the pesticide?
- Will the pesticide eliminate a prey base or food source for some other wildlife species?
- Will damage by this one insect adversely affect the whole plant?
- Is there a more wildlife-friendly approach?

Try to completely avoid weed-and-feed products, which combine water-soluble chemical fertilizers with weed-killers. These products spread toxic chemicals over a large area. People, pets, and wildlife are exposed to the weed-killers, then rains may send the chemicals flowing into local waterways. These chemicals can also harm your soil, kill earthworms, and make your lawn easy prey for disease, drought damage, and weeds.

Precautions to be followed when using any pesticide in the landscape include the following:

- Read and follow the directions on the label every time. Don't rely on memory.
- Use small hand-held sprayers or brushes. Sprayers that attach to the end of the hose tend to leak, and it is difficult to control their outflow.
- Don't mix any more of the chemical than you need. If you end up with extra chemical in the sprayer, continue spraying the area in accordance with the directions until you have dispersed all the mix, instead of dumping the mix on the ground or down a drain, toilet, or sewer.
- Don't store unused mix in any container.
- Don't spray on a windy day.

PART 2

PACIFIC NORTHWEST WILDLIFE IN THE LANDSCAPE

CHAPTER 5

MAMMALS

Mammals are furry, four-legged animals that nourish their young with milk. Squirrels, raccoons, and some other mammals are born blind, furless, and quite helpless. Others, such as deer and hares, are ready to move about soon after birth. Most mammals have an excellent sense of smell, sight, or hearing, and will detect your presence long before you detect theirs.

Mammals tend to be inconspicuous and elusive, so people are often unaware of their presence. Many times their trails, droppings, or nests are all that can be seen. Although more than a hundred kinds of mammals live in the Pacific Northwest, you're most likely to see only the ones that have adapted to living near humans.

Squirrels, raccoons, opossums, skunks, mice, and some bats have adjusted particularly well to urban living. They do so by eating whatever is around and/or coming out only after dark. Travel corridors such as railway lines, power-line rights-of-way, thickly vegetated ravines, and hedgerows help these mammals move around and remain camouflaged in urban areas. For information on bats, see Chapter 13.

When Mammals Are Active

To avoid being eaten, most small mammals, including mice, voles, moles, and shrews, are active only at night. The coyotes, foxes, skunks, domestic cats, and owls that prey on these smaller mammals are also partially or totally nocturnal. Larger mammals, such as deer, bears, cougars, coyotes, and raccoons, may be active both day and night.

In order to conserve energy during winter, many mammals enter a state of reduced metabolic activity that resembles sleep. However, few mammals actually hibernate throughout the entire winter.

Mammals on Your Property

If your property and the areas around it have few places where mammals can find food and shelter, chances are slim that many mammals will live around you. To enhance your property for mammals:

- Include plants that provide berries, seeds, and nuts at different times of the year. These will be eaten while on the plants and after they have fallen.
- Leave plant litter in which small mammals can find food and shelter, and prunings for others to gnaw on.

Mammals and Rabies

Rabies is a disease that can affect all mammals. People become infected with rabies if they are bitten or scratched, or if saliva from an infected animal enters open skin or mucous membranes (nose, eyes, mouth). Rabies in humans is uncommon in the Pacific Northwest.

Here are some measures for preventing rabies infection: Avoid contact with wild animals, especially bats, foxes, raccoons, and skunks. Make sure your pets are vaccinated against rabies. This includes your cats as well as dogs, since more cats than dogs are reported rabid. Although it is not possible to determine by simple observation if an animal is infected, signs which should lead you to suspect that it may be rabid include nervousness, aggressiveness, excessive drooling and foaming at the mouth, or abnormal behavior (e.g., wild animals losing their fear of people or animals normally active at night being seen in the daytime).

If you suspect that an animal has rabies, notify your local animal control division or health department. Do not attempt to capture the animal yourself. If you are bitten or scratched by an animal, wash the wound thoroughly with soap and water as soon as possible and notify the health authorities immediately. If you come into contact with a bat (e.g., awake to find one in your room or see one near an unattended child), contact the health authorities immediately.

- Preserve some brambles and thickets for deer to browse and for small mammals to use as shelter.
- Leave dead or dying trees (snags) alone when possible. These provide homes and food-storage sites for flying squirrels and other mammals. Logs provide homes and attract food sources, such as insects for shrews.
- Install bat houses and flying-squirrel boxes in suitable locations. (See Chapter 13 and

Table 1. Food plants commonly used by deer

(Adapted with permission from the Woodland Fish and Wildlife Project)

Mule deer	Black-tailed deer	White-tailed deer (Westside)	White-tailed deer (Eastside)
Trees and Shrubs	***Trees and Shrubs***	***Trees and Shrubs***	***Trees and Shrubs***
Bitter cherry	Bitterbrush	Blackberry	Bitterbrush
Bitterbrush	Buckbrush (*Ceanothus* spp.)	Elderberry	Buckbrush (*Ceanothus* spp.)
Currant	Deer brush (*Ceanothus* spp.)	Red-twig dogwood	Deer brush (*Ceanothus* spp.)
New growth of Douglas-fir	New growth of Douglas-fir	Salal	Evergreen ceanothus
Mock-orange	Red huckleberry	Snowberry	New growth of Douglas-fir
Ninebark	Red alder	Snowbush (*Ceanothus* spp.)	Red-cedar
Deer brush (*Ceanothus* spp.)	Red-cedar		Sagebrush
Red-twig dogwood	Salal		Serviceberry
Serviceberry	Salmonberry		Bitter cherry
Snowberry	Snowbush (*Ceanothus* spp.)		Willow
Snowbush (*Ceanothus* spp.)	Trailing blackberry		
Thimbleberry	Vine maple		
Wild rose	Willow		
Willow			
Forbs and Legumes	***Forbs and Legumes***	***Forbs and Legumes***	***Forbs and Legumes***
Alfalfa	Alfalfa	Bulrush	Alfalfa
Balsamroot	Balsamroot	Clover	Burnet
Bluebells	Cat's ear	Pearly everlasting	Clover
Burnet	Clover	Sword fern	Dandelion
Clover	Deer fern		
Dandelion	Deer vetch		
Hawkweed	Fireweed		
Prickly lettuce	Pearly everlasting		
Trefoil	Plantain		
Twinflower	Vetch		
Grasses and Others	***Grasses and Others***	***Grasses and Others***	***Grasses and Others***
Bluegrass	Bluegrass	Bulrush	Fescue
Cheatgrass	Wheat	Sawgrass	Orchard grass
Oats	Oats		Wheatgrass
Wheat	Lichen		Lichen
Lichen	Mushrooms		Mushrooms
Mushrooms			

Appendix D for placement and management.)

- Leave part of your lawn unmowed or pasture ungrazed, to attract field mice and other small mammals. These may become food for hawks, owls, snakes, and other wildlife.
- Build a pond or ground-level birdbath to attract the wide assortment of animals that will visit a water source.
- Build a brush and/or rock shelter.
- Keep domestic dogs, cats, and other pets indoors or fenced.

Feeding Mammals

The best way to provide food for mammals is to maintain a variety of natural food sources on your property. Artificially feeding mammals using human or pet food can create undesirable situations for you, your children, your pets, and the animals. Mammals that have been artificially fed often lose their fear of humans and pets, and develop a territorial attitude that may lead to aggressive behavior. Artificial feeding also tends to concentrate animals in a small area and may raise that area above its normal carrying capacity. (See Chapter 1.) Overcrowding can result in increased diseases or parasites, which may be transmitted to pets or humans.

Artificial feeding may also negatively affect your property and other wildlife. For instance, if the local deer population is larger than it would be naturally because you're feeding them, they will eat more twigs, branches, and bark from nearby shrubs and trees. This heavy browsing may kill the trees and shrubs, reducing the number of homes available for birds and other wildlife.

Some people feed deer and other mammals during the winter. However, winter is one of nature's ways of balancing the size of wildlife populations with the available habitat.

Watching Mammals

Most mammals are shy, quiet, dull-colored, and nocturnal, so they are not easy to see. You'll have a better chance of spotting most mammals during early morning and evening twilight hours.

You don't always have to see mammals to enjoy them, however. Animals often leave clues about their movements and daily routines. Among these are the following:

- Little piles of scales from pine and fir cones on stumps, logs, and rocks left by Douglas squirrels and chipmunks.
- Trails that mark routes to food, shelter, or water. These include the narrow runways (1–2 inches wide) in the grass made by meadow mice.
- Nibbled ends of twigs on shrubs and small trees made by deer.
- Gnaw-marks on bark and branches on the ground in

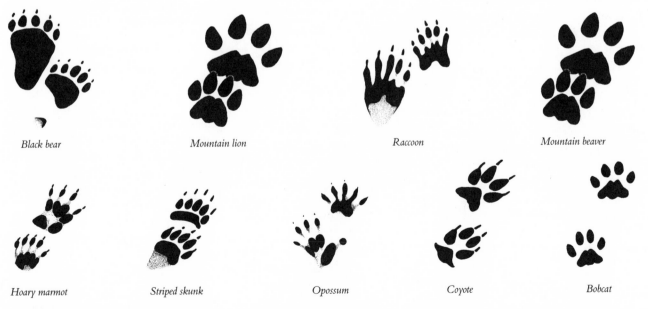

Black bear   Mountain lion   Raccoon   Mountain beaver

Hoary marmot   Striped skunk   Opossum   Coyote   Bobcat

Figure 1. Animal tracks may be something you concern yourself with only when you stumble upon them. On the other hand, you may find yourself eagerly anticipating the snow season because tracks are so much more plentiful then. In the absence of snow, seek out damp sand or soft dirt near a water source such as a river, stream, wetland, or pond. The tracks here have been adapted from Animal Tracks of the Pacific Northwest, *by Karen Pandell and Chris Stall. The Mountaineers, Seattle, 1981.*

winter made by mice and rabbits.

- Large ovals or circles of matted-down vegetation where deer have bedded down.
- Rub-marks on tree trunks and branches made by male deer polishing their antlers in the fall.
- Browse-lines in stands of trees caused by many deer feeding in the area.
- Fern fronds bitten off and laid to wilt on logs. Mountain beavers take this "hay" into their burrows to use for food and making nests.
- Long claw marks on tree trunks made by bears marking territory boundaries.
- Scat (animal droppings). Many droppings will contain seeds in summer and fall; contents of scat provide clues to diet.
- Bits of fur or hair on the ground or caught in a fence or among vegetation.
- Animal tracks in snow, in loose dirt, in soft ground, or near water (Figure 1). There are books that can help you learn to identify and interpret animal tracks.
- Teeth marks at the base of trees at the water's edge made by beavers.
- Sounds. Listen, for instance, for the characteristic "slap" of water made as a beaver dives below the surface; the sound is made by the beaver's tail striking the water and serves as a warning to others of potential danger.

Common Mammals Found in the Pacific Northwest
(from smallest to largest)

Mammals that are less than 12 inches long including tail

Bats

More than 12 species: see Chapter 13.

Shrews

(see Plate 1) 10 different species: 3–6 inches long

The smallest mammal is a shrew. Most have sharply pointed snouts and bead-like eyes. Shrews have very high metabolic rates, and they eat their own weight or more in food each day. Most live in moist leaf litter, grasses, logs, thickets, and stumps, and they are not as common in hot, dry areas. Like bats and swallows, shrews eat a lot of insects (also worms and larvae), and can be harmed by pesticides. They are active day or night and seldom live more than one year.

Voles

(see Plate 1) 10 different species: 5–9 inches long

Voles look like a cross between a mouse and a mole. Most voles have stout, chunky bodies, brownish-gray fur, relatively short tails, black, bead-like eyes, and blunt noses. They live in a variety of habitats, primarily meadows and marshes. Voles eat grasses, bulbs, roots, tree bark, and various seeds and berries. They nest in the ground under a mat of grass or decaying wood and are eaten by owls and hawks, coyotes and foxes. Voles are active day or night; some

species are good swimmers, other species seldom come above ground.

Mice

More than 12 species: 6–9 inches long

The common, introduced house mouse has a grayish-brown back and belly, beady eyes, and a naked tail. Although generally found indoors, house mice are sometimes seen in parks and gardens. They nest in buildings, breed year-round, and eat almost anything. The native deer mouse has fur ranging from pale gray to dark reddish-brown with a white belly; the tail is furry, white below, dark above. More of a "country" mouse, the deer mouse is found in forests and grasslands, also in rural homes and cabins throughout the Pacific Northwest. Deer mice eat seeds, nuts, berries, and insects, and they nest in burrows in the ground, cavities in trees, and buildings. Mice are eaten by dogs, cats, hawks, owls, snakes, weasels, raccoons, foxes, coyotes, and skunks.

Pocket gophers

Six species: 6–9 inches long

Pocket gophers have grayish to buff fur, large, yellowish incisor teeth, large hand-like front claws, and small eyes and ears. Pocket gophers get their name from their fur-lined cheek pouches or pockets. Their soft fur rubs in both directions to allow them to move through a tunnel forward or backward. Pocket gophers are found in a variety of habitats and dig underground tunnels to eat roots, tubers, and bulbs. They are seldom seen above ground and are always solitary; each burrowing system has only one gopher. They make extensive tunnels and continue burrowing in soil or snow in the winter. Their presence is indicated in summer by mounds of earth, which they push out, and in winter and early spring by ropes of earth lying on the ground. Predators include coyotes and foxes, which can dig them out of their burrows, and hawks, owls, and snakes.

Moles

Four species: 6–9 inches long

Moles have chunky bodies with pointed snouts, short legs, and naked to sparsely haired short tails. **Townsend's moles** (the largest moles in the United States) are nearly 9 inches long. They are well-adapted to underground life, with very small eyes and big feet for digging. Moles have an excellent sense of smell and touch, with a sensory organ located on their naked or nearly naked snouts. They eat earthworms, insects, larvae, and slugs. As moles burrow underground to find food, shelter, and a nesting place, they create dirt mounds and tunnels. They push dirt up from the center of the mound to form a dirt volcano, unlike gophers, which form piles on one side of their burrow entrance. Although moles don't eat plants, plant damage may be caused when moles push plants around while burrowing or when other creatures, which do eat plants, use their tunnels. Predators include coyotes and foxes, which can dig moles out of their burrows, and some hawks, owls, and snakes.

Chipmunks

(see Plate 1) Six species: 6–12 inches long

Chipmunks have a series of alternating light and dark stripes over their backs and upper sides running from the shoulders to the rump. The sides of their heads are striped and their bellies are white or buff-colored. They are found in coniferous woods throughout the Pacific Northwest and are often seen scurrying along the ground to hide behind a tree or rock pile. They seldom climb trees. Chipmunks mostly eat seeds, but may also eat fruits, leaves, stems, corn, some fungi, and insects. In fall, they begin storing food in a den, usually covering the cache with grass. They can nest under stumps, logs, or brush or rock piles, and their predators include cats, dogs, foxes, weasels, snakes, hawks, and owls.

Flying squirrels

(see Plate 1) 9–11 inches long

Flying squirrels have glossy olive-brown fur above, a white belly, and a folded layer of loose skin along each side of the body, from front to hind leg. They have very large, dark eyes. They live in forests throughout the Pacific Northwest, are nocturnal, and are only seen during daylight hours if frightened from their nests. They eat seeds, nuts, fungi, berries, fruits, and blossoms, and they will visit bird feeders at night. They also eat insects, and occasionally small birds and birds' eggs. Flying squirrels don't actually fly, but glide downward through the air using their skin flaps. For every two feet that the animal glides forward, it drops about a foot. Small groups of flying squirrels sleep together in dens found in holes in mature trees, usually lined with leaves and shredded bark. They will also use nest boxes (See Appendix D). They are eaten by owls.

Douglas squirrel (Chickaree)

(see Plate 1) 10–12 inches long

Upper parts of the Douglas squirrel

Encounters with Bears and Cougars

Bears

Unlike their rare cousin the grizzly bear, black bears are abundant in the Pacific Northwest. Black bears will usually avoid people, but their size, strength, and surpassing speed make them a potential danger. Most confrontations with bears are a result of surprise encounters at close range, so take precautions to avoid startling a bear. Remember, because of their keen sense of smell, black bears are attracted to food and odors.

If you live in or near wooded black bear habitat, you can reduce the potential for a black bear conflict by taking these precautions:

- Don't leave food out that bears can get into. Keep pet foods and livestock feed (including chicken feed) indoors.
- Use garbage cans with tight-fitting lids, and store cans in your garage or a shed until pick-up day.
- Wash barbeque grills immediately after use, and keep any fish parts and meat waste in your freezer until they can be disposed of properly.
- Enclose any beehives and fruit trees in chain-link or electric fencing, where practical, to prevent depredation.

Black bears tend to avoid humans, but should you come in close contact with one, here are some tips:

- Stay calm and avoid direct eye contact, which could elicit a charge. Because bears are nearsighted, if one has not caught your scent, it could

continued on next page

mistake you for prey. Try to stay upwind and identify yourself as a human by standing up, waving your hands above your head, and talking. The bear will probably leave you alone.

- Do not approach the bear, especially if cubs are around. Give the bear plenty of room and slowly back away. Leave the bear an escape route at all times. If you are too close, a black bear may "bluff" charge, although it is highly unlikely the bear would touch you. Because black bears can reach speeds of 30 mph or more, running away is not a wise decision. Running may also stimulate the bear's instinct to chase.
- If you cannot safely move away from the bear and the bear doesn't flee, then try to scare it away by aggressively clapping your hands or yelling. Black bears can climb trees, so fleeing up a tree is not a safe option.
- In the unlikely event a black bear attacks you (meaning he thinks you are prey), fight back aggressively using your bare hands or any object you can reach. As a last resort, should the attack continue, protect yourself by curling into a ball or lying flat on the ground on your stomach and playing dead. Do not look up or move until you are certain the bear is gone.

are dark, dusky-brown or olive; the belly is yellowish or rusty, separated from the upper sides by a black line. The tail is blackish-brown, edged with white or yellow. Douglas squirrels occur mostly in wooded areas west of the Cascade mountains. Noisy, shy little squirrels of evergreen forests, they nest in tree cavities or on tree branches. Douglas squirrels eat mainly fir seeds, and store cones in the fall under fir trees. Look for piles of cone scale on logs and stumps. The loud, sharp *peeoo, peeoo, peeoo* call is distinctive. They are eaten by hawks, owls, and house cats.

Mammals that are 12–24 inches long including tail

Red squirrel

(see Plate 1) About 12 inches long

Red squirrels have dark-reddish or brownish upper parts, with the upper sides separated from the white lower sides by a black line. They have grayish-white bellies and dark tails. They are found east of the Cascade mountains and inhabit evergreen forests and semi-open woods. They eat a variety of foods, including seeds, nuts, and mushrooms, and usually have a favorite feeding stump where remains of conifer cones accumulate in piles. Red squirrels nest in tree cavities or in the branches of trees. They are active throughout the year and will even tunnel in snow. Their predators include large hawks and owls.

Columbian ground squirrel

(see Plate 1) About 12 inches long

Columbian ground squirrels have large, bushy tails with mottled upper parts, and dark-reddish feet. They live mostly in burrows in meadows, open woods, and cultivated fields east of the Cascade mountains and are seen mostly in spring and early summer months during daylight hours. They go dormant from July–August to February–March. Columbian ground squirrels eat a wide variety of foods, but mostly green vegetation. They offer startling chirps for conversation and hurried chatter when alarmed. They are eaten by snakes, hawks (particularly red-tailed hawks), weasels, badgers, and coyotes.

Eastern gray squirrel

(see Plate 1) 12–18 inches long

Eastern gray squirrels' upper parts and sides are grayish (more reddish in summer), and under parts are whitish. Their tails are long and bushy with long, white-tipped hairs. Some Eastern gray squirrels are totally black. Eastern gray squirrels were introduced from the eastern United States into public parks in the Pacific Northwest in the early 1900s. They are aggressive and are now spreading beyond urban areas, crowding out native squirrels. They eat fruits, bulbs, seeds, refuse, and nuts stored in the ground. They seldom venture far from trees, and would rather climb than run. They are notorious for raiding bird feeders and scaring off small birds. They are adept beggars, ingenious at obtaining food, and they adapt to humans easily. They nest in tree holes or build leaf nests in branches. In urban areas they have few predators.

Mountain beaver

12–18 inches long

Mountain beavers look like tailless muskrats. They have dark brown fur, very short tails, and beady eyes. They're not really beavers, but were so named by trappers to enhance the marketability of their fur. Mountain beavers are the oldest known living rodents. They occur mostly west of the Cascade mountains in moist lowland forests, sometimes on densely tangled slopes and damp ravines in urban areas. They have very poor eyesight and seldom enter cleared ground, where they could be an easy meal. Mountain beavers are strictly vegetarian; they prefer shoots of native plants including sword and bracken ferns, nettles, salal, huckleberry, and vine maple. They also eat rhododendrons and other ornamental shrubs. They den in burrows with tunnels and chambers, and their musky odor may be noticeable. Mountain beavers may whine when annoyed and are usually nocturnal.

Rabbits and Hares

(see Plate 1) Eight species: most commonly 12–18 inches long

Rabbits are born naked, blind, and helpless in a nest lined with fur. Hares are born with fur and with eyes open, in crude nests, and are able to run within a few minutes after birth. Rabbits and hares are active day and night, never far from dense cover. They are eaten by coyotes, owls, and large hawks.

The common introduced **Eastern cottontail rabbits** are most likely to be found near people throughout the Pacific Northwest. They have brownish-gray bodies, short, white cottony tails, long ears, and long hind legs. They live in dense shrubbery, brush piles, burrows in the ground, and debris. They eat green vegetation when available, and in winter they eat bark and twigs.

Some people have released pet **European rabbits**. These have black, white, or brown fur, or a combination.

Snowshoe hares are found in timbered areas, swamps, and thickets—not normally around people and not normally in arid areas. **Jackrabbits** (really hares) live in arid sagebrush areas, and are not often seen close to humans.

Yellow-bellied marmots

18–24 inches long

Yellow-bellied marmots have dark-grayish upper parts, strongly marked black and white faces, and yellow bellies. The sides of their necks have conspicuous buffy patches. They are found throughout the Cascade mountains and east of them on talus slopes, rock fills along highways, and other rocky places. They are also found in old log piles, under buildings, and in burrows in cut banks. They eat green vegetation including dandelions, alfalfa, clover, and garden vegetables; also bark, seeds, and fleshy roots. Marmots will generally gorge themselves before winter sleep. They whistle or chirp an alarm when danger is sighted, and are one of the prominent guardians of rocky places. Young marmots are preyed on by hawks.

Skunks

18–24 inches long

Striped skunks are about the size of house cats, with two white stripes down their backs. **Spotted skunks** are smaller, with broken

Cougars

As cougar numbers increase and habitat dwindles in the Pacific Northwest, the more likely you are to encounter a lion. Young, newly independent cougars of one or two years of age, presumably having difficulty finding food for themselves, account for the majority of the cougar/human interactions reported in the Pacific Northwest.

If you live in cougar country:

- Keep pets indoors or in secure kennels at night for safety.
- If practical, bring farm animals into enclosed sheds or barns at night, especially during calving or lambing seasons.
- Do not leave pet food or food scraps outside.
- Store garbage in cans with tight-fitting lids so odors do not attract small mammals.
- When children are playing outdoors, closely supervise them and be sure they are indoors by dusk.
- Provide light along walkways in the landscape around your house.

Relatively few people will catch a glimpse of a cougar, much less confront one. If you do come face to face with one, your actions can either help or hinder a quick retreat by the animal. Here are some tips:

- Stop, stand tall and don't run. Pick up small children immediately. Running and rapid movements may trigger an attack. Remember, a cougar's instinct is to chase.
- Face the cougar, talk to it firmly, and slowly back away. Always leave the animal an escape route.
- Try to appear larger than the cougar by getting above it. (e.g., stepping up onto a stump). If

continued on next page

wearing a jacket, hold it open to further increase your size.
- Do not take your eyes off the animal or turn your back. Do not crouch down or try to hide.
- Never approach the animal, especially if it is near a kill or with kittens. Never corner the animal or offer it food.
- If the animal does not flee and shows signs of aggression (crouches with ears back, teeth bared, hissing, tail twitching, and hind feet pumping in preparation to jump), be more assertive. Shout, wave your arms and throw rocks. The idea is to convince the cougar that you are not prey, but a potential danger.
- If the cougar attacks, fight back aggressively and try to stay on your feet. Cougars have been driven away by people who have fought back using anything within reach, including sticks, rocks, shovels, backpacks, and clothing—even your bare hands. Generally, if you are aggressive enough, a cougar will flee, realizing it has made a mistake.

white spots on their backs and sides. Both inhabit open woods, prairies, farm land, brushy streamsides, and marshes in all but the hottest areas and the Cascade mountains. They eat mice, amphibians, eggs, insects, carrion, acorns, berries, and garbage. Their dens are in the ground, and they burrow under old buildings, woodpiles, culverts, and porches. Skunks are generally nocturnal. They are docile, but when alarmed can fire a pungent musk 10 feet or more. Their glands can shoot five to eight times, totaling roughly two tablespoons of musk. Coyotes, red foxes, great horned owls, bobcats, and other meat-eaters will eat skunks, but pass them by when other food is available.

Mammals that are 2–3 feet long including tail

Porcupine

2–2½ feet long

Porcupines have blackish or yellowish coarse fur, heavy bodies, and short legs. They're slow-moving, with long quills interspersed through their hair, especially on their backs and tails. Porcupines are found in forested areas throughout the Pacific Northwest. They eat buds, leaves, and the inner bark of trees; also grasses, berries, alfalfa, and other crops. They are also fond of salt. Porcupines den in hollow trees or rock caves and do not hibernate. They are mostly nocturnal and spend the day under cover or high in trees, which they climb with ease. Awkward on the ground, they are easily hit by vehicles and harassed by dogs. Young porcupines are eaten by mountain lions, coyotes, and bobcats. Quills are not thrown, but they dislodge

easily from the porcupine's body, and they are barbed. Quills embedded in an animal's lips, tongue, or palate make eating impossible and may cause eventual starvation.

Opossum

2–3 feet long

Opossums have white or grayish-white fur, with pointed white faces, black ears and feet, and long, naked, rat-like tails. These are the only marsupials (mammals that carry their young in pouches) in the Pacific Northwest. They were introduced to the Pacific Northwest in the early 1900s and survive by constantly adapting to change in their habitat. Their range seems limited only by the severity of winters they can withstand. They are abundant, especially in urban areas. They're slow moving and often hit by cars. They can kill rats and snakes but prefer insects, fruit, carrion, nuts, grains, reptiles, birds, and birds' eggs, and they love to eat snails. They use their tails mainly for grasping. Their dens are made in hollow logs or trees, rock piles, brush piles, and spaces under buildings and decks; also in abandoned burrows of other animals. Opossums are usually nocturnal and sleep for long periods of time in winter but do not truly hibernate. When frightened, opossums frequently bare their teeth (they have 50 sharp teeth, more than any other mammal), or imitate death ("play possum") by falling on their sides, closing their eyes (not always), and sticking out their tongues. Small opossums are eaten by dogs, foxes, large hawks, and bobcats.

Raccoon

(see Plate 1) 2–3½ feet long

Raccoons are easily recognized by their black face masks and ringed tails. They are found throughout most of the Pacific Northwest and are most abundant in urban areas. Cubs are born in dens in hollow logs, culverts, or large snags; also under large rocks, decks, or buildings. By early winter the young are generally on their own. Raccoons are good swimmers and climbers. They're mainly nocturnal and don't hibernate but can become much less active in the winter. Raccoons eat a wide range of foods (crayfish, frogs, freshwater mussels) that they gather along the water or below the surface with considerable handling. They also eat berries, insects, birds' eggs, garbage, and pet food. Raccoons prefer to soak food before eating it. Dogs may kill adult raccoons, and bobcats, coyotes, and great horned owls may eat young cubs.

Raccoons can carry a parasitic roundworm that can cause serious health problems in children. This worm is excreted with raccoon feces and sticks to the raccoon's fur, from which it can be transferred to humans by handling. Like other wild animals, raccoons should not be kept as pets because of potential health and safety problems.

Mammals that are 3–4 feet long with tail

Red fox

3–3½ feet long

Red foxes have reddish-yellow fur, white bellies, and bushy tails with white tips. They are found in mixed woods near open fields, including urban areas. The introduced species common in some lowland areas escaped or was let loose from fur farms and has multiplied. The native red fox occurs high in the Cascade mountains. Red foxes eat insects, rabbits, rodents, birds, fruits, berries, and grasses. They will store the excess, especially if there is snow. Red foxes are most active at night, although they often will be active during the day. Their den is an abandoned burrow, culvert, or other secretive spot. Young foxes may be preyed upon by eagles and great horned owls, while adult foxes fall prey to bobcats, domestic dogs, and coyotes.

River otter

3–4 feet long

River otters have large, weasel-like bodies with brown fur above and a silvery sheen below. They have broad snouts with distinctive whiskers, webbed feet, long, thick, tapering tails, and an awkward gait. River otters live near fresh or salt water, particularly lake shores and larger streams throughout the Pacific Northwest. They're always seen in or near water. River otters eat fish, frogs, and aquatic invertebrates and den in banks with the entrance under water. River otters will bask on lakeshore docks and are often seen in groups. They're great wanderers and can travel 50–60 miles along rivers in a single year or 5–6 miles on land. They are very playful and like to slide on their bellies on mud- or snow-covered slopes into the water.

Coyote

(see Plate 1) 3–4 feet long

Coyotes have gray to reddish-brown fur, erect ears, long, slender noses, and tall, thin legs. Their bushy tails are held down as they

Sprayed by a Skunk?

A skunk can spray up to 10 feet with great accuracy from the two ducts located under its tail. However, it's a placid creature, and will spray only when it feels threatened. It provides advance warning by stamping its front feet, raising its tail, and turning its rear end toward the offender.

To remove skunk spray, flush the area with large quantities of clean water. The painful irritation that occurs when the spray gets into the eyes will soon pass. If a pet or other animal has been sprayed, wash it thoroughly in tomato juice or vinegar, followed by a thorough shampoo. This will make the smell tolerable. It will take several washings and time to bring total relief.

If a non-living object has been sprayed, use chlorine bleach or ammonia if possible. If human skin has been sprayed, mix a quart of 3 percent hydrogen peroxide, ¼ cup of baking soda (sodium bicarbonate), and 1 teaspoon liquid soap. Use this to wash the area.

run. They're found throughout the Pacific Northwest on the edges of forests, blackberry thickets, farm areas, and urban greenbelts. True scavengers, they eat rodents, rabbits, domestic cats, small domestic dogs, chickens, insects, fallen orchard fruit, and most berries. Several coyotes may cooperate in chasing down deer or other prey. Coyote dens are holes in banks, culverts, uprooted trees, or under buildings or brushy areas. They're secretive and mostly nocturnal, but are sometimes seen hunting during daylight. Their howl is frequently heard during the night, or in response to a siren during the day. Their exceptional sense of smell, vision, and hearing, coupled with their evasiveness, enables them to survive both in the wild and in suburban areas. In urban areas, coyotes are less likely to fear people and more likely to associate them with an easy, dependable food source.

Beaver

(see Plate 1) 3–4 feet long

Beavers have brown fur and paddle-like tails. Near extinction in the mid-1800s from unregulated trapping, beavers are now common throughout the Pacific Northwest, even around urban areas. Beavers eat twigs, branches, and bark of young deciduous trees and shrubs; also aquatic and flowering plants. In midsummer they add protein-rich algae to their diets. Faced with starvation, beavers will eat coniferous trees. When all the available food is gone, beavers move to another location.

Beavers build dams and canals to create deep water near their food source, for safety, and to facilitate the transport of tree limbs. Beavers will build dams out of whatever materials are at hand: wood, stones, mud, or crop residue. Sometimes they will create burrows in banks by streams, creeks, or lakes, also under stumps or docks. Lodge entrances are always underwater. Beavers are social and are often seen in family groups. They are generally nocturnal and don't hibernate.

Beavers create new wetland habitat for wildlife. Raccoons and herons hunt frogs and other prey along marshy edges of beaver ponds. Fish, reptiles, and amphibians overwinter in deep beaver ponds. Dead trees, with their populations of insects, support woodpeckers whose holes provide homes for many other species.

Mammals that are 5–6 feet long with tail

Black bear

5-6 feet long

Black bears are the smallest of the North American bears. They can be black, cinnamon-colored, or even lighter. Black bears are found throughout the Pacific Northwest in forested areas with streams and wetlands and in rural gardens. Mostly vegetarian, black bears eat berries, nuts, fruits, grasses, seeds, tubers, and honey. They may also eat insects, rodents, salmon, carrion, and occasionally deer fawns and elk calves in early summer. The black bear's den is under a tree or in a hollow log, snag, cave, or large brush pile. Mainly nocturnal, black bears are often seen during the day in spring and summer. They don't fully hibernate, but doze and wake to feed. Black bears have poor eyesight, but an excellent sense of smell and hearing.

Mammals that are 6 feet or longer including tail

Cougar (mountain lion)

6–8 feet long

Cougars have tawny to grayish backs with lighter undersides. Their fur is short, and their tails may be one-third of the total body length and tipped with dark brown or black. Cougars live throughout most of the Pacific Northwest and are found in mountain forests and in broken, semi-wooded canyon areas where deer are found. A cougar needs a territory of about 50 square miles, and will sometimes enter urban areas along greenbelts and large stream corridors. Cougars eat mostly deer, also elk, rabbits, beavers, raccoons, grouse, and occasionally livestock and domestic dogs and cats. Cougar dens are located in hidden, sheltered spots, in slash or log piles or caves. Cougars are very shy and solitary, and are most active at dusk and dawn. However, they will hunt day or night in all seasons. They're good climbers and, when pursued, often take to trees.

Mammals that are 3–5 feet tall to the shoulder

Deer

3–3½ feet tall

Deer are most active in mornings and evenings, and on moonlit nights. Adult deer are preyed upon by cougar, bear, bobcat, coyotes, and dogs. Young fawns are eaten by black bears, coyotes, and bobcats.

White-tailed deer have grayish hair in winter and reddish-brown hair in summer. Their tails are large, a distinctive flag when the deer runs. White-tailed deer live primarily in northeast Washington but are also found in the north-central part of the state and along the valley bottoms of southeast Washington.

Columbian white-tailed deer, an endangered species, are found in western Washington along the Columbia River near Cathlamet. Northwest whitetails are common in rural areas of eastern Washington, especially around edges of agricultural crops where cover is nearby. They eat leaves and twigs of shrubs, trees, grasses, and herbaceous plants.

Mule and black-tailed deer have grayish coats in winter and reddish-brown coats in summer. Their tails are small and black-tipped (mule deer) or black on top (black-tailed deer). Mule deer are larger than black-tails but where ranges overlap near the Cascade crest the two races interbreed. Mule deer are larger and have more massive antlers than do black-tailed deer. Mule deer live east of the Cascade crest while black-tails live on the west side. Both eat leaves and twigs of shrubs, trees, grasses, and herbaceous plants.

Elk

4–5 feet tall

Elk have reddish-brown hair, a pale yellow rump patch, and a small tail. The male elk (bull) has large spreading antlers. There are

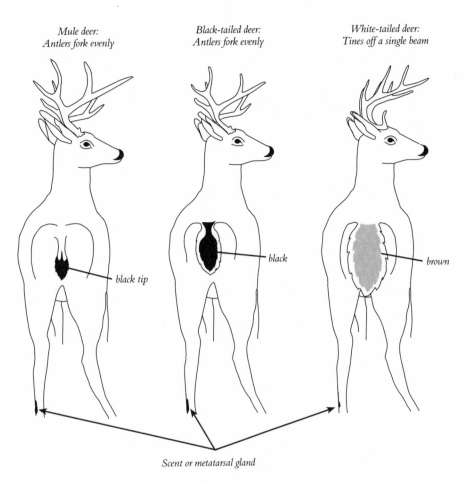

Figure 2. Characterictics used to identify deer.

two subspecies of elk in the Pacific Northwest: **Roosevelt elk** in the forests west of the Cascade mountains, and **Rocky Mountain elk** east of the Cascade crest. Genetically mixed populations of Roosevelt and Rocky Mountain elk are found east of Interstate 5 and west of the Cascade crest. Elk are social animals and usually form small herds of 6 to 100 animals. Migrating herds are led by an older cow elk. Elk eat mainly grasses, sedges, and herbs. The male has a high-pitched bugling call that starts with a high pitch whistle and ends with a few low-toned grunts; this is heard during the rutting season in August and September. Adult elk are preyed upon by cougar, and elk with young calves are especially vulnerable to black bear.

CHAPTER 6

BIRDS

Birds are the most visible wildlife in your yard. You can't help but notice their songs and calls, colorful plumage, and daytime activities, such as nesting, feeding, and flocking. Almost 300 species of birds are native to the Pacific Northwest. Many of them could call your yard home for at least part of the year, depending on what you provide for them.

The diversity of bird species and their visibility in your yard can be increased by strategically planting sources of food and cover, and by including simple enhancement features like birdbaths, feeders, nest boxes, and perching places.

A variety of birds means more than just greater viewing enjoyment. Many bird species help control insect populations around your property. Owls and hawks help control mice, gophers, and other small mammals on larger properties.

For information on hummingbirds, see Chapter 17.

Bird Behavior

Migration

Seed-eating birds such as finches and sparrows are able to spend winters in the Pacific Northwest because natural sources of seeds are available throughout the year. Insect-eating birds, including chickadees and nuthatches, add some seeds to their diet and glean enough insects from leaves, twigs, and crevices in the bark of trees to make it through the winter. Towhees and robins that overwinter scratch for bugs in leaf litter.

However, many insect-eating birds can't find enough food here to keep their metabolic furnaces stoked, so they fly south. Avoiding cold temperatures is actually a less important reason for leaving than not having an adequate food source.

The seasonal migration of birds from the Pacific Northwest to warm tropical zones is one of the wonders of the natural world.

Each autumn, almost half the bird species that breed in the Pacific Northwest migrate south to tropical Central and South America. These birds, called "neotropical migrants," spend six to nine months in these warm areas before returning north in spring to mate and rear young. Most vireos and warblers winter in western Mexico and northern Central America, as do western tanagers, black-headed grosbeaks, and orioles. Bank and cliff swallows go to the Amazon Basin. Some barn swallows head to Tierra del Fuego, but rarely make it all the way.

Nesting

As the days become longer in the spring, a bird's activity turns to nesting. Migratory birds return from their southern wintering grounds. Other birds, like varied thrushes and mountain bluebirds, may leave the lowlands for higher elevations. Some chickadees and jays leave suburban areas and re-turn to more rural areas to nest.

Male birds typically arrive on the nesting grounds a few days to a few weeks before the females arrive. They usually select the general location where they want to nest and then drive all rival males from the area. When the females begin arriving, the males try to entice them to remain and mate. Birds generally choose just one mate for a nesting season. Some male birds, such as grouse and pheasant, regularly mate with several females. Canada geese, eagles, and crows mate for life.

Nest-building can take a lot of time and energy. A pair of woodpeckers, for example, may work several hours a day for nearly a month to carve out a hole. Most birds will spend from one to two weeks making a nest, although some take only a day or so.

Many birds will desert a nearly-built nest if disturbed. Avoid pruning trees, shrubs, brambles, and other likely nesting spots in the spring and early summer when birds are nesting. If you must prune, mow, or weed-eat at this time, carefully examine the area for nests before you begin, or wait to see if birds are flying into the area carrying nesting materials.

Some birds prefer to nest in cavities of trees, either ones they've excavated themselves (primary cavity nesters, like woodpeckers) or already existing ones (secondary cavity nesters, like bluebirds). In areas where older trees with cavities are scarce, many cavity-nesting bird species will readily accept nest boxes.

Parent birds will aggressively fly at and dive-bomb any perceived threats to their nest or young. This includes cats and other domestic pets, as well as humans. Although these attacks can be intimidating,

Table 1. Some nest materials used by birds

Barn and cliff swallows	Light-colored feathers, mud
Bluebirds	Light-colored feathers
Bushtits	Moss, small twigs, spiderweb
Chickadees	String, yarn, moss, fur, horse and other hair, light-colored feathers, cotton, bulrush down
Crows	Moss, twigs, strips of cedar and birch bark
Goldfinches	Thistle and dandelion down, grass
House finches	Fine straw, string, yarn, small twigs, cotton, bulrush down
House wrens	Cotton, bulrush down
Hummingbirds	Lichen, cotton, lint, bulrush down, bark shreds, spider webs
Dark-eyed juncos	Fine straw, hair
Kinglets	Rootlets, light-colored feathers
Mallards	Reeds, grass, sticks, feathers
Orioles	String, yarn, hair
Pine siskins	Rootlets, small twigs
Robins	Coarse straw, string, yarn, mud, fur
Steller's jays	Sticks, mud, grass, pine needles

few birds (crows being one exception) ever actually strike the target of their attack. Try to stay away from nesting areas until the young are raised (three weeks or so) and ask others to do the same. If you must walk there, wave your arms slowly overhead to keep the birds at a distance. If you're frightened, wear a hat.

Roosting

Birds sometimes seem to suddenly disappear, especially at dusk. You know they haven't gone far because by morning they're back to their usual activity in all their accustomed places.

They've gone to "roost" or "rest" in secure places, usually in trees, shrubs, or large brush piles. They may also use nest boxes or natural cavities. These roost sites are places where birds go at the end of the day, either in groups or alone, to rest, stay warm, and spend the night.

Perching

Hawks and owls commonly use tall dead trees and branches as places to rest and look for prey. The tree's height provides the birds with a wide visual range, easy takeoff, and greater attack speed when hunting. Where tall snags or dead branches don't exist or can't remain because of safety constraints, perch poles can be installed to provide "lookout posts" for owls and hawks. You can add extensions to fence posts, or crossbars to sprinkler risers in grassy fields, orchards, or woodlots. Remember that a perch is only as useful as the surrounding habitat. Management of open habitat is essential if breeding and hunting grounds for hawks and owls are to be maintained or created.

Small birds need places to perch, also; these are often used as places to preen, or vantage points from which to survey a feeder or pond before flying to it. A perch

Supplementing Birds' Nest Materials

Supplementing the supply of nest materials is a small way to enhance your property for nesting birds. It's best to place nest material where birds can easily see it, such as near a feeder. Provide natural materials if you can, otherwise use short lengths (no longer than 3 inches) of string and yarn to avoid the risk of a bird becoming entangled. Nesting material should also be placed away from cats, perhaps in the middle of the lawn or high in shrubbery. An empty wire-mesh suet feeder hung from a branch is an excellent place to offer nesting materials; it's convenient for the birds, and the wind can't scatter the pieces.

You can provide mud for robins, phoebes, barn and cliff swallows, and other birds that use mud to build their nests. A tray (an inverted garbage can lid works well) of mud recessed into the ground is easily kept moist and can be placed near the nesting site in an area where it doesn't set birds up for ambush by house cats. The soil should be like a mud-pie. Clay-like soil with some humus is a good building material because the plant fibers provide strength to the nest. Have mud available from early April through mid-June.

Children often enjoy providing nest materials. The activity lets kids observe animal behavior that we usually don't notice and might allow them to discover where some secretive birds are nesting.

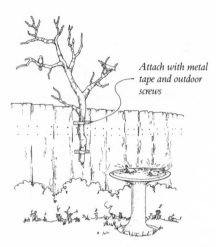

Attach with metal tape and outdoor screws

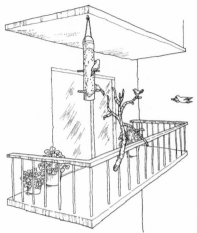

Girdling   *Girdling*

Figure 1. If there are no natural perching areas, a branch can be attached to a fence or deck railing. A perch can also be created by girdling a limb on a living tree or large shrub.

Roost Boxes

The need for protected roost sites, which is often overlooked, offers another opportunity to attract a diversity of birds. Although birds sometimes use nest boxes as roosts during the non-nesting season, you can build or make special roost boxes that are even better for this purpose (see Appendix D). A nest box is generally designed with an entrance at the top, space for nest construction below, and ventilation for hot spring and summer days. However, since excess heat is usually not a problem at night or during winter, a roost box needs no ventilation holes. For the same reason, a roost box is best built with the entrance at the bottom, helping to trap birds' body heat inside the box. Place the roost box in a sheltered site with the entrance facing away from the prevailing wind.

For species that roost alone, such as woodpeckers, smaller boxes are better because the bird can warm the interior quickly with its own

continued on next page

may be a dead branch in a tree or shrub, a wire fence, or a utility line. Most songbirds perch near some protective cover where they can easily fly to evade predators such as hawks.

Perches for Hawks and Owls

To replace missing habitat features such as snags and open trees, tall posts with crossbeams or large limbs placed along the edge of fields can approximate the location and function of natural perches. In addition, a suitable tree near open habitat (at the edge of a woodland) can be made into a natural-looking perch (see Figures 3 and 4 in Chapter 16). Try one perch pole every 200 feet around a field or per acre to provide birds with enough perches for optimum hunting.

If erecting poles, it's best to use a 4 by 4 inch or thicker timber 16–20 feet long, buried 3–4 feet in the ground. Treated wood set in concrete will stand firm. If concrete isn't appropriate, you can brace the pole with struts. The crossbeam at the top should be no less than 2 by 4 inch to ensure adequate hold for the birds. It can be supported by diagonal struts. Orient the crossbar so that birds can land on it against the prevailing wind. Where wind directions vary, use two crossbars.

It is best to situate a perch in

the area of a brush pile or other area which provides a home for small mammals and other prey. Perches that are being used by hawks and owls can be identified by the presence of pellets and "whitewashed" vegetation at the base of the post.

Perches for Songbirds

If there are no perching branches in your yard, you can build some. Popular sites would be near a birdbath or other water source, next to a feeder, or in a sunny edge of a wooded area.

A quick and simple way to make a perch is to attach a dead limb to a fence post or other structure using nails, wire, or a clamp (see Figure 1). A slower technique is to girdle the bark on a branch of a tree or large shrub. This will kill the branch but not the whole plant, and provide the leafless tangle that attracts songbirds.

Smaller, portable perches can be placed around your property and moved to accommodate farm machinery or a lawn mower. A 6-foot stake with a sharpened end can easily be tapped into the ground. Bluebirds and flycatchers will even use a 3-foot-tall perch placed by their tree cavity or nest box.

Managing Your Property for Birds

Although you need to consider several things when managing for birds around your property, the most important is to protect undisturbed wild areas. Examples include any size wetland, areas of tangled vegetation, unmowed grassy areas, and any place that contains snags.

Other important things you can do to manage your property for birds are:

- Provide natural sources of food and shelter by preserving or planting additional layers of vegetation with trees, shrubs, ground covers, and flowers. Pay particular attention to varying the fruiting season of your landscape plants. Try to achieve a roughly even mix of evergreen and deciduous plants.
- Leave some of your "forest floor" in open soil, or mulched with leaf litter, to provide for ground-foraging forest dwellers such as thrushes.
- If you are especially ambitious and have no real open spots in your landscape, consider creating an opening by removing a few trees. If you do decide to remove a tree to create an opening, consider leaving 6–15 feet of trunk as a wildlife tree. (See Chapter 16 for information on how to safely do this.)
- Provide access to a source of clean water in the form of a birdbath or a pond with a gently sloped edge that creates a beach. Include a dust bath in a place where it won't be ambushed by house cats.
- Carefully manage a variety of feeders, nest boxes, and roosting boxes for different bird species.

- Construct a brush pile for shelter from winds and predators.
- Avoid using pesticides, especially insecticides. About 70 percent of the breeding birds in the Pacific Northwest eat mostly insects. Most young birds are fed insects and spiders, snails, and worms. Even seed-eaters such as finches get this diet as babies because it contains more protein, calcium, other nutrients, and fluids needed by growing bodies.
- Leave some of your vegetable garden for birds during winter. Broccoli, carrot, squash, fennel, and parsley seeds will be eaten by goldfinches, chickadees, juncos, and house finches. Also, plant a portion of the garden with sunflowers and millet; these are popular seed sources in winter.
- Leave hedges unclipped, or prune them naturally by selective branch removal. Restrict your pruning to winter, if possible, after any fruit that shrubs produce has been eaten, and when birds are not nesting.
- Leave potential perches for hawks, owls, and songbirds. Add perch poles where natural perches don't exist.
- Provide supplemental nesting material, such as grass and small shrub cuttings, in early spring.
- Keep your cats indoors and discourage other cats from visiting your property. Keep dogs and other pets confined, too, especially during nesting season.

body heat. Since woodpeckers sleep clinging to a wall of their roost site, roughen the surface of the wood on the inside of the roost box.

Without loose bark or other similar cavities in which to roost, a brown creeper may nestle into a depression in rough bark near the base of a tree, sleeping upright with bill tucked into back feathers and tail serving as a prop. You can create an artificial roost site for this bird by fastening a large piece of bark, canvas, tar paper, or shingle to a tree trunk in tent-like fashion with the opening facing downward. The roost should be in a protected site no more than 6–8 feet up on the trunk of a large, rough-barked tree. The roost "tent" should be at least 6 inches deep, and should hug the tree trunk so there is a crevice of only about 2 inches for the birds to slip into. Bats also may use this roost area if their needs are met (see Figure 2 in Chapter 13).

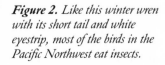

N. Yarbrough

Figure 2. *Like this winter wren with its short tail and white eyestrip, most of the birds in the Pacific Northwest eat insects.*

Water Developments for Birds in Arid Areas

There are many areas in the drier regions of the Pacific Northwest that have all the necessary habitat features to support good coveys of quail, chukar, grouse, or other birds, but water is lacking during the hot summer months in many of these locations. A small amount of water is sufficient if it is reliable. Most birds utilize less than an ounce of water per day.

The availability of well-spaced, permanent summer water is particularly important to young birds. Adults can travel considerable distances for drinking water and to feed on succulent late-summer annual vegetation, but chicks must have open water within a reasonable distance of their nest. Supplementary water can ensure that more young will survive.

A water development is not just a simple matter of collecting water and making it available. The development needs to be well planned to be safe and to achieve its purpose. Water developments planned for wildlife use should do the following:

- Maintain or provide adequate cover around the watering area, either by saving the natural cover or by plantings and brush piles.
- Provide at least one escape route to and from the water. Take advantage of the natural terrain and vegetation where possible.
- Be covered to reduce evaporation if necessary.
- Be large enough to ensure that the basin will retain sufficient flow from the seep and leave open water available at all times.
- Be fenced from livestock. Fences can serve the purposes of preserving the water source

and protecting food and cover needed for small species of wildlife. Protective fences should be negotiable by wildlife except where trampling or wallowing by big game will damage the spring source.

- Provide safety from wildlife drowning by construction of gentle basin slopes or ramps in tanks and basins.
- Provide, where applicable, an information sign to inform the public as to the purpose of the development.

The simplest type of water development is the excavation of a small spring or seep. Here available moisture can be collected in a permanent drinking basin. A basin may be easily constructed of concrete (see Chapter 11) or a synthetic liner (see Chapter 10).

Even a seep of very small volume can be collected in a drinking basin in sufficient quantities to maintain a relatively large population of birds, since most species drink only once or twice per day. This gives the basin a chance to refill during the night and between the morning and the afternoon drinking period. Practically all seeps run considerably more at night than during the day.

Stock troughs (Figure 3) can be designed to be wildlife-friendly.

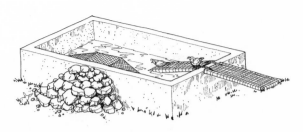

Figure 3. *A stock trough can serve as a water source for birds and other wildlife when entries and exits are well designed.*

Drinking ramps and escape ramps can be installed on them for adult upland game birds and for their young after they are able to fly or jump to the edge of the trough. Birds can walk down the ramp to the water level and drink without falling into the water. Also, if birds fall in the water, there is a good chance they will make their way out up the ramp.

The gallinaceous guzzler (Figure 4) was developed to provide water for all kinds of birds in very arid climates. It is a hole which has been dug in the ground and lined with concrete, a synthetic liner or fiberglass, or steel. A collecting cover, usually made of corrugated plastic, metal, or other impervious surface, is then placed over the top. This is sloped to catch any moisture and drain it into the reservoir. If properly constructed, it will store water for long periods of time. Contact your local Fish and Wildlife office for information on a design for your area.

Figure 4. *Gallinaceous guzzlers require minimal maintenance: occasional cleaning and inspection for leaks. A small ramp leading from the surface down into the hole should be included.*

Watching Birds in the Landscape

One of the greatest joys of providing feeders, nest boxes, birdbaths, and dust baths is watching the birds that are attracted to them. Viewing opportunities are best when birds are given the space they need to feel comfortable and secure. Winter is a good time for watching birds. Because the leaves are off many trees and shrubs, birds are easily seen. Birds also cover more area in search of food in winter, and some songbirds gather in mixed flocks that congregate around feeders and birdbaths. Winter is also a good time to look in trees and large shrubs for nests made during the previous nesting season.

In spring, during the courtship period, most birds are in their prime. Male birds try to make themselves as conspicuous as possible, both by songs and plumage displays. Males have various ways of announcing their presence to females: woodpeckers produce a loud noise by hammering their bills on a tree, the side of house, or other structure; the ruffed grouse perches on a drumming log and beats its wings in an accelerating drumroll; the male bushtit builds an intricately woven, sock-like nest which hangs from a branch. During winter you may find a bushtit nest that was built remarkably close to human activity.

Birds leave other clues as well. You might find a few feathers in a shallow hole where a bird took a dust bath. "Whitewash" (droppings) beneath trees indicates the presence of a roost, nest, or colony of birds. Owls can't digest the fur and bones of the small mammals they eat. These leftovers are regurgitated in pellet form. Look for them around the bases of large trees. Woodpeckers and sapsuckers drill holes in trees. Kingfishers and bank swallows excavate burrows in stream banks. Orioles weave large bag-shaped nests. Nighthawks cry at night, and owls hoot.

The color of a bird's plumage generally indicates the kind of plant cover it seeks for protection. For instance, the greenish yellow of many warblers is difficult to distinguish from the sun-dappled foliage of the trees they inhabit. The dusky browns of thrushes and sparrows blend well with the piles of leaves and grasses they frequent under shrubs and thickets close to the ground.

Females of nearly all bird species have plumage that blends with their environment. Many male birds, such as the yellow warbler and the American goldfinch, exchange their bright breeding colors for duller ones in winter.

A good way to learn more about local birds is to join a local chapter of the National Audubon Society. Field trips, sharing ideas and sightings with others, and access to local bird experts are some benefits.

A field guide is helpful for positive identification of a species, and many excellent ones are available. A good field guide should be lightweight, strongly bound to withstand frequent use, and have a water-resistant cover. Illustrations should be in full color and show distinguishing traits. Clear, concise descriptions are necessary for identifying a bird in its various forms, including male in breeding and non-breeding plumage, female, juvenile, and geographical variations. Some popular field guides are listed in Appendix E.

Learning the calls and behaviors of species will help you to identify them, learn more about them, and stay in tune with their activity. The best way to do this is simply to listen and observe birds at different times of the day.

Learning bird distress calls and signals could help you to prevent tragedies if the young begin to fly in an area frequented by house cats. Some distress signals used by birds include:

- A "broken-wing" display.
- Circling repeatedly.
- Crying out from a position overhead.
- Diving at intruders.

Some species may have their own distinctive distress signals: ducks pump their heads; young owls sway from side to side and"pop" their beaks; Canada geese hiss and charge.

Stay clear of nests, as parent birds may abandon a nest with eggs if they are repeatedly disturbed. When disturbed, birds also give visual clues that may lead predators to a nest; predators also may follow human scents to nests.

Common Birds Found in the Pacific Northwest

You may want to purchase one of the many excellent field guides to birds, not only to identify the rare visitor but also to more positively identify the ones you see every day. The photographs and short descriptions in this book will start you on the way, but are not intended to substitute for a field guide. In many cases the brief descriptions are of adult male birds in breeding plumage. Females, juveniles, and males in non-breeding plumage look quite different, and some species are differentiated from closely related ones or by traits too subtle to treat here in depth.

Introduced Birds

Common birds that have been introduced to the Pacific Northwest are the European starling, the house (English) sparrow, and the rock dove or pigeon. All have a long history of association with and adaptation to human settlements. Starlings and house sparrows were released in New York in the 19th century and have since spread throughout most of North America. Pigeons were brought to North America from Europe in 1606. Their native habitat is coastal cliff areas, but they now use the exposed beams, ledges, and artificial cavities of buildings for nesting.

The food sources for these birds are associated with human activity and include bread and other handouts, seed in bird feeders, and discarded food. These birds all use paved surfaces for feeding, and starlings forage for grubs in lawn areas. Easily procured food allows sick and injured birds to survive much longer in urban areas than they would in natural conditions.

Starlings and house sparrows take food intended for native birds and compete with them for nest sites. These species don't migrate and thus are able to occupy the preferred habitat of migratory birds before they return. House sparrows and starlings have driven birds like the bluebird and purple martin from areas where they were once common. (For information on how to deal with problems associated with these species, see Chapters 12 and 14.)

Key

Eastside = Areas east of the Cascade crest in Oregon and Washington and east of the Coastal mountains in southern British Columbia.

Westside = Areas west of the Cascade crest in Oregon and Washington and west of the Coastal mountains in southern British Columbia.

Resident = Seen year-round in the area.

Migrant = Migrates to the area from somewhere outside the Pacific Northwest or between lower and higher elevations within the Pacific Northwest.

Neotropical migrant = Spends winter in tropical Central or South America

N. Yarbrough

Figure 5. *The European starling spends a lot of time in lawn areas where it forages for grubs.*

Water and Shore Birds

Hooded merganser

Eastside and westside resident

Hooded mergansers are small (18 inches long), handsome ducks that can be seen diving for fish along wooded streams and wooded shorelines of lakes and wetlands. The male has a black head with a large white hood and brown sides. The female has an orange or rust-colored crest. They nest in tree cavities and may use a nest box if natural cavities are not available. Other cavity-nesting ducks include the **wood duck, common merganser, common goldeneye**, and the **Barrow's goldeneye**.

Wood duck

(see Plate 2) Eastside summer resident, westside resident

Male wood ducks are unmistakable with their multicolored sides, sleek green crest, and red bill. Wood ducks inhabit wooded areas near large ponds, lakes, sloughs, or other still water. Food and cover are provided by waterside trees and shrubs that overhang the water. They nest in tree cavities, roost on large branches, and will use a nest box where natural cavities are not available.

Mallard

Eastside and westside resident

Male mallards are easily spotted because of their shiny green head and white neck ring. Mallards are common in both salt and fresh water and are often seen tipping their rear ends straight up while looking for food on the bottom of a pond, lake, or slough. They nest on the ground or sometimes in a tree or elevated, open-topped nest box near water. They eat seeds (including acorns and grain), grass, and insects.

Canada goose

Eastside and westside resident

These black and white geese are commonly found in open, grassy areas such as city parks. They usually nest on the ground near water but will use any available elevated structure that is safe from predators. They mate for life but will remate if a partner dies. The goslings remain with their parents until the next breeding season.

Great blue heron

(see Plate 2) Eastside and westside resident

Great blue herons are large (40 inches tall), prehistoric-looking, gray-colored birds that fly slowly with their feet trailing behind and neck tucked in. Their call is a deep, rough *croak*. They are most often seen at the edge of open-water areas such as lakes, rivers, mudflats, and salt marshes, but will visit almost any pond or other small pool that contains fish. In addition to fish and other aquatic life, they eat small rodents found in fields.

Osprey
Eastside and westside neotropical migrant

These large (24 inches tall) fish-eating birds may be seen diving in rivers, lakes, reservoirs, and fresh- and salt-water marshes. Identified in flight by white underparts and long, downward-bent, narrow wings, they are commonly confused with bald eagles. They nest along rivers and lakes and migrate south to Mexico each fall.

Doves, Pigeon, Quail, Pheasant, and Grouse
Mourning dove
(see Plate 2) Eastside and westside resident

Mourning doves are a bit smaller than city pigeons (rock doves) and have a long pointed tail, large dark eye, and a dark bill. They are seen in farmlands, backyards with scattered trees and shrubs, grass-lands, and clearcuts. They often appear in urban and suburban gardens, particularly those close to open fields. Mourning doves are especially attracted to water in dry areas, frequently visiting streams, pools, and other drinking and bathing spots. Their call is a mournful *who-ooh, who-who-who*. They will visit certain types of feeders.

Band-tailed pigeon
(see Plate 2) Westside migrant

These native pigeons have a purplish head and breast, a dark-tipped yellow bill, and a small white crescent on top of the neck. They have a more rapid flight than do rock doves; the call is an owl-like *whoo-whoooo*, repeated several times. They are found in urban areas with tall trees, especially oaks, and will visit a seed feeder.

Rock dove (Pigeon)
Eastside and westside resident

This species of pigeon was brought over from Europe in the 1600s by early settlers. Breeders have created many different colors; most are gray with iridescent feathers around the neck. They have a strutting walk and their call is a low *coo*. The original preferred habitats of rock doves were high cliffs, ledges, and caves near the sea. They have adapted to the "urban cliffs" of tall city buildings and bridges, and are common in city and suburban parks and gardens, as well as farmlands.

California quail
(see Plate 2) Eastside and westside resident

Valley quail have a distinctive black head plume that grows straight up on the female and tips forward on the male. The call is a loud *look right here* or *Chicago Chicago*. They inhabit wooded urban areas where there is brushy growth, orchards, areas along streamsides, and rangelands where there is some cover. They are nearly always located close to a source of water such as a stream, pool, or irrigation ditch. In the landscape, a combination of dense shrubby cover and mixed herbaceous borders is especially inviting, and a source of drinking water on

the ground will be popular. Valley quail don't fly much, especially in winter when they need to conserve energy; at night they roost in low tree branches 15–25 feet off the ground. They will use certain feeders and also constructed brush shelters. (See Chapter 15.)

Ring-necked pheasant
Eastside and westside resident

The ring-necked pheasant was brought to the Pacific Northwest from China in 1881. The male stands about 24 inches tall, has a very long tail, reddish-brown body, and a white neck ring. The female is brown with a short tail. Pheasants are usually seen walking on the ground near a thick cover of woody plants and thorny shrubs. The call is a loud double *squak*, heard often in spring. They will visit a ground feeder.

Ruffed grouse
Eastside and westside resident

Ruffed grouse are named for the black feathers ("ruffs") on the side of the male's neck. They are common in mixed forests, thickets, and brushy streambanks. In spring the male perches on drumming logs and beats its wings in an accelerating drumroll to attract hens. The nest is a bowl scraped at the foot of a tree, or stump lined with leaves and some feathers. They will visit a ground feeder.

Eagles and Hawks
Bald eagle
Eastside and westside resident

Adult bald eagles are very large (36 inches length) with a dark brown body. After they reach the age of five, the feathers on their heads and tails turn white; the white head earned them the name "bald eagle." They are more common on the westside, but are found on the eastside near large bodies of water. Bald eagles perch near rivers, salt water, or lakes where they find fish and water birds to eat. In winter they also eat dead deer, elk, and livestock, as well as waterfowl and spawning salmon. Large open-topped conifers or cottonwoods are used for nesting, which usually occurs on shorelines but may be 5–7 miles inland.

Red-tailed hawk
(see Plate 2) Eastside and westside resident

Red-tailed hawks are the most commonly seen hawk in urban areas on the westside. The adults have a tail that is rusty-red on top, a paler red if viewed from underneath. They can be seen soaring in wide circles over fields and other open spaces, or perched on light poles or fences in search of small rodents. They also eat insects, birds, frogs, snakes, and lizards. Other large hawks that may be seen include the **Swainson's hawk** and the **Northern harrier** (marsh hawk). The **rough-legged hawk** is common on the eastside during the winter months.

Sharp-shinned hawk
(see Plate 2) Eastside and westside resident

Sharp-shinned hawks are jay-sized with a short neck and a squared-off tail. They're seen in urban areas, especially during winter months, where they zip through underbrush and tangles in search of small birds, which they often catch in flight. They may be attracted to feeders that have a lot of feeding activity. **Coopers's hawks** are eastside and westside residents that are slightly larger than sharp-shinned hawks but hunt in a similar fashion. **American kestrels**

(formerly sparrow hawks) are also eastside and westside residents. They prefer open country such as farmlands and grasslands dotted with a few trees or adjacent to a woodland. Kestrels will use nest boxes.

Owls and Nighthawks
Barn owl
(see Plate 2) Eastside and westside resident

Barn owls are slender owls about 16 inches high. They have a distinct white, heart-shaped face and white underparts. The voice is a hissing or raspy scream. They inhabit agricultural land, other open areas, and urban grasslands with tall thickets and trees nearby. They eat mostly rodents and may use nest boxes. The **barred owl** is an eastside and westside resident and is similar in size. Barred owls are often mistaken for spotted owls.

Great horned owl
(see Plate 2) Eastside and westside resident

Great horned owls are large owls (24 inches tall) that are mostly brown with a white collar; tufts on each side of the head look like ears or "horns." The voice is a low, hooting *hoo-hu-hu-hu-hoo*. They inhabit undeveloped forested areas, urban areas with mature trees, and open country including sage and grassland areas. They eat gophers, mice, rabbits, and rats; they are so large and powerful they can also eat cats, skunks, porcupines, and other owls. They will use a nest box.

Saw-whet owl
Eastside and westside resident

Saw-whet owls are only about 7 inches high. They are difficult to spot but easily approached. During the breeding season they can sometimes be seen near nest

holes during the day. Their call is a series of *took* notes, unevenly spaced and repeated. They may use nest boxes if natural cavities are not available. **Screech owls** are eastside and westside residents and are occasional visitors to rural and suburban yards and gardens that adjoin open areas where they can hunt. They drink and bathe frequently and appreciate a source of water.

Common nighthawk
Eastside neotropical migrants

Nighthawks are about 10 inches long with long, pointed wings that contain white patches. They are usually seen at dusk swooping for gnats, midges, flying ants, and other insects. The male produces a buzzing sound as air passes over its wings when it dives. They usually nest in open, gravelly areas but may nest on flat rooftops of downtown buildings. They winter in South America.

Swallows
Violet-green swallow
(see Plate 2) Eastside and westside neotropical migrant

Violet-green swallows have a purple-green back, white belly, and a white patch over the rump. They fly with rapid wing beats alternating with brief glides. They eat insects that they catch in flight. These swallows are common in urban areas during summer and spend the winter months in Central America. They readily take to nest boxes.

Barn swallow
(see Plate 2) Eastside and westside neotropical migrant

Barn swallows have a dark blue back, rust-colored underparts, and two long tail streamers. They are agile fliers, often catching insects

Mammals
(Plate 1)

Shrew — Jim Pruske

Vole — Jim Pruske

Chipmunk — Tom Boyden

Flying Squirrel — Russell Link

Douglas Squirrel (Chickaree) — Jim Pruske

Red Squirrel — Tom Boyden

Columbian Ground Squirrel — Tom Boyden

Eastern Gray Squirrel — Jim Pruske

Eastern Cottontail Rabbit — Jim Pruske

Raccoon — Tom Boyden

Coyote — Tom Boyden

Beaver — Joy Spurr

Birds
(Plate 2)

♂ ♀ ♀ Wood Duck

Jim Pruske

Great Blue Heron

Jim Pruske

Mourning Dove

Jim Pruske

Band-tailed Pigeon

Jim Pruske

♂ California Quail

Jim Pruske

Red-tailed Hawk

Jim Pruske

Sharp-shinned Hawk

Dennis Paulson

Barn Owl

Idie Ulsh

Great Horned Owl

Jim Pruske

Violet-green Swallow

Jim Pruske

Barn Swallow

Jim Pruske

♂ Red-winged Blackbird

Jim Pruske

Birds
(Plate 3)

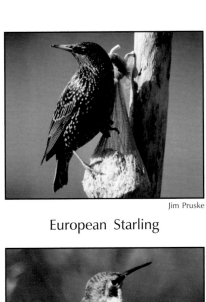

Jim Pruske

European Starling

Jim Pruske

Steller's Jay

Jim Pruske

Cedar Waxwing

David Hutchinson

♂ Rufous Hummingbird

David Hutchinson

♀ Anna's Hummingbird

Dennis Paulson

♂ Calliope Hummingbird

Jim Pruske

♂ Northern Flicker

Idie Ulsh

♂ Downy Woodpecker

Jim Pruske

Red-breasted Nuthatch

Jim Pruske

Bewick's Wren

Jim Pruske

Black-capped Chickadee

Dennis Paulson

Western Meadowlark

Birds

(Plate 4)

Jim Pruske
♂ Varied Thrush

Jim Pruske
♂ Western Bluebird

Dennis Paulson
♂ Western Tanager

Jim Pruske
♂ ♀ Evening Grosbeak

Jim Pruske
♂ Black-headed Grosbeak

Jim Pruske
White-crowned Sparrow

Jim Pruske
♂ House Sparrow

Jim Pruske
Pine Siskin

Jim Pruske
♂ House Finch

Jim Pruske
♂ American Goldfinch

Jim Pruske
♂ Dark-eyed Junco

Jim Pruske
♂ Spotted Towhee

Reptiles and Amphibians

(Plate 5)

Dennis Paulson
Common Garter Snake

Jim Pruske
Gopher (Bull) Snake

Jim Pruske
Northern Alligator Lizard

Kelly McAllister
Western Fence Lizard

Kelly McAllister
Painted Turtle

Jim Pruske
Red-eared Slider

Jim Pruske
Long-toed Salamander

Kelly McAllister
Rough-skinned Newt

Jim Pruske
Pacific Treefrog

Jim Pruske
Bullfrog

Jim Pruske
Red-legged Frog

Dennis Paulson
Western Toad

Butterflies
(Plate 6)

Anise Swallowtail — Tom Boyden

Western Tiger Swallowtail — Tom Boyden

Pale Swallowtail — Tom Boyden

Clodius Parnassian — Tom Boyden

♂ Pine White — Idie Ulsh

Orange Sulphur — Dave Nunnalee

♂ Cabbage White — Idie Ulsh

♂ Sara Orangetip — Dave Nunnalee

Brown Elfin — Dave Nunnalee

♀ Purplish Copper — Idie Ulsh

Spring Azure — Dave Nunnalee

Silvery Blue — Dave Nunnalee

Butterflies and Moths
(Plate 7)

Lorquin's Admiral

Tom Boyden

Red Admiral

Idie Ulsh

Painted Lady

Idie Ulsh

Mourning Cloak

Tom Boyden

Milbert's Tortoiseshell

Idie Ulsh

Mylitta Crescent

Idie Ulsh

Satyr Comma

Idie Ulsh

Common Ringlet

Dave Nunnalee

Woodland Skipper

Idie Ulsh

Clear-winged sphinx moth
(a bumblebee mimic)

Tom Boyden

Polyphemus moth

Idie Ulsh

Ceanothus silk moth

Tom Boyden

Miscellaneous Insects
(Plate 8)

Tom Boyden
Anise Swallowtail Larva

Tom Boyden
Western Tiger Swallowtail Larva

Jim Pruske
♀ Western Pondhawk

Jim Pruske
Eight-spotted Skimmer

Tom Boyden
♂ Cardinal Meadowhawk

Jim Pruske
♂ Paddle-tailed Darner

Idie Ulsh
♂ Northern Bluet

Dave Pehling
Orchard Mason Bee

Dave Nunnallee
Bumble Bee

Dave Pehling
Hover Fly

Dave Pehling
Green Lacewing

Dave Pehling
Ground Beetle

within inches of the ground. Barn swallows can be found in nearly any open country, often but not always close to water. Meadows, golf courses, parks, large lawns, pastures, and agricultural fields are favorite haunts. They build a mud nest on beams and ledges of buildings, bridges, culverts, and man-made structures. They are common in urban areas in the summer and migrate to South America for the winter,

Like the barn swallow, the **cliff swallow** needs open country over which to hunt on the wing, usually near bodies of water and the cliffs, dams, and large man-made structures upon which it nests.

Tree swallow
Eastside and westside neotropical migrant

Like violet-green swallows, tree swallows also have white underparts and metallic-like, greenish-blue upper parts, but they lack the white rump patch. They are often the first swallow to return to the north from Central America. They are less common than violet-green swallows in urban areas and usually found near water, including wet meadows, marshes, and the shores of lakes, ponds, and streams. They readily take to nest boxes. Another swallow, the **purple martin,** is a very rare westside neotropical migrant.

Blackbirds, Ravens, Crows, and Magpies
Red-winged blackbird
(see Plate 2) Eastside and westside resident

Male red-winged blackbirds are black except for a bright red-and-yellow shoulder patch. Their call is a loud *konk-a-ree*. They breed in wetlands of many kinds, especially those with thick stands of cattails. They also nest in nearby upland, brushy fields and pastures, and they forage in flocks in agricultural fields. During non-breeding season they form flocks, visiting all kinds of open land. They will use a variety of feeders. The male **Brewer's blackbird** has yellow eyes and a shimmering, iridescent dark head. These blackbirds walk while foraging on the ground and are common in agricultural as well as urban areas and parks.

European starling
(see Plate 3) Eastside and westside resident

Breeding males are all black with a large, sharp, yellow bill; in winter they are spotted, with a darker bill. The call is a rapid series of repeated whistles, squeaks, and bubbling noises, often including imitations of other bird species. Starlings prefer open country, including farmlands, orchards, and urban and suburban areas. They avoid woodlands, but are extremely adaptable to a wide variety of habitats. Thousands of individuals roost communally, especially in fall and winter, in buildings and trees. They were introduced to the United States. For more information, see the sidebar "Introduced Birds."

American/Northwest crow
Eastside and westside resident

American/Northwest crows are smaller than ravens and have rounded tails. They aggressively defend their nests, which are cone-shaped, about 2 feet wide and deep, and located toward the top center of a tree. Human alteration of the landscape has encouraged the proliferation of crows, as some of their favorite habitats include farmlands, roadsides, highways, suburban areas, thinned or logged-over forests, and orchards. Crows are very adaptable and live essentially wherever there are trees. The **common raven** is an eastside and westside resident. It is larger and heavier than a crow and has a wedge-shaped tail. Both will come to feeders and birdbaths.

Black-billed magpie
Eastside resident

Magpies have striking black-and-white plumage and a long tail. They are common around agricultural and open, shrubby areas and are often seen along highways. They nest mostly in wetlands with willow trees but also use trees around farms and other sites. Their abandoned bulky nests are occupied by other birds, including long-eared owls and Swainson's hawks, which expand the magpie nests to suit their needs.

Jays, Waxwings, Orioles, and Buntings
Steller's jay
(see Plate 3) Eastside and westside resident

Steller's jays have a tall black crest. Their calls include a loud *shack* and a *keeler* similar to that of a red-tailed hawk. They are mostly a bird of deep coniferous forests, including higher elevations, but they also inhabit mixed coniferous-deciduous forests. In fall and winter they frequently move into the lowlands. They will use a feeder and often drive other birds away. **Scrub jays** are local eastside and westside residents. They have no crest and are more common in drier areas, especially near oaks.

Cedar waxwing
(see Plate 3) Resident

Cedar waxwings are sparrow-sized birds with a black mask, brown upper parts, and a small crest. The males have some yellow under the tail. Their call is a high, thin *zee*. Waxwings rove quite a bit, and may appear anywhere there are abundant fleshy fruits on trees and shrubs. Their favorite habitats in the non-breeding season include orchards, parks, forest edges, and second-growth woodlands along streams and rivers. In the breeding season, pairs seek out nesting territories near farm ponds and open woodlands. They will use feeders. **Bohemian waxwings** are winter residents and are larger and more common on the eastside than are cedar waxwings.

Bullock's oriole
Eastside neotropical migrant

Bullock's orioles are conspicuous robin-sized birds. Males have bright orange eyebrows, cheeks, and underparts. They are noisy birds, most often found around tall shade trees with shrubby undergrowth. They can also be found in open woodlands. Bullock's orioles are adaptable and have learned to nest in street trees in suburban areas. The nest is a woven pouch. They winter in Mexico or Central America and will visit a fruit feeder.

Lazuli bunting
Eastside neotropical migrant

Lazuli buntings are uncommon summer residents. The male is a beautiful, turquoise blue with a rust-colored breast. The song is a lively *sweet-sweet chew-chew*. Buntings inhabit open, brushy fields and pastures, forest edges, and brushy clearings, favoring areas along the margins of rivers and streams. They occasionally use feeders.

Hummingbirds
see Plate 3 and Chapter 17

Woodpeckers
Northern flicker
(see Plate 3) Eastside and westside resident

Northern flickers are large woodpeckers with salmon-colored wing undersides distinctly visible during their slow, bouncy flight. The call is a *cheer, flicka, flicka*. They're common in urban areas and are often seen hopping on lawns and fields where they dig in the ground, tear up anthills, and catch prey with their sticky tongues. They nest in holes that they excavate in trees. They use feeders (primarily suet feeders) and nest boxes.

Pileated woodpecker
Eastside and westside resident

Pileated woodpeckers are very large (nearly crow-size) with a bright red crest and a raucous call. They often excavate dead wood at the base of a tree, leaving a rectangular hole. They inhabit densely wooded areas and forested urban areas. Pileated woodpeckers nest in cavities in large trees and require a large territory containing large snags. They will visit a suet feeder.

Downy woodpecker
(see Plate 3) Eastside and westside resident

Downy woodpeckers are small (about robin-size) and have a short bill. Males have a red patch on the back of their head and spots on their outer tail feathers. They inhabit orchards and areas containing large deciduous and coniferous trees. A similar but less common species, the **hairy woodpecker**, is slightly larger and more common in areas dominated by conifers. Both species may use nest boxes and definitely use suet feeders. The **acorn woodpecker** is seen in oak and pine–oak woodlands in

Oregon. These birds have a bright red cap that's more extensive on the male.

Nuthatches, Wrens, and Chickadees
Red-breasted nuthatch
(see Plate 3) Eastside and westside resident

Red-breasted nuthatches are small birds with a white stripe above the eye, a black cap, and a red breast. Their long toes enable them to walk upside down or sideways down tree trunks where they dig for insects. The call is a nasal, rapid *ank ank ank*. They inhabit areas with many coniferous trees or where there is a mix of conifers and mature deciduous trees, as well as orchards. They use nest boxes and feeders (especially in winter). The **white-breasted nuthatch** has a black crown and nape, white face and underparts, and is more often seen on the eastside.

Bewick's wren
(see Plate 3) Westside resident

Bewick's wrens are sparrow-sized with a distinct eye stripe and a cocked-up tail. The call is two to three high notes, dropping lower, ending on a thin trill. They inhabit areas with many trees and are also found near thickets, brush piles, gardens, and farmlands. The **house wren** is similar in appearance but is an uncommon summer resident in urban areas. Both will use nest boxes and suet and seed feeders. The **winter wren** is very small, with a short straight-up tail and rusty-brown upper parts. The call is a staccato *timp* sung from a high stump or a branch in the moist understory of tall trees.

Black-capped chickadee

(see Plate 3) Eastside and
westside resident

Black-capped chickadees are very
active and conspicuous birds with
large white cheeks and a black cap
and bib. Their short, pointed bill
helps them dig out or pick up
insects from branches. The call is
a *chick-a-dee-dee-dee* or *dee-dee-dee*.
They inhabit forest edges, thickets,
and urban areas with mature trees.
The **chestnut-backed chickadee**
has rust-colored sides and a dusky-
brown crown and bib. Both will
use feeders and nest boxes.

Bushtits, Creepers, and Meadowlarks

Bushtit

Occasional eastside resident,
westside resident

Bushtits are tiny gray-brown birds
that fly in roving flocks during
winter and after breeding season.
They inhabit thickets and woodland edges, and they glean enormous numbers of aphids, scale
insects, leaf hoppers, and other
insects from the bark and leaves
of trees and other plants. They
are attracted by sources of drinking and bathing water and by suet
feeders in the winter. Their sack-
like nest is hung in a small tree or
shrub, often in clear view.

Brown creeper

Eastside and westside resident

Brown creepers are sparrow-sized
birds that often fly to the bottom
of a tree and spiral up the trunk,
head first, probing the bark for
insects. When they reach the top,
they fly to the bottom of the next
tree and start over. The call is a
high-pitched *trees, trees, trees, beautiful trees*. They inhabit forests with
tall trees and build a nest behind a
ridge of loose bark.

Western meadowlark

(see Plate 3) Eastside and
westside resident

Meadowlarks are robin-sized birds
that have a bright-yellow breast
with a black V below their neck.
They are found (more often heard)
in pastures, cultivated fields, and
grasslands. The song has several
clear, whistled introductory notes
followed by a rapid cascade of bubbling, rich, flute-like notes. It is
the state bird of Oregon.

Kingbirds, Robins, Thrushes, and Bluebirds

Western kingbird

Eastside neotropical migrant

Kingbirds are robin-sized but are
more streamlined in appearance.
They have a greenish-gray back,
pale-yellow belly, and a heavy
bill. They are open-country birds
found most often in agricultural
areas. The call is a sharp *kit* or
whit. They generally nest in large
bushes or on the crosspiece of
utility poles. **Eastern kingbirds**
are more common in some areas,
particularly British Columbia.

American robin

Eastside and westside resident

Robins are common in residential
areas and parks, woodland edges,
and open country. They need
open ground on which to forage
for food and some woods or at least
a few scattered trees and shrubs for
nesting and roosting. They cock
their heads to the side as they hunt
for worms. In winter they eat a lot of
berries. The calls include a soft *tut...
tut...tut* and a loud *wink*. They will
nest on platforms (see Appendix D).

Varied thrush

(see Plate 4) Eastside and
westside resident

Varied thrushes are similar in shape
and size to robins, but are sleeker

and shyer, and are colored orange
and black. Adult males are blue-
gray above, orange below, with an
orange stripe behind the eye. Their
call is a long soft *buzz* and a soft
took. They inhabit areas with trees
and dense vegetation and sometimes forage on lawns for worms.
They nest in the mountains and
winter in lower elevations. They
will use a platform feeder, especially in winter.

Western bluebird

(see Plate 4) Eastside and
westside resident

The adult male western bluebird
has purplish-blue upper parts and
a rusty-colored breast. The females
have duller, brownish upperparts.
The call is a short *pew* or *mew*, also
a loud chattering note. Western
bluebirds inhabit woodland clearings, pastures, orchards, and fields
with scattered trees and/or fence
posts. They are now rare in urban
areas and largely dependent on nest
boxes for breeding. They occasionally visit feeders. The male
mountain bluebird is sky-blue
with a lighter breast. On the female the blue is most noticeable
on the tail and rump. They nest
in the mountains and move to
lower elevations in the winter.

Tanagers and Grosbeaks

Western tanager

(see Plate 4) Eastside and westside
neotropical migrant

Western tanagers are unmistakable
birds. The adult males are bright
yellow with a strawberry red head.
The back and tail are black. The
females are dull greenish-yellow
above and yellow below. Tanagers
are forest birds, preferring areas
with dense, mature trees. They
will use feeders. They winter in
Mexico and Central America.

Evening grosbeak

(see Plate 4) Eastside and westside resident

Evening grosbeaks are robin-sized with a huge cone-shaped bill. The males are brightly colored with a yellow forehead, black crown and tail, and a yellow belly. The females are mostly gray with black and white on their wings. Evening grosbeaks prefer dense coniferous and mixed deciduous-coniferous forests for nesting, year-round when food is plentiful. They roam in flocks extending over all kinds of habitats outside the breeding season and will use feeders. **Black-headed grosbeaks** (see Plate 4) are an eastside and westside neotropical migrant. They have a large, thick bill and a black head with a reddish collar. Their call is a loud *spik* or soft, whistled *whee*. They're usually secretive and nest in open woods, along streams, and at edges of fields. They winter south of the United States and will use feeders in summer.

Sparrows, House "Sparrows," Pine Siskins, and Finches

White-crowned sparrow

(see Plate 4) Eastside and westside neotropical migrant

These are large sparrows with a black-and-white-striped crown. They are generally seen individually in a variety of urban habitats, including parking lots at malls. They also inhabit grassy and rocky open places with shrub areas. They commonly nest in brush piles and frequent birdbaths and seed under feeders.

Song sparrow

Eastside and westside resident

Song sparrows have a brown head with a gray stripe on the crown. Their song contains many notes and begins with a *tweet, tweet, tweet* sound. They inhabit brushy thickets near water, also drier habitats of fence rows, old fields, and open groves with dense undergrowth. Higher perches from which to sing are essential to the male during the breeding season. Song sparrows frequently use birdbaths and seed under feeders.

House sparrow

(see Plate 4) Eastside and westside resident

The house sparrow is brown with a black bib and is often seen around fast-food restaurants, stores, and people's yards. It is actually a finch and not a sparrow. They are common at seed feeders and birdbaths. For more information on this bird, see the sidebar "Introduced Birds."

Pine siskin

(see Plate 4) Eastside and westside resident

Adult pine siskins are light brown or tan above, with dark streaks. Their wings have a yellow band across the base. They prefer nesting in the edges of coniferous forests and in second-growth forest clearings in the mountains. In the winter they roam about in large flocks, and can be found just about anywhere that food is available. They are common at feeders.

House finch

(see Plate 4) Eastside and westside resident

Male house finches have a bright red head, breast, and rump; the females are all brown with streaks on their sides and belly. These finches inhabit a wide variety of areas and are particulary common where bird feeders are regularly provided in urban areas.

American goldfinch

(see Plate 4) Eastside and westside resident

In summer adult male goldfinches are bright yellow with a black forehead and cap; females are brownish-olive above and paler yellow below. They inhabit weedy fields, orchards, meadows with a few scattered trees, openings in forests, and brushy old fields. In winter they roam about in nearly all types of open terrain. They frequent both birdbaths and feeders and are the state bird of Washington.

Juncos and Towhees

Dark-eyed junco

(see Plate 4) Eastside and westside resident

These are sparrow-sized birds with a black head that make them look as though they're wearing a hood. In flight, white bars on the outside of their tails are easily seen. Most populations spend the winter months in lowlands, and spring and summer higher up in the mountains. They are often seen in forest edges, brushy fields, hedges, parks, gardens, and along roadsides. They are common around a variety of different feeders.

Spotted towhee

(see Plate 4) Eastside and westside resident

These are robin-sized birds with rusty coloration on the sides and a distinct red eye. They inhabit shrubby forest edges, thickets, brush piles, and vegetation along streams and are often seen foraging on the ground, scratching with both feet. They nest on the ground or low in a shrub, vine, or small tree and will eat seeds found under feeders.

CHAPTER 7

REPTILES AND AMPHIBIANS

Reptiles and amphibians are collectively called herps from the same root as *herpetology*, the study of reptiles and amphibians. Some 50 species of herps are native to the Pacific Northwest; about half are amphibians (salamanders, newts, frogs, and toads) and half are reptiles (turtles, lizards, and snakes). Herps feed on large numbers of insects and other invertebrates, including slugs, sowbugs, earwigs, earthworms, and cutworms.

Because of their habitat requirements, it's difficult to maintain wild herps in most yards. Cats, dogs, children, lawn mowers, and cars all take their toll. Also, the loss of wetlands and natural ponds eliminates breeding areas, pesticides poison them and their food, and landscape maintenance can damage their shelters.

The chance of seeing and hosting reptiles and amphibians increases if your property adjoins an undeveloped area, like a greenbelt or other wild area, or if it is next to a stormwater retention pond or other fresh water area. Your property may then become part of a herp's home range if you establish and maintain a natural landscape. Such areas can support alligator lizards, fence lizards, garter snakes, treefrogs, red-legged frogs, and perhaps redback and long-toed salamanders.

If your yard is surrounded by concrete and highly maintained landscapes, chances are slim that it will be visited by these interesting animals.

Reptile and Amphibian Biology

The body temperature of reptiles and amphibians is directly regulated by the environment and their behavior. As their bodies cool, their movements decrease. When body temperature is not within a comfortable range, the animal will move somewhere else. To get warmer, a lizard or snake will move onto or beside a rock pile or concrete walkway, or it will lie on a paved road at night to obtain the stored heat. To cool off, it may retreat into the shade of a brush pile or under a log. Similarly, a turtle will bask on a log to absorb the sun's heat or retreat into the water to cool off. Most herps hibernate in the winter and some will also aestivate (hibernate in summer) in hot areas.

Most amphibians start their lives as totally aquatic animals with gills and a pronounced tail fin; this is familiar to many people as the tadpole stage. Over a period of time, dependent on food and water temperature, legs develop, the tail and gills are absorbed, and the amphibian becomes a terrestrial, air-breathing animal.

Since most species lay eggs in water, most amphibians require open water to survive. Species that lay their eggs on land do so in moist locations such as the inside of a rotting log, under a rock or a woodpile, or in the splash zone along a stream.

Amphibians' skin is not waterproof the way ours is. Moisture

Life Cycle of a Pacific Treefrog (Chorus Frog)

The treefrog is the most widespread and abundant of the Northwest frogs (see Plate 5). It averages 3¼ inches and may be green, brown, tan, or gray. A black stripe runs along its nose, eye, and shoulder. It can turn from a light color to a darker one in ten minutes. This frog is especially musical at rainy times, with a highly pitched *krrreck* or musical chirping or chorusing, the sound most people associate with frogs.

Like most amphibians, treefrogs spend part of their lives on land and part in water. During the non-breeding season adult treefrogs live in the brush or in woods and sometimes in wet meadows or next to a body of water. During dry periods they often will spend the day in a rock or log crevice, rodent burrow, or in some other protected place and come out at night.

Treefrogs have skin glands that produce a waxy coating. This makes it possible for them to survive quite dry conditions. In early spring, treefrogs move into breeding areas, such as ponds, where the males begin chorusing to attract females. The female lays eggs in small, jellylike masses that are about 2 inches long. Egg masses are often attached to sticks or vegetation below the water surface or are located on the bottom of a shallow pond or even a large puddle. Individual eggs are tan to gray-brown above and yellow-gold to cream below. The eggs develop into tiny hatchlings that are golden-tan with dark eyes. These quickly turn into tadpoles with short, round bodies and eyes that poke out at the sides of their heads. The tadpoles turn into air-breathing adults that climb onto land but return, like their parents, to breed in water.

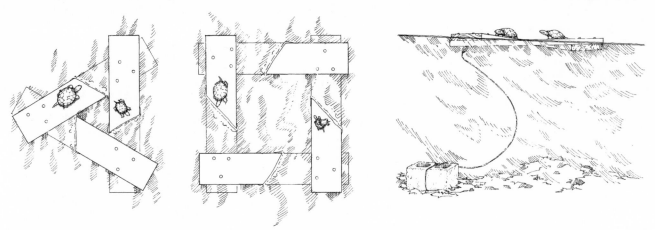

Figure 1. Simple platforms can be built to provide turtles with basking areas.

How to Build a Basking Platform for Turtles

Turtles bask on fallen tree limbs, logs, rocks, and many other objects that protrude above the surface of ponds and slow-moving water. If this debris is removed, turtles are forced to bask on the shore and they become easy prey. However, you can provide them with a safe basking platform. Two patterns for a constructed platform are shown in Figure 1. Note that in the plans the 2 by 8 by 48 inch boards are overlapped. This allows some of the planking to be submerged so that even a small turtle can easily crawl out of the water. Be sure to set the material far enough away from shore to prevent pondside predators from gaining access.

Attach the anchor through an eyebolt set into one corner. Two opposing anchors will prevent the platform from swinging in the wind and ending up close to shore, which can happen easily in small ponds. The finished platform should be set to catch the morning sun. Two feet or more of water with vegetation or limbs providing cover, yet allowing sunlight through, is ideal.

and air pass through their skin. Amphibians must maintain a certain level of skin moisture in order to survive, which is why you'll encounter most of them in moist places. (Toads are an exception; they also occur in dry areas.)

Having skin that allows water to pass through does offer some advantages. For example, amphibians can breathe through their skin. This allows them to remain under water for long periods of time and permits some species to hibernate at the bottom of ponds and lakes, or underground in moist soil. But this permeable skin also makes them unusually sensitive to

handling and to environmental toxins.

Reptiles represent a further evolution away from water. Their scaly skin is waterproof—it keeps body fluids in and water out. As a result, all reptiles breathe principally through lungs. Turtles, for example, must regularly surface for air. (Some turtles can absorb oxygen from water, which is useful during hibernation when the turtle is immobile.)

Reptile eggs have a leathery shell that retards water loss, but they are still laid in moist areas to conserve water. The eggs hatch into juveniles that look like small adults.

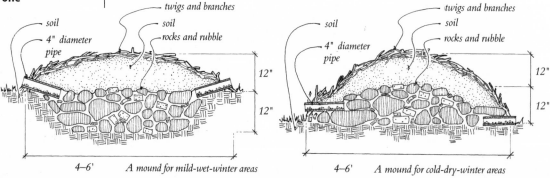

A mound for mild-wet-winter areas

A mound for cold-dry-winter areas

Figure 2. Herp hibernation mounds for mild-wet-winter areas, and cold-dry-winter areas can be made from common materials. Alligator lizards, redback salamanders, Northwestern salamanders, and gopher snakes are some examples of herps that will overwinter (or escape from the heat) in a specially prepared mound. Locate the mound in a well-drained area next to existing cover and/or water. Keep rocks and rubble open enough to allow animals to enter into the mound.

Herps Around Your Property

Herps tend to be inconspicuous and elusive. Except for garter snakes and some frogs, they are rarely seen or heard. However, habitat features such as floating basking sites in ponds may be used by turtles and frogs that bask in the sun, making them more obvious. Garter snakes can often be seen in a sunny spot near escape cover. Piles of rocks, broken concrete and old firewood, logs, large pieces of plywood, and stumps all serve as shelter and basking sites for herps.

Avoid destroying natural habitat in one area by removing stumps, logs, and rocks to create another area. However, sites to be cleared for development or that have been logged are excellent places to salvage habitat materials such as rotting stumps and other large woody debris. Boulders can sometimes be obtained from gravel operations. Before scavenging, always seek permission from the landowner.

These things can protect herps and their habitat:

- Protect buffer areas next to streams, lakes, or ponds. North-facing slopes are productive areas for amphibians because they tend to be cooler and moister.
- Preserve the leaf litter under trees and shrubs.
- Consider building a small, fish-free (fish eat all stages of amphibians) pond for frogs, toads, turtles, and salamanders. Even a pond that dries out in early summer provides important breeding habitat for treefrogs and some salamanders.
- Add waterside plants to ponds to create suitable habitat. Plants such as bulrush provide cover for herps, food for birds, and perching spots for dragonflies and other aquatic insects.
- Protect small-mammal burrows next to ponds and other water sources for herps to find shelter in. Build a herp hibernation mound next to your pond (see Figure 2).
- Avoid using pesticides and herbicides. Herps can be poisoned by them directly or indirectly through their food, for example slugs and snails.
- Allow herps to safely return to their breeding areas. If possible, protect known migration paths from automobiles and lawn mowers.
- Leave a portion of your grass unmowed, especially in areas that adjoin a wet area, forest edge, or any other distinct habitat. If the grass in these areas has to be cut, set the mower blades as high as possible, or use a weed-eater and leave grass 6 inches high.
- Survey the area and move or corral any herps to a safe location prior to mowing. Mowing in the morning, before the heat of the day, may prevent killing snakes and lizards that are basking.
- Discourage cats and dogs from using your yard. They are effective hunters and can severely impact herp populations in and out of the water.
- Prevent the introduction of bullfrogs anywhere. They are not native to the Pacific Northwest and will eat all other herps and anything else they can get in their mouths. (For more information on bullfrogs, see "The Problem Bullfrog.")
- Encourage your friends and neighbors to preserve wildlife habitat on their property.
- Support public acquisition of greenbelts, remnant forests, and other wild areas in your community.

Amphibian Ponds

Many low-elevation amphibians (red-legged frog, Pacific treefrog, western toad, long-toed salamander) use ponds for breeding during several weeks or a few months in late winter and early spring (mid-January to mid-March). To attract and maintain a breeding population of frogs, toads, or salamanders your pond must meet the following requirements:

Location: Because most pond-breeding species spend the majority of the year in forests, shrubs, and vegetated landscapes, your pond should be located amongst some wild habitat. In order to attract amphibians, the pond must be within ½ mile of a pond or wetland that already has breeding amphibians and be connected to this area by a vegetated corridor of woodlands or other "natural" area.

Exposure: Provide both sun and shade. Tree, shrub, and other cover is important in maintaining cool water in ponds during late spring and summer, and it is especially important along the edges of small, shallow ponds, which tend to warm up more quickly and have deficiencies in dissolved oxygen.

Pond depth: The pond can be from 4–24 inches deep. Ideally water levels should not fluctuate drastically from February through June, with shallow water remaining through August. For most species water levels can become higher, but if water levels go down, eggs may become dehydrated and die. Frog eggs eventually float on the surface, so depths can fluctuate

Rattlesnakes and Rattlesnake Bites

The Western rattlesnake ranges from 18 inches to 4 feet long when mature. Color patterns differ with habitat, but all have distinctive dark, diamond-shaped blotches along the middle of the back. It's also possible to identify the snake by its broad, triangular head, which is much wider than the neck. The number of segments on its rattle does not indicate the true age of the snake since rattlesnakes lose portions of their rattles as they age. The Western rattlesnake is found in most of eastern Washington and scattered throughout Oregon (rare in the Willamette Valley). In interior British Columbia it's found in Ponderosa pine/bunchgrass habitats and sandy deserts. This snake is most common near den areas, which are on south-facing, rocky hillsides, in rock crevices not shaded by vegetation. It will den with other snake species. It's most active at night and at dusk, and it may be active during the day if the weather is cool. It eats mice, wood rats, ground squirrels, and small rabbits; it sometimes eats nesting birds and eggs.

A rattlesnake bite seldom kills people, although painful swelling and discoloration near the bite may occur. Rattlesnakes calibrate the amount of venom delivered, and since the primary purpose of venom is to stun or kill prey, most bites are not fatal to humans, unless fatalities are caused by improper treatment. When exploring in rattlesnake country, know how to treat their bite.

If a human or pet is bitten by a rattlesnake:

1. Treat the victim for shock and temperature extremes only, and

continued on next page

Table 1. Spawning vegetation used by some Pacific Northwest amphibians

Pacific treefrog and long-toed salamander
Creeping spike-rush, *Eleocharis palustris*
Small-fruited bulrush, *Scirpus microcarpus*
Taper-tipped rush, *Juncus acuminatus*

Red-legged frog and Northwestern salamander
Beaked sedge, *Carex rostrata*
Slough sedge, *Carex obnupta*
Toad rush, *Juncus bufonius*
Water-parsley, *Oenanthe sarmentosa*

late in the season if egg masses are originally spawned in deep water.

Vegetation: Flexible, herbaceous, thin-stemmed emergent plants are ideal for many pond-breeding species. Rigid, woody, wide-stemmed species, including spirea and cattail, should be avoided.

Most egg-laying takes place in either open water next to vegetation or in open water above short, new spring vegetation. Fifty percent open water and 50 percent vegetation cover may be ideal. Both 100 percent cover and completely open water are undesirable.

New ponds can be improved until vegetation is established by adding thin-stemmed material, including dead spirea, conifer, and other woody vegetation.

Food and cover: Protect and keep (or bring in) large, coarse, woody debris, including logs, logs with root wads, and stumps, in areas next to the pond and in corridors. This provides cover (burying places), food (snails, slugs, insects), and shelter (moist, damp areas) for migrating amphibians. Material should ideally be larger than 2 feet wide and 6 feet long to retain moisture during summer

droughts. Areas of leaf and twig litter for adult amphibians are also important. Live and dead aquatic plants will supply the necessary snails, insects, and other food sources for their larvae.

Protection from predators: Control pets, as they will kill or play with amphibians and in turn can be harmed by the noxious chemicals that amphibians use for defense. Plant cover is important so that larvae can escape predation.

The Dangers of Herps

Most herps are harmless animals that prefer to avoid encounters with people. Although snakes are often seen as threatening, they will hiss or strike only if they're cornered. The Western rattlesnake is the only truly venomous snake in the Pacific Northwest.

Like all wild animals, snakes should be left alone. Except for a rattlesnake that poses an immediate danger to people or pets, no snake should ever be killed.

Less obvious dangers are presented by the mild poisons in amphibians' skin glands. These are probably defensive weapons for protection against predators.

Handling or eating amphibians, or allowing pets to bite, lick, or play with them, can cause illness in humans and pets.

Watching Reptiles and Amphibians

Herps may be difficult to track down. Yet, like most wildlife, they leave telltale signs. Snakes shed their skins between rocks and in the grass. At night your flashlight will cause the eyes of frogs and toads to glow. Listen at night in spring for the peeps and croaks of frogs and toads. Turtle eggs are often buried along the shoreline—look for signs of digging.

Amphibians tend to be more active at lower temperatures than reptiles are. As a result, amphibians may be observed in the fall and early spring when most reptiles are still hibernating. On warm, rainy nights in spring, search trails, roads, and other openings for migrating adult amphibians, and on cool, rainy evenings in late fall search these same areas for dispersing young amphibians. Examine ponds and wetlands from February through April for breeding adults and eggs, and later in spring and early summer (through July) for tadpoles and juveniles.

Handling amphibians can be hazardous to their health. Their permeable skin could absorb harmful chemicals from your hands, such as lotion or bug repellent. Amphibians can die from contact with the higher temperature of your hand and from moisture loss, known as dessication. Handling them increases this risk.

Observe reptiles and amphibians, like all wild animals, from a respectful distance. Please don't attempt to capture them, and don't keep wild ones as pets.

Collecting and Releasing Herps

Amphibians and reptiles are difficult to attract to a landscape so people sometimes consider capturing them and relocating them to their property. This is environmentally unsound and illegal. Many of these animals are rare, threatened, or of unknown status, and there is no justification for moving them to areas where they may well not survive.

If a particular animal is not present in your yard, there is probably a good reason: the habitat may not be appropriate, the animal may never have occurred there, or it may be the wrong time of year for the species. If you enhance your yard for herps, and if they occur naturally in your general vicinity, sooner or later you'll probably enjoy them on your property. Meanwhile, work with the wildlife species that do visit your property and preserve nearby wild areas.

Another reason for not relocating these animals is their well-developed homing instincts. Many will immediately leave an unfamiliar area and, drawn by instinct, try to return to their place of origin. This usually results in their being killed on the roads or by predators they are exposed to on their journey, which is usually quite short.

Many species of herps are popular as pets. Unfortunately, owners often tire of them and release them to fend for themselves. Many of the pet-store species cannot survive our climate and will die if released. On the other hand, some species, such as the red-eared slider turtle and snapping turtle, often do survive.

get him/her to a doctor right away. Shock is responsible for more snakebite deaths than is the actual venom.
2. To treat for shock, keep the victim quiet and maintain body temperature. If the victim is cold, wrap him/her in a blanket; if hot, cool him/her off by fanning. If the victim's face looks pale, elevate the feet. If the face looks red and flushed, elevate the head.
3. Make sure the doctor who treats the victim knows how to treat snake bites and, if not, call the Poison Center:
British Columbia 1 (604) 682-5050
Oregon 1 (800) 452-7165
Washington 1 (800) 732-6985

Most important:
• Do not cut or suck the wound.
• Do not apply ice or cold packs to the wound.
• Never use a tourniquet.

The Problem Bullfrog

An adult bullfrog will eat any animal it can get into its mouth, including native frogs, turtles, crayfish, fish, small snakes, ducklings, mice, and tadpoles. Under no circumstances should you take or purchase bullfrog tadpoles for your home pond, transfer wild-caught bullfrogs, or in any way encourage them to expand their range.

The bullfrog is the largest frog in the Pacific Northwest and was introduced from eastern North America. It's pale underneath and green/olive on top, and has a green head and large "eardrum"— a drum-like circle of flesh. Males have a yellow throat (see Plate 5). The distinct booming call sounds like *kajoonk* or *jugarum*. It also squeaks when it jumps into the water to avoid an intruder. This frog requires permanent water year-round and is abundant in small impoundments, cattle-tramped banks, and ponds, also lakes and slow-moving streams. The tadpoles differ from most frog tadpoles by requiring two years to metamorphose; they can be huge (4–7 inches). These tadpoles are equipped with a chemical that makes them unappetizing to fish and birds.

The following information will help you to identify young bullfrogs:

Eggs: The thin gelatinous egg mass is 2 feet wide and floats on the surface of the water or rests on the bottom of the water body just before hatching. Eggs are very small, black on top, and white underneath.

Hatchling: The hatchlings are small, short, slender, gray-tan, and with short tails.

continued on next page

They may introduce new diseases to or be competitors with established colonies of native turtles.

The bullfrog is another non-native, introduced species that has successfully spread throughout the Pacific Northwest. Large populations of bullfrogs are believed to have contributed to the drastic decline of native frogs and turtles. If you are adding plants or water to a small pond, make sure you are not also adding bullfrog eggs or tadpoles. A humane way to euthanize adult or juvenile bullfrogs is to place them in the freezer.

Common Pacific Northwest Reptiles and Amphibians

Snakes

Garter snakes
(see Plate 5)

The three common garter snakes that occur in the Pacific Northwest are the **common garter snake**, the **Western terrestrial garter snake**, and the **Northwestern garter snake**. All have subspecies that vary in color and patterns.

These snakes are common in gardens, wet meadows, and areas with heavy underbrush and are usually, but not always, found close to water. Logs, rock piles, and plywood placed in a garden will provide hiding places. They eat amphibians, mice, voles, birds, fish, snails, and slugs. When disturbed, they will attempt to escape or will strike, bite, and smear anal secretions and fecal material. A bite from one of these non-venomous snakes may be alarming, but will rarely break the skin.

The **common garter snake** is found throughout the Pacific Northwest and grows to about 2 or 3 feet in length. It has brightly colored stripes (cream, yellow, green, blue) running lengthwise; the stomach is usually grayish.

The **Western terrestrial garter snake** is also common, except in coastal and high mountain areas. It's larger than the common garter snake, usually gray/brown or black with a dark checkered pattern between yellow stripes that run lengthwise. Darkly pigmented forms with brighter yellow stripes occur in some areas. Despite its name, the terrestrial garter snake spends a lot of time in water.

The **Northwestern garter snake** occurs only west of the Cascade mountains. It varies in color more than do the other garter snake species. It's usually dark gray or black with yellow or red stripes that run lengthwise. These are relatively small snakes, usually less than 2 feet long. This garter snake will forage in the rain, feeding mostly on earthworms, slugs, and snails, occasionally on salamanders and frogs.

Gopher snake (Bull Snake)
(see Plate 5)

A mature gopher snake can reach 3–4 feet in length. It's found throughout Oregon and in Washington in the hotter, dryer areas east of the Cascade mountains. It's tan with dark-brown blotches along its back. This snake mimics a rattlesnake by its coloration and by striking, hissing loudly, and rapidly vibrating its tail in grass and leaves, but it's not poisonous. It's a constrictor, so it kills prey by squeezing until the prey suffocates. It eats mostly mice and rats.

Racer

The racer is about 3 feet long, plain brown or olive above, with a pale yellow belly. It's thinner than a garter snake of comparable size. Racers are found in warm, dry, open or brushy country generally east of the Cascade mountains. This snake holds its head and neck above the ground when hunting, is very fast and an excellent climber, and eats lizards, grasshoppers, and crickets.

Lizards

Northern alligator lizard
(see Plate 5)

The northern alligator lizard is long-bodied, sometimes exceeding 10 inches. It has short legs and wiggles as it walks. With its large, platelike body scales and general shape, it resembles an alligator. It's found mostly west of the Cascade crest in Oregon and Washington, west of the Coast Range in British Columbia, and on the southern half of Vancouver Island. The alligator lizard prefers moist, forested areas, and is active at lower temperatures, unlike most lizard species. It's active day and night, although in winter it hibernates below ground in rock fissures, or beneath surface debris in rock and brush piles. As with most lizards, it can lose its tail to predators, and will sometimes bite if picked up or restrained. It feeds on beetles, grasshoppers, crickets, aphids, and spiders.

Western fence lizard
(see Plate 5)

This lizard ranges from 6 to about 8 inches long and is recognizable by the blue patches on its belly. The male is darker on top and more brightly colored underneath than is the female. The fence lizard is found throughout Oregon except on the northwest coast, and is more common east of the mountains in Washington and British Columbia. It occurs in a variety of habitats, but requires vertical surfaces such as rocks, logs, trees, fences, or building sides. Very watchable with binoculars, it can often be seen sitting in the open, relying on its speed to escape from predators. The Western fence lizard requires higher temperatures than does the alligator lizard, so this lizard is not found in deeply forested areas. The fence lizard readily loses its tail to predators and feeds on beetles, flies, caterpillars, ants, and spiders.

Western skink

A slender, shiny lizard, the western skink is often found in leaf litter or under logs in pine or oak woodlands. It is widespread east of the Cascade mountains in Oregon and Washington and can be found in western Oregon, but it is absent in western portions of Washington and British Columbia. Young lizards often have bright blue tails, apparently to divert predators when they are trying to escape. The western skink is probably more common than realized, due to its secretive, burrowing lifestyle. It feeds on beetles, grubs, ants, and spiders.

Sagebrush lizard

This lizard is slightly smaller than the western fence lizard, reaching about 6 inches. It is gray and brown on top with a light gray stripe on its back from head to tail. Blue belly patches are faint or absent on females. Usually the most common lizard in the sagebrush habitat east of the Cascade mountains, it's also found in open

Tadpole: The tadpoles have long bodies and pointed snouts, and their bellies may be yellow. They are dark green with black dots, with orange or bronze eyes. A two-year-old tadpole may be the length of your finger and half your palm.

Juveniles: Juveniles have a ridge around their large and conspicuous eardrum, are green to brown, and have orange or bronze eyes. Juveniles squeak when scared into water. Most will not survive to reach maturity, but those that escape predation and other hazards may live 15 years or more—longer than any other frog species, except toads.

forests with a brushy understory in southwest Oregon and is very watchable through binoculars. It's not much of a climber. Sagebrush lizards eat beetles, flies, butterflies, caterpillars, wasps, spiders, ticks, scorpions, and aphids.

Side-blotched lizard

This is a small, slim lizard that averages 2–5 inches in length. Found in dry, rocky areas east of the Cascade mountains, it's easily recognized by a blue-black spot behind each front leg. Its body becomes a golden-tan color during the breeding season in spring and early summer. It eats beetles, flies, ants, caterpillars, and small grasshoppers. It's often easy to observe this lizard in areas where it occurs.

Turtles

Painted turtle
(see Plate 5)

The painted turtle ranges in size from 8 to 14 inches. It's more common east of the Cascade crest, but is also found on the southern half of Vancouver Island. This turtle has red, yellow, and black markings on its underside and yellow stripes on its neck, head, and legs. The top of its shell is dark olive-green marked with thin, dark-red lines. It lives in marshy ponds, small lakes, and slow-moving streams and rivers, especially those with muddy bottoms and heavy aquatic vegetation. Painted turtles hibernate in the mud floor of a lake or pond and bask in sun offshore on rocks, fallen logs, or basking platforms. They're very watchable through binoculars. Painted turtles eat algae, moss, snails, mussels, dragonflies, crickets, flies, beetles, ants, and fish fry.

Red-eared slider
(see Plate 5)

The average length of the red-eared slider is 5–12 inches. It has a prominent yellow, orange, or red blotch or stripe behind its eyes. The shell is olive to brown, and its underside is yellow. This is the non-native "dime store" turtle that pet-owners release into ponds and lakes throughout the Pacific Northwest. It's not very cold-hardy, so most large populations occur in mild lowland areas west of the Cascade mountains in Washington and Oregon, and west of the Coast Range in British Columbia. Fond of basking, and often seen stacked one upon another on a favorite log, they're very watchable through binoculars. The young eat water insects, crustaceans, mollusks, and tadpoles, turning to a plant diet as they mature.

Salamanders and Newts

Northwestern salamander

This dark-brown salamander is 4–8 inches long with conspicuous raised glands on the sides of its head. It occurs in moist lowland forests west of the Cascade crest in Oregon and Washington and west of the Coast Range in British Columbia. Since it spends most of its life burrowing, it's seldom seen on land unless it's migrating to breeding sites. It's found under sword fern, logs, and bark in forests, and sometimes in rodent burrows during summer and winter. This is one of the few amphibians that continues to be found in large ponds and medium-sized lakes that have large numbers of introduced fishes and bullfrogs. In a defensive posture it stands on its legs, lowers its head, leans toward the disturbance, lashes its

tail, and exudes a white, poisonous secretion that can sicken even larger predators like skunks and raccoons. It eats soft-bodied invertebrates such as slugs and worms.

Long-toed salamander
(see Plate 5)

This salamander is a grayish color below, brown-black above, and has a gold-yellow or green back stripe. The stripe may be continuous or discontinuous. It's 2–5 inches long. The long-toed salamander is widespread and fairly common from sagelands to forests throughout the Pacific Northwest. The larvae require puddles or small ponds with no fish predators. The adult is especially active above ground during rainy periods. These salamanders eat spiders, crickets, larvae, and various other adult invertebrates.

Redback salamander

This salamander is slate-colored underneath, with a back stripe that is usually red, but sometimes tan or yellow. Some populations are black with small white dots. It's 2–4 inches long. The red-back salamander lives its entire life on land and has no need to return to water to breed. It's found under forest debris, wood, and brush piles in the damp coniferous forests and talus slopes west of the Cascade mountains in Oregon and Washington, and west of the Coast Range in British Columbia. It generally lays its eggs under logs, and the hatchlings develop directly into small salamanders. It eats spiders, crickets, larvae, and various other adult invertebrates.

Tiger salamander

The tiger salamander has a distinct pattern of olive or pale-yellow markings between black markings on its back. It's 4–12 inches long and common in some areas east of the Cascade mountains, in portions of the Columbia Basin, and northeast Washington. It spends most of the year burrowed underground and is active at the surface mainly at night and during or after spring rains. It eats a wide variety of insects, spiders, worms, and tadpoles.

Rough-skinned newt
(see Plate 5)

This 4–8 inch long newt is orange underneath and uniformly brown above, with rough skin (except on breeding males). It inhabits wet regions of the Pacific Northwest near many permanent bodies of water, and often is abundant in small ponds at low altitudes. It's active day and night, especially during rainy periods. During the day, you're likely to see this graceful swimmer in the water. Males may spend the entire year in the water, while females go to water only for breeding. This salamander swims by undulating its body and large, paddle-like tail. It will twist and thrash its tail when disturbed; its skin secretions are toxic if eaten. It eats a wide variety of soft-bodied invertebrates, like slugs and worms, as well as aquatic insects and tadpoles.

Frogs and Toads
Pacific Treefrog

See "Life Cycle of a Pacific Treefrog," and Plate 5.

Bullfrog

See "The Problem Bullfrog."

Spotted frog

The spotted frog is most common east of the mountains. It's reddish on the underside and brownish above with large, dark spots, and eyes that point upwards. The call is a quiet, low-pitched, hollow knocking sound. It inhabits marshy edges of colder, permanent ponds, lakes, and streams. In some places it is being (or has been) displaced by the introduced bullfrog. Spotted frogs eat insects, mollusks, and crustaceans.

Red-legged frog
(see Plate 5)

This colorful frog inhabits moist forests near ponds, lakes, and slow streams, especially where aquatic vegetation provides shade and cover. It grows to 4 inches in length and is best known for the red underside, especially the legs. The call is a quiet, low-pitched, throaty stuttering. The red-legged frog is found west of the Cascade mountains in Oregon and Washington, and west of the Coast Range in British Columbia. During non-breeding season this frog may be found several hundred yards away from permanent water. It eats mostly invertebrates.

Western toad
(see Plate 5)

The western toad is warty, brown and green above, with a light back stripe down the middle. Males grow to 5 inches in length; females may be larger. This toad is increasingly rare and nearly gone from the Puget Sound region. It's more common in other areas of the Pacific Northwest, with the exception of the highest mountain and driest desert areas. It inhabits a variety of forested, brushy, and mountain meadow areas, and is generally found near water east of the mountains. It's active in the day during wet weather but is primarily nocturnal. Adults retreat from dry or high temperatures to rodent burrows, or spaces under logs and rocks, and they are able to partially bury themselves. The warty skin on a toad is a collection of glands that secrete a milky fluid—harmless to humans, but an irritant to predators such as cats. (Toads do not give warts to people who touch them.) The call of adult males sounds like the peeping of baby chicks. Adult toads eat flying insects, spiders, crayfish, earthworms, and other invertebrates. Tadpoles feed mostly on algae and quickly turn into little toadlets.

Great Basin spadefoot

This toad is gray or olive-gray with light stripes on the sides and back, and with dark brown or reddish spots. It's 2–3 inches long. The spadefoot toad has a conspicuous black spade at the base of the first toe of each hind foot. It uses its spade feet for digging into the soil, where it spends eight months or more in dormancy. The call is a loud, low-pitched, nasal quacking that sounds like a slowed-down recording of ducks. It inhabits sagebrush, bunch-grass prairie, and open ponderosa pine forests, and it uses a variety of temporary waters including rain pools, roadside ditches, and small ponds. It's usually active above ground at night only in the moist months of spring (March–June). Skin secretions may cause humans to sneeze. It eats ants, beetles, grasshoppers, and flies.

FISH

Fish may not be the first thing that comes to mind when you're planning your landscape for wildlife. However, if a pond, lake, or creek occurs on your property, you will want to know something about the needs of fish.

Like all animals, fish need food, water, shelter, and space. One of the things that determines what fish are in an area is the quality of the water. For trout and salmon, the water needs to be clean, cold, and well-oxygenated. Goldfish and native minnows are not nearly so demanding and can tolerate less oxygen and warmer water.

Fish also need shelter from predators; this is particularly true of small fish, because more animals are capable of eating them. Depending on the species, fish can find shelter in areas between rocks, within aquatic vegetation, under tree and shrub limbs, or at the bottom of a deep pond. In a stream, important fish shelter includes logs and rootwads.

Shelter in a pond can include floating and sunken materials, which give fish a place to escape from herons, kingfishers, and mergansers. Sunken escape cover, such as a pipe with screened ends or a wire box with entries big enough for small fish but too small for otters or larger fish to enter, are worthwhile additions. A Christmas tree can also be effective cover for small fish when the tree is submerged in a large pond. In an even larger pond you can tie three or four trees together, attach a weight, and submerge them.

Restoring or improving fish habitat in and along creeks and streams is a developing technology. Many things can be done, but projects generally involve heavy machinery, an engineer, a biologist, and permits. If you or a neighborhood group want to improve a stream for fish, contact your local Fish and Wildlife office for technical assistance and examples of successful projects.

Protecting the area next to a stream, creek, or lake generally doesn't require a permit. This can involve planting a variety of native plants where few or none exist. Plants stabilize banks and provide shade to keep the water cool, a source of insect food, and cover in the form of fallen leaves, branches, limbs, and logs. For a list of streamside plants, see Appendix B. For information on planting around a pond, see Chapter 10.

Keep cows, sheep, and other domestic animals off the banks of streams, creeks, and even ditches. These animals cause erosion, eat vegetation, and pollute the water.

Fish Ponds

Nearly any size or style of pond can accommodate fish. A pond can be bathtub-sized and contain ornamental goldfish, or large enough to be called a lake and stocked with trout. For information on designing, planting, and maintaining ponds for fish and other wildlife, see Chapter 10.

Fish will have a profound effect on the pond's ecosystem. Too many fish can deplete the oxygen and add an unhealthy amount of ammonia from wastes. Fish also eat plants and animals, including frog and salamander eggs, tadpoles, and dragonfly larvae.

Do not add more fish to a pond than the pond can support. In a new pond, try to add no more than one inch of fish-length per square foot of pond surface area. Established ponds may have as many as two to three inches of fish per square foot. All ponds without an artificial supply of oxygen, such as a fountain or waterfall, should

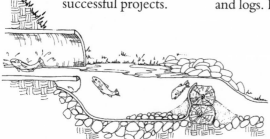

Figure 1. *Culvert with stream improvement to increase fish passage.*

be stocked sparingly so as not to starve fish of oxygen.

Individual fish tend to grow to the pond: the bigger the pond, the bigger each fish. A goldfish in a bowl never gets too big for the bowl, however much it is fed. But if it is moved to a pond, it will soon start growing.

A permit may be required if you are stocking your pond with game fish or grass carp. Consult your local Department of Fish and Wildlife office; they may be able to save you from having problems with the fish in your pond and direct you to a commercial source for stocking your pond.

Types of Fish for Fish Ponds

In addition to your goals for the pond, the choice of fish for your pond should be made based on pond size and depth, water supply and clarity, maximum temperature, amount of light, and water-level fluctuation.

The Department of Agriculture or your local Department of Fish and Wildlife office should have updated regulations, recommendations on what fish to use, and a list of commercial sources for non-ornamental fish, including those raised to feed either wildlife or humans. These fish will, among other things, have been inspected for the absence of disease organisms; this will ensure that you start out with healthy fish.

If your goal is to raise ornamental fish, a commercial aquarium supply or pond-supply store is the place to ask for advice.

Don't collect wild fish; it's illegal, and the likelihood of getting fish that will flourish in your pond is not good. In addition, never release fish from your pond or allow them to escape into the wild. Ensure that overflow from your pond cannot reach a creek, stream, or other water body. Crappie (pronounced crawpee) and bluegill are good choices for small ponds because they will remain small. Other fish that are easy to care for, can survive in a small pond (400 square feet of surface area and a depth of 3 feet) with good water quality, and are a suitable food source for an occasional heron, kingfisher, or raccoon include:

Brown bullhead	Perch
Carp	Pumpkinseed (Sunfish)
Goldfish	Rainbow trout
Catfish	Stickleback

Trout Ponds

To raise trout for eating, consider the following checklist of essentials for a trout pond:

- A surface area of approximately ¼ acre.
- A site free of sewage, excess sediment, and other pollutants, such as salt from roads or paths.
- Water free of heavy runoff or flooding.
- A dependable water supply that holds water without excessive fluctuation.
- Water temperature 50–65 degrees Fahrenheit, a minimum dissolved oxygen level of 5 ppm., and a pH of 6 to 8. (Devices to take these measurements are available from aquarium and biological equipment stores.)
- Complete independence from any other fish-bearing water. This is to prevent fish from escaping from your pond into the wild.

Fish Need to Breathe, Too

Like land animals, fish "breathe" oxygen, but with gills instead of lungs. The amount of oxygen in water can be decreased by pollutants such as sewage or lawn fertilizer. When these chemicals wash into a pond or stream, they absorb oxygen from the water, leaving less oxygen for fish.

Oxygen levels vary enormously over any 24-hour period. For example, levels may drop in cloudy weather because the poor light reduces photosynthesis, and the warmth of the water reduces the capacity to hold oxygen. (On a chemical level, warm water holds less oxygen than cold water does.)

The best way to judge oxygen levels in a pond is by watching the pond. A sudden change of water color from greenish to clear is one warning sign that planktonic plants are failing to produce adequate oxygen; the fish, at first lethargic, may soon be gulping for air at the surface of the water.

If some of the essential habitat components in the checklist are missing from your trout pond, contact your local Department of Fish or Wildlife office for advice. Take water temperatures at the surface and about 6 feet down during the hotter part of the summer. Check the amount of dissolved oxygen in the water at the same time.

Just because a pond is fed by springs, don't assume that it has enough dissolved oxygen to support fish. Artificial aeration of the pond waters may be necessary during the heat of the summer. In areas where ponds freeze and become covered with snow, it may be necessary to aerate the water in the winter as well.

Any artificial feeding should cease when trout become inactive as the weather grows colder. Start feeding again when the fish are seen darting about actively seeking food in the spring. A common problem in trout ponds is over-feeding. Give the fish only as much food as they will clear up in 20 minutes; otherwise wasted food acts as a fertilizer and algae will explode in growth. When the water warms up in midsummer, uneaten food may also cause the dissolved oxygen level to go down.

Preparing Your Fish Pond for Winter

After fish have been carefully fed to prepare them for winter, they can survive the winter months by living at the bottom of a pond. In a frozen pond, there is nearly always room beneath the ice where the fish can hibernate. The potential problem is not that the fish will freeze but that the ice may trap toxic gases so that the fish suffocate, though this does not often happen. You can cut a hole in the ice of a large frozen pond to release gases and let oxygen in. Never bang on the ice, as the sharp sound may give fish a concussion and kill them. Also, keep snow brushed off an iced pond to allow light to reach plants.

Ways to keep your pond from freezing completely include:

- Heat a small pond just enough to keep it from freezing. This will keep it open to wildlife that need water year-round for bathing and drinking. Use a thermostatically controlled submersible heater such as a birdbath deicer or a stock-tank heater available from a farm supply store.
- Turn the fountain or filter pump on to keep water moving in the pond. Raise the inlet to within a foot of the surface to avoid recycling the warmer water at the pool bottom, where the fish overwinter.
- Use a small, cheap aquarium pump designed to oxygenate water in a fish tank to create air bubbles. House the pump in a convenient shed, run a length of plastic piping to the pond, and fix the end about one foot below the water surface. Air bubbles will keep just enough of the surface clear of ice to let gases escape and to provide a couple of birds a place to have a quick bath.
- Float some small rubber balls, pieces of styrofoam, or wood in the water. Black objects will absorb more heat. This remedy may also help to prevent ice-expansion damage to a concrete pond.
- Cover part of the pond with either plywood or plastic (leave areas open for air circulation), leaves, or straw.

Fish Watching

Fish watching gives you the opportunity to slow down and tune in to your surroundings. You can watch fish in almost any body of water large enough to support fish. The greatest variety of wild fish are found in slow streams and sloughs that provide a variety of habitats, including pools and riffles, undercut banks and meanders, and fallen trees and aquatic plants. However, watching fish in a local public pond may be possible if you don't have access to another area.

Just because you cannot see fish, don't assume none are present. A small creek, 10 feet wide or so, may have as many as 100 juvenile trout and salmon per 100 feet of creek in the late summer. Experienced spawner survey-ors may see only 10–25 percent of the adult salmon actually in a creek. Even under the best condi-tions, you will only see a small fraction of the fish present.

When and Where to Look

Clear, sunny days are better for seeing into the water than are overcast days, and you can usually see better when the sun is high in the sky than you can in the late afternoon or early morning, when the sun is low on the horizon.

To seek out fish, begin by slowly walking along or creeping up to the water's edge until you find an angle from which you can see into the water. The trick to seeing into the water is to avoid surface glare. Keep still and look for smooth areas (areas sheltered from the wind) on the water's surface. Late evenings and early mornings are particularly good times to observe fish eating at the surface. If you see rings like those caused by raindrops, but it isn't raining, these may signal the presence of feeding fish. Fish feeding near the surface also sometimes produce a visible wake.

A high place—a bridge, overhanging bank or tree, or pier—may be a good location from which to view fish. Viewing spawning fish from this vantage will be a memorable experience.

A pair of binoculars can be useful. Another important tool for fish watchers is a pair of polarized sunglasses. These specially tinted glasses block some of the light rays from reaching your eyes, cutting down on surface glare and improving vision into the water. Polarized sunglasses cost about the same as ordinary sunglasses and can be purchased at most sporting-goods stores.

In shallow, cold water, you can use hip boots or wear chest-high waders with a face mask and snorkel. A face mask alone works in almost any body of water.

Regardless of age or size, fish are very wary. To improve viewing opportunities and prevent stressing the fish, conceal your presence and limit your movements. Wear clothing that blends into the surroundings, move slowly and deliberately, and do not crowd the fish. Remember that fish can "hear" vibrations in the water, so throwing rocks in the water or splashing through the stream is a sure-fire way to scare them off.

One way to conceal yourself is to use a fish blind, similar to the blinds used for duck hunting. A fish blind is a natural or artificial structure you can hide behind to avoid being seen by fish. It can be as simple as a strategically placed bush, or as complex as a plywood lean-to covered with camouflage netting. A sturdy blind that can

be left in place from year to year would be very effective at a point of known fish activity, such as a favorite spawning area. For information on how to construct blinds, see Chapter 20.

When looking for juvenile trout, salmon, or any small fish species, watch for movement. These 1–2 inch fish are camouflaged extremely well and, when motionless, are virtually impossible to detect. When they move, they do so quickly, then immediately become motionless again. If you come upon a pool with small moving fish you will see what appear to be shadows darting from one end of the pool to the other. Novices may mistake these as the shadows of insects moving over the water.

Identifying the Fish

Identifying fish is the most difficult part of fish watching. As an observer looking down at the fish, you are at an extreme disadvantage, as evolution has protected fish from aerial attacks by camouflaging them. If fish were easy to see, predators would find them too. Most people have to have a fish in the hand, or at least at a side view through a face mask, to identify them. If you are patient, however, you may be able to see some of a fish's distinctive marks when it turns in the water. Observing a fish's behavior can also give you a clue as to its species.

If you spend time looking at pictures in books and visiting museums and aquariums, you can become familiar with the fish in your area. A couple of excellent books on fish identification are listed in Appendix E.

Fish Watching Safety and Etiquette

- Always ask permission to enter private property.
- Be careful where you walk. Fish eggs are easily killed by a poorly placed foot.
- Avoid creating erosion to streambanks.
- If you are headed into the wild, always inform a responsible friend or family member about where you will be and when you plan to return.
- Wear clothing that will keep you warm and dry, and carry a raincoat even on "good weather" days.
- Remember that fast-moving water can be dangerous.

Common Fish of the Pacific Northwest

The Pacific Northwest is home to about 50 species of native freshwater fish. The best known are those that are recreationally fished, such as coho salmon, rainbow trout, and mountain whitefish. There are also a variety of native minnows, suckers, sculpins, and other fish.

Another two dozen non-native fish species were introduced to rivers and streams to boost sport fishing. These include brook, brown, and lake trout, as well as warm-water species such as bass, bluegill, catfish, crappie, and perch. Goldfish and koi are often introduced to park and garden ponds.

The following are some of the more commonly observed fish in the Pacific Northwest.

Coho Salmon

The coho is the salmon most likely to be observed in streams west of the Cascade mountains. The adult spawning coho may be a deep red on the sides with a dark head. Coho salmon are the most adaptable of all the salmonids and often spawn in small tributaries that flow only in winter months, underscoring the importance of even the smallest creeks. They have a tendency to migrate far upstream and easily surmount barriers a foot or

Chinook salmon

U.S. Department of Fish and Wildlife

Native Pacific Northwest Salmon

Name	Nickname
chinook	king
coho	silver
chum	dog
sockeye	red
pink	humpie

two in height. Coho spawn in the late fall or early winter, and the young will spend about a year in the creek before migrating to the ocean. Young coho prefer pools, particularly those with logs, root-wads, and other large rocks to hide in, under, and around.

Steelhead

The steelhead is simply a rainbow trout that went to the ocean while the rainbow trout stayed home. Generally, steelhead and rainbow trout are more heavily spotted than are coho. Steelhead are found in the same streams as are coho and

in larger streams. When found with coho, the steelhead will generally be in the faster water, such as riffles. Steelhead differ from coho in that the juveniles generally spend two summers in the stream before migrating to the ocean, they spawn in the spring, and they don't necessarily die after spawning.

Searun/Resident Cutthroat Trout

The cutthroat trout is similar to the rainbow in that it exists in both resident and searun forms. The cutthroats found west of the Cascade mountains are generally more heavily spotted than are rainbows. They also have two splashes of red on the membranes under the lower jaw, hence the name. They prefer stream velocities intermediate between those preferred by coho and rainbow. Also, they are often found in smaller streams than are the other two species. They spawn from February to May.

Lampreys

Lampreys are the eels commonly observed in streams on the west coast. Like salmon and trout, they generally spawn in fresh water and then make much of their growth in salt water. Lampreys are an ancient group of fish; they have no bones or scales; the adults have a sucking mouth and feed by attaching

U.S. Department of Fish and Wildlife

Steelhead trout

themselves to some other species of fish and consuming body fluids. Like the steelhead, lampreys spawn in the spring, often in the same general area as the trout. Young lampreys live in the mud of stream bottoms, where they feed on detritus filtered through their mouths. Despite appearances and some cultural biases, lampreys are a popular food item where they are abundant. As with the salmonids, lamprey numbers appear to be declining in some areas, probably due to changes in their habitat.

Suckers

Generally larger than native minnows, suckers are bottom-feeding fish that derive their name from their sucker mouth. They live entirely in fresh water and feed on the bottom of lakes and streams. Their habitat preferences, like those of the minnows, may include waters that are somewhat warmer than those preferred by trout and salmon. Still, many suckers share streams with salmonids, and smaller suckers are part of the prey base for larger trout.

Sculpins (Bullheads)

Commonly called bullheads, sculpins are generally small fish with huge heads and mouths. They usually live on the bottom among and under stones in both streams and lakes. Many sculpins make a downstream migration to spawn and then swim back upstream to spend the rest of the year. Unfortunately, they don't jump well. Consequently, even a 6- to 12-inch barrier in a creek, such as a poorly installed culvert, may, in the course of a few years, cause the extinction of the sculpin population that lived upstream of it. Sculpins are opportunistic preda-

tors; they eat whatever fits into their mouths. This often includes trout and salmon fry in the spring.

Sticklebacks and Other Minnows

These small fish inhabit swamps, lakes, slow sections of streams, and the brackish waters of estuaries. They may, in fact, be the world's most widely distributed fish species. Sticklebacks are small (a 3-incher is huge) fish with large eyes and strong spines on their backs. They are a common food item for larger fish where they are found. Often they are seen swimming slowly or in small groups and then darting rapidly for cover upon the approach of a possible enemy.

Two other small minnows are common to Pacific Northwest streams. The **longnose dace** (to 5 inches) can be recognized by its long, overhanging snout, slender body, and rounded, forked tail. It is found in riffles among bottom stones in summer but moves to deeper pools in winter. The plumper **red-side shiner,** recognized by the bright red splotch above the base of its pectoral fins, prefers lakes but can also be found in streams and irrigation ditches. All small minnow species may compete with juvenile salmonids for food, but they and their juveniles also serve as food for trout.

Bass and Sunfish

Bass and sunfish are introductions to the Pacific Northwest. They're generally found in warmer water than salmonids prefer and are more often found in lakes and ponds than in streams. These fish have a tremendous reproductive potential and are popular with anglers because they are abundant, aggressive, and good to eat. However,

they are able to survive in a wide range of environments, even where they may be in competition with native species. For this reason they should not be transplanted into bodies of water that contain native fish or into bodies of water from which they could escape into other bodies of water.

Carp

Carp are native only to Asia and were introduced to Europe as a potential food source. In the 1800s, they were imported into the United States for the same reason. Carp are now widely distributed in the Pacific Northwest. Carp eat all sorts of vegetable and animal matter, and in shallow water they may seriously compete with waterfowl for the available food. Though it is not uncommon for carp to attain weights of 35–50 pounds, the average is nearer 2 to 6 pounds. Wild carp resemble goldfish (a type of carp) without the vivid colors. The upper jaw has two whisker-like growths called barbels.

Goldfish

Goldfish may grow to 16 inches, but more commonly are in the 6- to 10-inch size class. Habits of goldfish are similar to those of wild carp, but they are generally smaller in size, less aggressive, and less prolific. However, under no circumstances should goldfish be released into natural waters.

CHAPTER 9

INSECTS AND THEIR RELATIVES

Insects can make your property more beautiful and interesting. The sight of a butterfly flitting from plant to plant or a beetle crawling along a leaf, or the sound of a cricket at night, can bring you closer to nature. What's more, once you realize the role that insects play in the wildlife web, you may develop a new respect for them.

If your property contains a variety of wildlife habitat, such as mature trees, infrequently mowed grassy areas, a pond or other permanent water source, it's more likely to support a wide range of insect species. However, almost any landscape will support insects if you aren't overly zealous about maintenance.

Insects are members of a larger group of animals called invertebrates, which are characterized by their lack of backbones. Many common invertebrates, such as spiders, millipedes, and sowbugs are often referred to as insects, although they're not "true insects."

A "true insect," such as a housefly, has a body composed of three parts: head, thorax, and abdomen. It also has three pairs of legs located on the thorax and generally two pairs of wings. For simplicity, however, throughout this chapter and book the term "insect" is used to refer to a wider range of invertebrates, including "true insects" as well as some of the insect-like invertebrates, such as spiders. Information about butterflies, moths, and caterpillars is provided in Chapter 18. Information on dragonflies and damselflies is provided in Chapter 10.

Birds aren't the only wildlife that depend on insects for food. Small mammals, including shrews, bats, and flying squirrels, are all major consumers, as are amphibians such as salamanders, frogs, and toads, and many snakes. Aquatic insects are a main source of food for many fish.

Insects are also major decomposers. Without decomposers dead leaves, feces, and carcasses would choke the earth.

In their own search for food, many insects pollinate flowers, ensuring the production of fruit, vegetables, and seeds. Native bees such as mason bees and bumblebees, and honeybees (introduced from Europe in the 17th century), accomplish the bulk of insect pollination. Butterflies, moths, ants, flies, and beetles also pollinate plants.

Insects also eat other insects that people consider "pests." For example, certain wasps and flies destroy tent caterpillars by laying eggs in or on them. When these eggs hatch, the immature wasps and flies feed on the caterpillars. Other effective predators include mud-dauber wasps, which feed on household spiders. (The blue mud dauber specializes in capturing black-widow spiders.) Ladybugs, lacewings, and dragonflies are well known for their ability to eat great numbers of aphids and mosquitoes.

Benefits of Insects

Insects are an essential link in the food web, providing an important source of protein to a variety of animals. For example, the majority of songbirds living on or passing through your property depend on insects for food. While some adult birds may eat seeds from your feeder, most newly hatched birds require a high-protein diet of fresh insects. It's not only cold weather and decreased day length that send birds south in the fall, but also their need for continual sources of insects to eat.

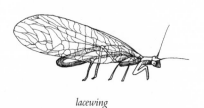
lacewing

ladybug

ground beetle

syrphid fly

Figure 1. *Some common predatory insects. A lacewing, a ladybug (also called a lady beetle), a ground beetle, and a hover fly (also called flower fly or syrphid fly).*

Watching Insects

Watching insects is a pleasant pastime that requires only a little know-how and a good eye. Just stand on the sunny side of a flowering shrub or small tree and observe the insect activity at eye level. Because many insects are more approachable than birds and other wild animals, their lifestyles can be studied very closely. They are considered by some people to be the most "watchable" of all wildlife.

Although many insects live in moist leaf litter, thick grass thatch, and cool shaded pools, a warm, sunny site is a good place to watch some species. Look at south-facing banks and bare spots where the sun reaches the ground for much of the day. Habitats for native bees include bare ground with sandy soil, bank sides, and beetle emergence holes in the dead limbs of trees or the broken stems of sumac, elderberry, roses, berry (*Rubus* spp.), and other plants. The heat generated by a warm compost pile may allow some insects to be active through the winter months.

To verify the presence of insects, look for egg cases glued to weeds and twigs, and swellings, called galls, created in plant stems. Check tree trunks and branches for tiny holes and tunnels chewed by wood eaters. You might even find an empty cocoon or a shed skin clinging to a twig. Because most insects are plant eaters, leaves and flowers nibbled by tiny mouths are also good indicators. Some larger insects leave footprints in dust or mud. Can you imagine what the trail of a millipede or cricket looks like?

Here are some other places to search for insects and suggestions on how to attract them:

On flowers. Plant a variety of plants to provide nectar and pollen for adult butterflies and other flying pollinators from early spring to late fall.

On feeble plants. Many insects thrive best on plants that are under stress. Although these plants look like poor examples of their kind, they may support more species of insects than do healthier plants nearby. Thus, these plants offer opportunities to sit and study the insect world.

In the vegetable garden. When faced with an insect problem, challenge yourself to try a wildlife-friendly control.

In a sunny weed patch. "Weeds" may provide food for butterfly larvae and shelter for beneficial insects.

In a grassland, particularly one that is mowed infrequently. When mowing, set the mower blades as high as possible, or use a weed-eater in order not to disturb insects living in the thatch. Postpone mowing part of your grassland until after the rains have come in the fall.

Along the edge and surface of a creek and pond, especially on or near plants. Peer below the surface with a diver's mask and look for caddisfly larvae or dragonfly nymphs creeping along the bottom. If there is no standing or running water nearby, consider building a small pond; even a seasonal damp area or spring will be used by many insects. For information on how to create a pond for dragonflies, see Chapter 10.

On or under fallen branches, fallen and standing trunks, dead branches in the crowns of trees, rotting heartwood in standing trees, and fallen twigs and fine branches. The richest insect populations tend

Insects by Habitat

Forests and other wooded areas: **June beetle, wood wasp, longhorn beetle, termite, yellow jacket, centipede, millipede, bark beetle, metallic wood borer, carpenter ant, alder flea beetle, March fly.**

Meadows and chemical-free lawns: **grasshopper, cricket, spittlebug, ground beetle, tiger beetle, katydid, snakefly, darkling beetle, robber fly, tortoise beetle, thatching ant, bumblebee, European crane fly, butterflies, orb weaver spiders.**

Wet places: **dragonfly, damselfly, mud-dauber wasp, caddisfly, mayfly, stonefly, water bug, dance fly, midge, alderfly, backswimmer, water boatman, diving beetle, whirligig beetle, mosquito, native crane fly.**

Ornamental gardens: **root weevil, aphids, slug, mason bee, rose sawfly, butterflies, cutworms, leafminers, potato flea beetle, soldier beetle, click beetle, ant lion, cabbage root maggot, scale, stinkbug, carrot rustfly, cherry sawfly, syrphid fly, garden spiders.**

See Table 2 for plants that attract bumblebees.

Bug Zappers

A widely used but ineffective method of insect control is the placement of electrocutors or "bug zappers" around the home. Although many insects may be attracted to these light traps (sometimes from considerable distances), not all fly into the trap. Those that remain nearby may actually increase the number of unwanted insects in your area. In addition, bug zappers attract and kill a wide variety of beneficial insects, such as the green lacewing, a predator of aphids and nymphs of scale insects, both of which harm plants. One study of 13,798 "zapped" insects showed that only 31 (0.22 percent) were biting flies. 13.5 percent were insect predators or parasites. So the net effect was probably to make the user's problems worse, not better.

Bug zappers do little to control mosquitoes. These devices don't attract all mosquito species. (Most mosquitoes that are out for blood are guided by the exhalation of their victims, not by light.) More than 95 percent of the insects they do kill are not mosquitoes.

Where insects cause the most annoyance, yellow light bulbs can be used; these attract fewer insects than do white incandescent lights or fluorescent bulbs. "Ultrasonic bug repellers" do not work in any way, shape, or form.

to be supported by wood located in partial shade. Deep shade may be too cold and damp for the development of some species, and full sun may be too hot.

On and under a rock pile. Incorporate rock, used brick, or recycled concrete in an unmortared retaining wall or other landscape project, or build a special rock pile. Use large rocks as accents in the garden.

On sap runs on trees. Sap runs come from an injury caused by gouges in the trunk that penetrate the bark. Sap runs that result from deep injury may run year after year. Generally, sap runs are short-lived, of little or no consequence to the tree, and healed by the end of summer. The insects attracted by sap runs can attract woodpeckers and other insect-eating birds that are enjoyable to watch.

Near an outdoor light at night. Outdoor lights (except red or yellow ones) are bad for insects. Outdoor lights should be minimized and kept off except when actually needed. Do not use a bug zapper.

Looking at Insects Close-up

Without a magnifying lens, activities of some insects are often unintelligible or completely overlooked. The naked eye cannot adequately observe a spider pulling thread from its body to weave a web. A hand-held lens helps you observe a butterfly's compound eyes and wing scales.

Since hand lenses are relatively inexpensive, you might want to consider carrying two kinds. The first and most important is the small but relatively powerful 10-power lens. It can show detail on an insect's eye or note the texture of pollen grains on the leg of a bumblebee. The diameter of this lens is usually less than an inch, making it easy to carry everywhere.

To use a hand lens properly, remember that the lens goes right up next to the eye and is held steady there. To focus, move the object that is being viewed, not the hand lens. (If this would disturb a live specimen, move the lens to the insect.) Hold the specimen in a good light or you'll not be able to make out the really interesting fine detail. Tie some brightly colored wool to your lens in case you drop it outside.

Some people prefer a loupe lens (the kind you hold in your eye socket) over a hand-held lens, because it allows you to leave both hands free.

In addition to a small, powerful lens, a larger magnifying glass, from 2 to 4 inches in diameter and with a handle, is very handy. Although usually only 3-power, this type of magnifying glass works well when you are watching insects in action.

For purposes of the backyard naturalist, a modestly priced dissecting scope (6–50 power magnification) is adequate. Very powerful microscopes (with 100x and up magnification) are not necessary and can have drawbacks, since more often than not you will want to use your dissecting scope to watch large organisms as they interact within their world. Avoid dissecting scopes (and microscopes) sold as toys. They often have poor lenses that distort the image or ring it with a rainbow halo. Check the larger horticultural tool and supply catalogs for dissecting scopes suitable for backyard use. (For a list of suppliers, look under "General Insects" in Appendix E)

Mason Bees and Mason Bee Boxes

Unlike the introduced honeybee, the mason bee is a Pacific Northwest native that nests singly and has no queen, hive, or honey. It can be easily encouraged to colonize the landscape to the benefit of the gardener, orchardist, homeowner, and nature lover. All it requires is a dry site for its nest, some damp soil nearby, and a nectar and pollen supply.

The shiny, dark-blue mason bee is about two-thirds the size of a honeybee (see Plate 8). It usually is seen flying, landing only momentarily to look for holes in which to lay its eggs. Homeowners sometimes become concerned when they see this bee entering cavities under shake siding or investigating nail holes. However, since they do not excavate holes in the wood, mason bees do not cause damage. In the wild, mason bees use old beetle holes in snags, broken hollow stems, and similar holes as nests. The bees can be seen working at any temperature above 50 degrees Fahrenheit; on chilly, drizzly days, they are often busy pollinating when honeybees remain inside the hive.

Mason bees are totally safe, even around children and pets. They're nonaggressive, although they can sting if trapped under clothing or are swatted, or otherwise mistreated. The sting is less painful than a honeybee sting and the stinger doesn't remain in the skin.

The Life Cycle

One of best things about having mason bees around is being able to observe their life cycle. The female bee first places a mud plug at the back end of a hole; this may be a hole in a nest block you provide (see Figure 2). She then brings in 15 or 20 loads of nectar and pollen, which she collects from nearby flowers. (You can safely watch the bee enter the nest block with pollen on the undersides of her abdomen.) When she has provided a sufficient supply of food for the larva, she lays an egg and then seals the cell with a mud plug. She then brings in more nectar and pollen, continuing the process until the hole is nearly full with a series of cells, each with an egg inside. Finally the bee plasters a thick mud plug at the entrance.

In each cell the larva hatches from the egg after a few days and begins to eat its provisions. After about ten days the pollen/nectar mass is all eaten and the larva spins a cocoon and pupates within the cell. In fall the bee transforms to the adult stage but remains in the protected cocoon throughout the winter. In the spring when the weather has warmed up sufficiently, the bees begin to chew their way out of the cocoons and the mud plugs. Females mate soon after emerging, then begin the egg-laying process in three to four days.

Not all holes may show the rough muddy closure. Holes that are smoothly stoppered may be created by potter wasps; those with a greenish sheen to the plug may be made by leaf cutting bees.

Pollinating with Mason Bees

Mason bees are ideal pollinators for apples, pears, cherries, plums, and related crops. They prefer blossoms of these trees and will work them exclusively when the blooms are available. One to three active females can adequately pollinate an entire mature tree. The bees will also forage a variety of other garden and landscape flowers for their nectar supply. These include rosemary, lavender and many other herbs, sunflower, Oregon-grape, and dandelion. Mason-bee nests need only be within a hundred yards or so of the nearest flowering tree, although the closer they are to the blossoms, the better. It's more important to place the block in a warm, dry, protected spot and not to disturb it than it is to move it close to the blossoms.

Building Homes for Mason Bees

If you wish to develop populations of mason bees to pollinate a home orchard, or just to have around to observe, set out one or more nesting blocks to attract them (Figure 2). A nesting block can be made by drilling holes 5⁄16 inch in diameter, 3 inches deep, and about ½ inch apart in any untreated 4 x 4 about 8–12 inches long. For mounting, drill a hole in the back of the block or attach the block to a piece of 1 x 4 that is 2 inches longer than the 4 x 4, and place it over a nail.

Attach the nesting block to a house or other structure, out of the wind and rain, and preferably in a place that receives morning sunlight. If one spot doesn't work, try another. Put the nest up before the bees begin nesting in February or March. Do not disturb the block, as the developing bees are very sensitive.

Mud is a necessary building material for mason bee nests and is why they are called "mason" bees. If no natural mud source is available near the nesting block (this is generally not a problem west of the Cascade mountains), dig a shallow hole, line it with plastic, and keep it filled with moist soil.

Over several seasons the holes in the block may become fouled with debris, and diseases and parasites may build up. The block can lose its attractiveness and mason bees may "go away." Blocks can be cleaned out manually with a 5 percent vinegar and water solution, or the nest holes can be fitted with paper straw inserts that are replaced each year to give the bees a clean nest cavity. Replaceable inserts are available commercially. It is important to monitor the bees' activity in early spring when blossoms appear so the block can be replaced, cleaned, or fitted with inserts.

You can design your nest block to make it easier to clean each year by drilling the holes completely through the block and providing a removable back. After the bees have left the block in spring, remove the back and clean out the individual holes.

Bumblebees and Bumblebee Boxes

Over a dozen species of bumblebees live in the Pacific Northwest (see Plate 8). Some have red tail-ends, some have white, some have black. Some are almost entirely a pale yellow.

Almost all species are very gentle and nonaggressive; however, several species will defend their nests vigorously. Most species are slow to sting, but will do so if you disturb a nest by digging in or pulling grass from the area around it.

Young queens emerge from hibernation in spring and begin a search for a suitable nest site, such as an old bird nest, mouse nest, or hole under a rotten log. The young are raised in clusters of wax cells, and a small amount of honey and pollen is stored nearby in wax honey-pots. Adults eat nectar and pollen. Workers occasionally chew small holes in blossoms to reach nectar deep inside. Bumblebees are important pollinators of many native plants and fruit crops.

Build a Bumblebee Box

A supply of dry insulation material in a dark undisturbed area is the bumblebee's basic requirement (see Figure 3.) Upholstery cotton is relatively inexpensive and available at upholstery shops. Fiberglass insulation also works. Two chambers are helpful because this provides bees with an area where they can defecate and not contaminate their nest area. Other features that add to the attractiveness of a bumblebee box are a tunnel entrance so the colony can better protect itself, protection from the weather, ventilation, and paint around the entrance hole to provide a color contrast. A couple of colors they seem to like are pale yellow and mauve. A landing surface at the entrance is also an attraction, particularly for young bees that are inexperienced foragers. Nest boxes must be sterilized, and contaminated items should be disposed of each year.

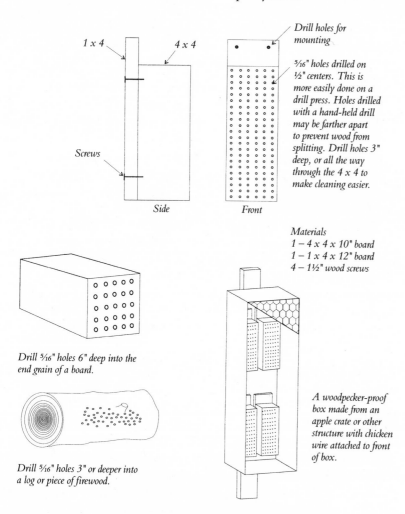

1 x 4   4 x 4

Screws

Side

Drill holes for mounting

5⁄16" holes drilled on ½" centers. This is more easily done on a drill press. Holes drilled with a hand-held drill may be farther apart to prevent wood from splitting. Drill holes 3" deep, or all the way through the 4 x 4 to make cleaning easier.

Front

Materials
1 – 4 x 4 x 10" board
1 – 1 x 4 x 12" board
4 – 1½" wood screws

Drill 5⁄16" holes 6" deep into the end grain of a board.

Drill 5⁄16" holes 3" or deeper into a log or piece of firewood.

A woodpecker-proof box made from an apple crate or other structure with chicken wire attached to front of box.

Figure 2. *Examples of nest structures for mason bees. You may want to include some ⅛" holes for other native bees and wasps.*

Table 2. Common garden and landscape plants that attract bumblebees

Aster	Huckleberry
Bee balm	Hyssop
Blackberry	Lavender
Blueberry	Penstemon
Chive	Raspberry
Clover	Rhododendron
Cotoneaster	Rose
Fireweed	Rosemary
Foxglove	Salal
Germander	Sedum
Globe-thistle	Spirea
Goldenrod	Sunflower
Gooseberry	Willow
Heather	

Nest chambers should be about 6 by 6 inches or slightly larger. The ideal entrance diameter is ⅝ inch but no larger than ¾ inch. The inside divider needs to be rough so bees can get a grip when they climb over it.

Let your nest box "weather" outside for a while before you put it into use. This will allow any paint or caulking odors to dissi-

pate. Set the nest box out in early spring when the first flowers, such as willow, are starting to bloom. Bumblebees may search all day long for up to three weeks to find an "ideal" nest site. Place the box on the ground upon a flat rock or a couple of bricks to keep it off the damp earth. Put it in a shaded place on the north side of a building or behind shrubbery. If you live in a rainy climate, it's not a bad idea to place an old cookie sheet or a piece of roofing paper over the top with another brick on top to hold it down. If you live where skunks or other honey-loving creatures live, you might want to elevate the bees to atop the woodpile or on a fence.

Bumblebees are very fickle. They may use your nest one year but not the next. Also, if they find a better location before they are really settled in or if they are too badly disturbed, they will move. If you get a box occupied, consider yourself lucky. Do not take the cover off your nest or the bees will probably vacate.

Common Insects Seen in the Pacific Northwest

You may want to purchase one of the many excellent field guides to insects, to identify not only the rare visitor but also the insects you see every day (see Figure 1).

Dragonfly and Damselfly
(see Plate 8)

These insects with bright metallic bodies and large gauzy wings are often seen darting over ponds, lakes, and streams in pursuit of prey. They fly with their six legs bent in front of them to form a "catching basket" for small flying insects. Their long, spiny legs are also adapted for grasping a plant to rest. Prey is consumed in flight or while the insect perches. A dragonfly's preferred meal is a mosquito or a midge, while a damselfly prefers aphids.

Dragonflies and damselflies regularly fly long distances in search of wet habitats, and if you have space to install a pond, you will probably attract them. However, they also feed and roost in fields and yards a mile or more away from ponds, for days at a time.

The nymph feeds on microscopic water organisms and small insects, including mosquito larvae. These nymphs themselves are important in an aquatic food web, and are eaten by fish, turtles, frogs and salamanders. For information on how to make your pond attractive to dragonflies, see Chapter 10.

Ladybug

Adult ladybugs come in various combinations of red, orange, black, and white (see Figure 1). The larvae are long and narrow,

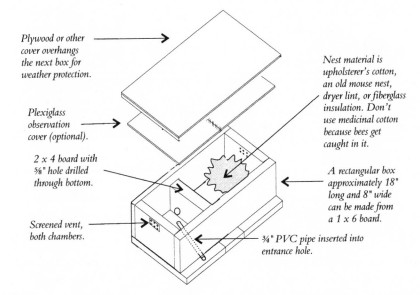

Plywood or other cover overhangs the next box for weather protection.

Plexiglass observation cover (optional).

2 x 4 board with ⅝" hole drilled through bottom.

Screened vent, both chambers.

Nest material is upholsterer's cotton, an old mouse nest, dryer lint, or fiberglass insulation. Don't use medicinal cotton because bees get caught in it.

A rectangular box approximately 18" long and 8" wide can be made from a 1 x 6 board.

¾" PVC pipe inserted into entrance hole.

Figure 3. *A bumblebee box.*

covered with lumps and spines, and sometimes spotted or banded with bright colors such as salmon and blue. They may resemble tiny alligators. The eggs are yellow to orange, spindle-shaped, and attached to plant stems or foliage in compact clusters. Ladybug adults and larvae feed on aphids and other small, soft-bodied insects, including mealybugs, scale insects, and spider mites. Each larva can eat 25 aphids a day; each adult can consume nearly 60 aphids daily.

Adults migrate to protected areas (sometimes houses) in large masses to overwinter. They can be swept up and deposited outside in a protected area, such as under leaves or in a large rock pile. They can also be stored in a refrigerator and released in spring. Adults may occasionally nip people but cause no harm. People often collect ladybugs in masses from sheltered overwintering areas in the mountains and sell them for controlling aphids. Unfortunately, ladybugs fly away after being released and are not effective when used in this way. It's better to encourage and protect local species already active in your garden.

Ground beetle

(see Plate 8)

Ground beetles range from ¼ inch to more than an inch in length (see Figure 1). There are many different species; most are dark and shiny. Adult ground beetles are found in many habitats and may hide under rocks, logs, or duff during the day. If you want to look at ground beetles, look at night, especially near the edge of standing water. Ground beetles actively hunt for prey any time of day during mild or warm weather. They eat caterpillars, cutworms,

and housefly maggots. A common species found in gardens eats small slugs and slug eggs. The narrow head of this species is an adaption to eating snails from the shell. Ground beetles are very sensitive to pesticides.

Water strider

The common water strider is dark gray, black, or brown on top and silvery underneath. It is from ½ to nearly an inch in length, has a long, narrow body, and can be either winged or wingless. They are found on the surface of still or slow-moving fresh water. Here they "skate" over the surface searching for prey and mates, and locate prey by detecting ripples in the surface of the water. Water striders are able to walk on water by means of tiny "hairs" on the feet. They overwinter in the adult form under rocks and logs near the water and emerge with the first warm days of spring.

Hover fly

also called flower fly, syrphid fly (see Plate 8)

Hover flies come in many shapes and sizes (Figure 1). They average ⅜ inch in length, and are often metallic green with several yellow bands on the abdomen. They mimic bees and wasps but do not bite or sting. They're called hover flies or flower flies because they hover near and land on flowers to feed on pollen, nectar, and honeydew from aphids. In spring the pointed-head, ½-inch larvae crawl over vegetation and eat aphids, mealybugs, and other small soft-bodied insects. Like lacewings, larvae grab the aphid in their jaws and suck out the body juices. Black oily smears of excrement on plant foliage are typical signs of hover-fly feeding.

Field cricket

Crickets are closely related to grasshoppers. They are shiny black, and about 1 inch long. They're found in warm, dry areas out of the wind, such as under loose objects and inside cracks and crevices. On warm summer evenings their chirping produces a pleasant chorus. They eat a variety of plant parts and dying or dead insects. Crickets may be attracted to the warmth inside your house in the fall.

American red-legged grasshopper

Adult American red-legged grasshoppers are about 1 inch long and gray-green to yellow-brown on top with a yellow underside. Their wings often show a flash of red or yellow during their rapid flights of 10 to 30 feet. They're found in open, grassy areas and open woods, and they crawl, hop, and fly only on sunny days. They eat grasses, weeds, crops, and ornamentals. Grass-hoppers are eaten by many wildlife species, including spiders, snakes, hawks, and owls.

Spiders

There are now more than 830 spider species known in Washington State alone. All spiders are predators, and are estimated to eat approximately half of all insects. Spiders are common along forest edges, in tall grassy patches, on rotten logs, under low branches around young conifers, in leaf litter, and along wetland edges.

Spiders have several methods of catching prey: some use webs, some are fast on their legs, and some stand very still, blending into the background until prey wanders by. Spiders attack and subdue their prey by biting with their fangs to inject venom.

PART 3

SPECIAL FEATURES FOR WILDLIFE LANDSCAPES

CHAPTER 10

PONDS

Of all habitat that you might consider creating, a pond can be the most satisfying. A well-managed pond adds a spot of beauty and tranquility to your landscape and provides food, cover, and water for an amazing array of creatures. Rich communities of plants and animals can exist in and around the smallest garden pond.

Planning a Wildlife Pond

Constructing a pond of any size can be expensive; however, the highest cost is for a pond that hasn't been carefully planned. A poorly planned pond can damage surrounding habitat, require frequent maintenance, and endanger people and wildlife.

The first priority in creating a pond is to have a clear idea of why you want it. Pond management can succeed only where there is a sense of purpose, of knowing not only what you are doing but why you are doing it. Ponds can be created for amphibians, ducks, dragonflies, songbirds, and fish. You can have a general pond for your wildlife sanctuary, an ornamental pond with a waterfall, a farm pond for livestock to drink from, or an irrigation pond for agriculture or orchards. For information on stormwater detention and retention ponds, see "Wetlands and Wetland Gardens" in Appendix A.

Also, before you select your pond site, observe the habitat that is already there. If a lot of site disturbance would be required in order to install a pond, you may decide the existing habitat is more valuable to wildlife than a new pond would be.

Permits and Advice

Before constructing even a small backyard pond, you should first check with your local planning office for permit requirements. Also, check with your insurance company, which may have additional safety requirements. Any pond project that involves a stream or creek should be dicussed with your local Department of Fish and Wildlife.

For a one-time consultation or help with design and construction, look in the Yellow Pages under "Ponds," "Landscape Contractors," or "Nurseries" that specialize in water gardens. If a pond exists nearby, ask the owner for information on how it was constructed, who did the work, what permits were required, and what problems were encountered. Also, ask pond owners what they would do differently next time. If you decide to use a contractor, find one with experience building ponds and request a list of references. Visit one of the contractor's ponds to see if you would like to have one like it on your property.

Pond Location

When you assess locations for the pond, consider all underground utilities and other potential obstacles, including tree roots, which can make excavation difficult. You may also require supplemental water to keep the pond full, and electricity to run a pump for a filter or waterfall.

A healthy pond needs daily exposure to at least five hours of sunlight during the growing season (spring through fall). However, a shallow pond that would heat up quickly may benefit from more shade.

Most likely, you'll want to locate your pond where you can enjoy watching the wildlife that use it. A pond that can be seen from the house is especially important if small children play in the area. If you want a closer view of pond wildlife, consider building a simple blind. A garden bench or viewing blind placed nearby also can increase the pleasure of pond watching. For information on viewing blinds, see Chapter 20.

An obvious place to locate a pond is in a low area where water naturally collects. However, because a high water table will cause a synthetic liner to "bubble up" or a concrete one to crack, it is better to locate these types of ponds above the highwater line. Plants that indicate water is close to the soil surface include willow, hardhack spirea, cattail, sedge, and horsetail. In winter you can dig a test hole the same depth as your proposed pond and observe it for 24 hours for signs of water. If the hole fills with water on a no-rain day, consider your water table to be high.

Generally a section of the edge of your pond will be slightly lower than the rest so that excess water is directed into an overflow area. You'll need to be aware of where the overflow from the pond will go when it is full. Wetland-type plants can be placed in this area but may need supplemental water in the

Safety

Families with young children need to consider the hazards of a pond and perhaps postpone construction until children are old enough to understand the danger associated with water. If you have a pond, contact your neighbors with small children to educate them about safety.

Access points such as a large, sturdy flat rock or platform at the pond's edge can make visiting safer for children and the elderly. Any rocks at or near an access point should be able to easily support the weight of an adult. Deep areas can be made inaccessible by closely planting shrubs or other vegetation at the edge of the pond.

Stringently follow safety guidelines and avoid electrocution by having electrical outlets near your pond installed with a ground fault circuit interrupter (GFCI), which prevents any shorting out or similar problems associated with outdoor electricity.

Seasonal Ponds

Many wildlife species find nothing wrong with a pond that regularly dries up. Such ponds come in a wide variety of shapes and sizes. Some large, natural temporary ponds are marked by a ring of wetland plants; others are barely seen in the middle of a woodland or grassland. Any small pool of water that's full for a few weeks to several months each year will be beneficial for some wildlife. Treefrogs, toads, and some salamanders have been known to breed in a pond that is 4 feet wide, 6 inches deep, and full of water only until late June. Butterflies and other invertebrates will congregate in the moist mud left on the bottom after the water evaporates.

Management for a seasonal pond may involve keeping aggressive plants from dominating the area and keeping dogs and domestic animals out.

summer when the water level lowers.

Bear in mind, however, that any pond constructed with a liner will normally have dry soil surrounding much of the pond because there is no natural transition zone from pond to dry land. To remedy this you can create a mini-wetland next to the pond. You can design a mini-wetland next to your new pond by figuring in additional length to your pond liner, or you can make it independent from your pond and locate it somewhere else. For information on how to create a mini-wetland, see "Wetlands and Wetland Gardens" in Appendix A.

Determining Shape, Size, and Depth

A natural appearance can be achieved if the general shape of the pond and its slopes are varied. Experiment with different shapes and sizes by using rope, a garden hose, or wooden stakes to create a pond outline. Varying the slope (see Figures 1 through 3) will allow for a mixture of plants and provide the different water depths sought by insects, frogs, turtles, and bathing birds.

A gradual slope (a drop of 6 inches for every 3 horizontal feet) on at least 50 percent of the shoreline is optimum. This is easier to create in a large pond, but may also be achieved in a small pond using the techniques shown in Figure 2. Coarse-textured material, such as sand or small rocks, should be used to create traction on slopes created with a synthetic liner. The covering creates a natural-looking surface and allows wildlife to move in and out of the pond more easily.

A beach-like area on these gradual slopes is extremely important. Many songbirds, including

robins, chickadees, and warblers, use the shallow (¼-inch to 1-inch) water at the beach for drinking and bathing. Mud in this area is also used as nesting material by cliff and barn swallows; a variety of insects use mud for basking and nesting. Tadpoles and other aquatic creatures escape predation by fish and other animals that may eat them by inhabiting the 1- to 3-inch level of water around the shoreline.

No matter how big you make your pond, after a while you'll probably wish you had made it bigger. An advantage of larger ponds is that the more gallons they hold, the larger the buffer for environmental changes. For example, a larger pond warms/cools more slowly, allowing the fish time to adapt. Generally it's not much more time-consuming to take care of a large pond than to take care of a small one.

A pond can be any depth. Even a 1-foot-deep pond can contain a variety of aquatic life if kept full and in partial shade in the summer. A more stable pond would have to be at least 24 inches deep, and 36 inches is preferable.

Pond Liners

If your new pond doesn't hold water naturally, you'll want to make it watertight by using a liner. The liner may be concrete, earth (clay), or a flexible, synthetic material manufactured for ponds. Prefabricated shells are both durable and easy to install. However, they are available only in limited sizes and shapes, and their slopes can be steep and slippery, features that are not child-friendly or hospitable to some wildlife. Also, some swimming pool liners and children's play pools are treated with chemicals to combat algae, which will leach into the pond and kill plants, fish, and possibly other wildlife that use the pond.

For a small pond, a flexible, synthetic liner is recommended. Available at landscape supply centers or aquatic plant nurseries, flexible liners allow you to easily shape the contours of the pond to your specifications. Furthermore, a flexible liner is guaranteed, impervious to freezing, nontoxic to plants and wildlife, and not too difficult to install. A minimum liner thickness of 30 mil is suggested, with 45 mil being preferred.

Filters

In an ideal pond ecosystem, plants absorb carbon dioxide and release oxygen into the water, while aquatic life forms, such as fish, frogs, salamanders, turtles, and insects, breathe dissolved oxygen and exhale carbon dioxide into the water. In addition the earth absorbs ammonia from fish wastes. Reaching and maintaining a balance in a small pond with an artificial liner can be difficult, especially if it contains fish. A filter will be necessary if you want to view fish in an artificially lined pond. A pond without fish may not require a filter.

Filters work in two ways: mechanically, by physically screening out particles in the water, and biologically, by converting toxic ammonia and nitrites from fish wastes and other material into material that can be utilized elsewhere. A biological filter can be bulky, so plan for its location early in the design stages. Many of the books listed in Appendix E describe how to construct a biological filter.

Waterfalls and Cascades

In ornamental garden ponds, waterfalls and cascades give an extra dimension in the liveliness of movement. Waterfalls and cascades can be practical as well as aesthetically pleasing. If water is circulated by one means or another, it will be constantly replenished with oxygen, to the greater benefit of pond life. Not all plants appreciate moving water; waterlilies, for example, prefer still water. But if the cascade is a trickle rather than a torrent and is carefully sited, there will be quiet backwaters for the lilies.

A waterfall can be especially important, especially in summer, in a pond that heats up easily or contains a lot of fish, because warm water holds less dissolved oxygen than cold water does. Many pond dwellers are weakened or may suffocate when the oxygen levels drop.

A simple waterfall can be created with stones at one end of the pond. Put a liner behind and underneath the stones to prevent leakage.

A typical cascade system is a series of small ponds or pools excavated into a natural slope in a descending chain, with water dropping from one level to the next over a lowered section of the pool or through pipes. These can be made very simply or they can be marvels of engineering skill involving elaborate designs, powerful pumps, and tons of beautiful boulders.

Waterfalls and cascades usually rely on small electronic pumps that fall into one of two categories: submersible or surface. Submersible pumps are cheap and simple but not as powerful as surface pumps. The submersible pump is placed in the pond, and its capacity to generate a flow is governed by the volume of water in the pond and the height to which the water will be lifted.

Surface pumps are needed for bigger ponds. The pump unit is placed close to the pond and must be housed in a weatherproof container. The electrical connection can be made by a competent amateur, and then it is simply a matter of running one polyethylene tube between the pond and pump and another between pump and waterfall or upper pool.

A commercial dealer for pumps will be able to help you with general design questions for your waterfall, cascade, or filter. Check the pump frequently for smooth operation and to avoid having a motor burn up from overwork.

Small Pond Construction with a Flexible Liner

Constructing a small pond using a flexible liner is an economical project you can complete with relative ease. Here are step-by-step procedures for constructing this type of pond.

Prepare the Hole

1. Dig a hole in the shape you have designed. Dig deep enough so as to create a pond of the depth you want plus 2–4 inches (to be filled with sand or washed gravel), but don't over-excavate. The hole should be smooth and free of sharp stones, and the sides should be packed down to prevent loose soil from slipping later.
2. Carefully dig out shelves for plants or rocks (Figure 2). Don't try to build up shelves with soil; they are apt to collapse later.

How to Use a Water Level

A water level is a useful tool to use when constructing a wide pond (one that is too wide for a level on top of a straight board). Buy (available from most hardware stores) or borrow some transparent flexible tubing long enough to lay down along the longest part of the pond. Fill the tubing with water, ensuring that there are no air bubbles. Lay the tubing across the hole for the pond and hold both ends vertically. If the perimeter of the surface of the pond is level at this point, the water in the tubing will be at the same level at both ends. Move one end of the tubing to several points around the proposed site to assure the pond is level.

3. Place a level on a straight board and lay it across the pond to make sure the perimeter surface is level within a quarter-inch. (See sidebar on how to use a water level.) Remember that the overflow will leave the pond through the lowest part of the edge.

4. Apply a layer of sand or similar protection (layered newspaper, sawdust, old carpets, or carpet pads) to create a buffer between the weight of the water pressing down and anything that might work its way up under the liner.

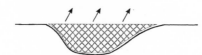

(A) Excavate to the final pond depth with consideration given to added material, such as the protective lining and soil.

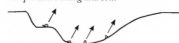

(B) Dig planting shelves out and remove sharp stones and roots.

(C) Lay in protective lining.

(D) Lay in liner and temporarily secure it with stones.

(E) Install gravel or soil if desired over liner and fill with water.

(F) Add final rocks. Trim excess liner and protective lining.

Figure 1. *Steps to follow when installing a pond liner.*

Estimate the Amount of Liner Material

The following technique will leave ample liner material for you to create a mini-wetland next to the pond.

1. If the pond hasn't been dug, start with its proposed length.
2. Next, take the intended depth of the pond and double it. Add this measurement to the length and then to the width of the pond.
3. Now add 4 feet to each of the above to allow for the apron that will extend out from the edge of the pond.

Sample:
Length = 10 feet;
width = 15 feet;
depth = 3 feet.
Two times the depth of pond (6 feet) added to the length = 16 feet.
Two times the depth of the pond (6 feet) added to the width = 21 feet.
Add length of the apron (4 feet) to each of the above to get the final dimensions of the liner you'll need = 20 x 25 feet.

Note: If the pond has been excavated, measure the length and width of the liner you'll need by running a measuring tape down one bank, along the bottom, and up the other side, and add enough for the apron.

Install the Liner

1. Place the liner over the protective layer in the pond and allow it to sag in.
2. Work the liner into the shape, folding it over at the corners and making sure that there is plenty extending all around the pond edge. Wrinkles and folds won't

weaken the liner. Avoid stepping on the liner with anything but bare feet or tennis shoes.

3. Temporarily secure the liner in place by laying large, smooth stones on the apron around the pond.

4. Add a 2- to 4-inch layer of sand or washed gravel over the liner. If the sides of the pond aren't too steep, the material won't settle into the bottom. This provides shelter for small organisms, creates a natural appearance, increases the surface area for growth of bacteria that break down fish wastes and other organic matter, and gives the fish something to root in for food (algae).

Complete the Installation

1. Fill slowly with water. The water will mold the liner to the shape of the hole. Soil may be added or removed under the liner's edge to adjust the level and overflow point.

2. Lay rocks directly on the liner around the pond edge to help conceal the liner. For safety, rocks at the edge should be able to stay in place on their own. See Figure 2 for edge options.

3. Trim the excess liner.

Finally, microscopic life is an important component of the ecosystem you want to create in your pond. One way to introduce microscopic life forms is to add a bucketful of water and mud from a local pond that supports a healthy amount of aquatic animal and plant life. Many tiny invertebrates, protozoa, and algae may be included in this mixture, creating a healthy diversity in your pond.

Pond Vegetation

Plants provide food, oxygen, shelter, hiding places, and platforms on which wildlife rest, live, lay eggs, and metamorphose. Plants also stabilize the pond shoreline, hide the pond liner, and shade the surface of the water to limit algae growth and keep the water cool in summer.

A new pond will need a year or so of plant growth before it will look natural and begin to appeal to a variety of wildlife. Plants nearby will colonize on their own, but adding your own will speed the process along and assure you get what you want. As elsewhere in your wildlife landscape, the aim is to provide a variety of habitats for wildlife. One way to do this is to offer a range of plant types. Never dump aquarium plants into your pond. Many are aggressive growers and can quickly take over.

Your pond should have no more than 65 percent of its surface covered with plants during the summer months. Oxygen enters the pool where water and air meet, and sunlight needs to reach submerged plants, algae, fish, and amphibian eggs.

Algae are free-floating microscopic plants without true roots, flowers, or leaves. They are an essential food for fish, tadpoles, ducks, and snails, as well as providers of dissolved oxygen for all aquatic creatures. In a balanced pond system, algae growth is controlled, creating at most a moss-like coating on the surface of the liner, which gives it a natural look.

An overabundance of nutrients (decayed vegetation, fish wastes, fertilizer runoff) will increase algae and color the water brown, yellow, pea-soup green, or even red. When this occurs, the pond is said to be "blooming." Algae blooms also occur in new ponds and in the spring before pond plants get big enough to shade the water adequately. The following are the types of plants you can include in your pond. Aquatic plant nurseries are good places to view these and other species not listed here.

Submerged plants are rooted or free-floating plants that grow completely underwater. They grow in 1–4 feet of water and are extremely important because they release all their oxygen into the water rather than into the air. They also provide egg-laying sites and hiding places for fish, frogs, and other aquatic animals. The seeds and leafy stems may be eaten by ducks.

Native Submerged Plants
Coontail, *Ceratophyllum demersum*
Elodae, *Elodea canadensis*

Floating leaf plants float either on or raised slightly above the pond surface. Their roots are generally 1–3 feet below the water's surface. Leaves of water lilies and other floating water plants provide shade for fish, resting places for frogs and dragonflies, breeding places for water beetles and snails, and attachment sites for other aquatic animals including caddis-flies and midges. Ducks, shorebirds, and muskrats eat the plants and the aquatic insects that live with them. Deer eat the leaves, stems, and flowers of pond lily; beavers eat the rhizomes. Floating plants, such as water fern and duckweed, can spread very quickly.

Native Floating Leaf Plants
Water fern, *Azolla mexicana*
Watershield, *Brasenia schreberi*
Duckweed, *Lemna minor*
White water lily, *Nymphaea odorata* (not native but naturalized)

Yellow pond lily, *Nuphar lutea* ssp.
 polysepala
Pondweed, *Potamogeton natans*

Marginal plants create the
immediate habitat surrounding
your pond and thrive in 6–12
inches of water. These plants help
camouflage the edges of a pond
constructed with an artificial liner.
Most can be grown in containers.
In an earthen pond, they help
strengthen the banks by preventing shoreline erosion. Marginal
plants are used as habitat by birds,
mammals, amphibians, and reptiles. Floating plants, such as spike
rush and cattail, can spread very
quickly.

Native Marginal Plants
Great water-plantain, *Alisma
 plantago-aquatica*
Inflated sedge, *Carex vesicaria*
Spike rush, *Eleocharis palustris*
Wapato (duck potato, arrow-
 head), *Sagittaria latifolia*
Hardstem bulrush, *Scirpus acutus*
Wool grass, *Scirpus cyperinus*
Small-fruited bulrush, *Scirpus
 microcarpus*
Soft-stem bulrush, *Scirpus validus*
Cattail, *Typha latifolia*

Pond Plants in Containers

In a pond lined with a flexible
liner, the best way to grow many
plants, especially marginal plants,
is in plastic containers. Containers
facilitate removing plants from the
pond for thinning, replanting, and
protection in cold weather. Containers also help keep aggressive
growers in check. Containers that
extend above the water surface
serve as small islands in your pond
and are places for frogs to rest and
hunt. A nice size for a container is
12 inches wide and 8 inches deep.
The containers can rest on the

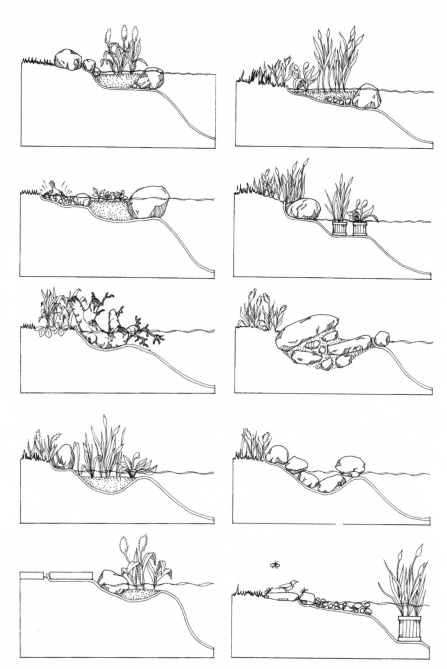

Figure 2. *Examples of ways to make the edge of a pond more wildlife-friendly.*

shelves you construct around
the inner edge of the pond (see
Figure 2).

 To plant containers, fill
them with a mix of one-quarter
sand, one-half garden loam, and
one-quarter compost. Be careful
not to use materials that float, such
as vermiculite and perlite. Also, be
sure there is nothing sharp on the
bottom of the container that could

wear a hole in the liner. An extra
piece of liner or carpet remnant
placed under the container will
provide added protection from
wear and tear.

 You can "mulch" the plants
in containers with an inch or two
of pea gravel and coarse sand to
prevent the soil from clouding
the pond. If you use only soil or
fine sand, fish—especially koi—

Table 1. Plants to use with caution or not at all

Aggressive growers to be grown in containers and used with caution

Yellow iris, *Iris pseudacorus* (non-native)

Yellow pond lily, *Nuphar lutea* ssp. *polysepala*

Common cattail, *Typha latifolia* (spreads vigorously)

Noxious weeds and/or very invasive plants not to use

Brazilian elodea, *Egeria densa*

Purple loosestrife, *Lythrum salicaria*

Eurasian water milfoil, *Myriophyllum spicatum*

Parrot's feather, *Myriophyllum aquaticum*

Reed-canary grass, *Phalaris arundinacea*

will root around in it and cloud the water.

Wildlife in and Around the New Pond

Animals living in a small pond are very vulnerable. A small pond can quickly warm up or freeze over, lose water or fill up. This places considerable stress on aquatic life forms, which cannot instantly move to a safer, more stable environment. Thus the life that colonizes a pond must be tolerant of a fluctuating environment or be able to adapt by mobility.

Even so, a surprising number of flying aquatic insects will colonize a new pond if there is another body of water within a half-mile. Water-boatmen, beetles, and dragonflies investigate new waters quickly and will stay

if conditions suit them. As your new pond begins to mature, other wildlife will visit and inhabit the area. Tiny aquatic mollusks and crustaceans will find their way from a nearby wetland on the feathers and feet of a visiting bird. Fish also may be introduced from eggs brought in by waterfowl. Larger ponds located in suitable habitat will attract frogs, newts, salamanders, toads, turtles, and snakes. Most of these will travel up to about a half-mile from their home pond or wetland, as long as there is adequate cover along the way.

Viewing Pond Life

You can easily observe small aquatic life in your pond using a few simple tools from your kitchen. With a measuring cup or meat baster, collect some water near a pondside plant. Put your sample in a white pan or deep white plate. A hand lens or magnifying glass will help you to see very small organisms. Sampling different places along the edge will net different creatures. Enjoy bug-watching, but ensure a steady diet for fish and other vertebrate life by putting your water sample and invertebrate organisms back where you got them.

Adding Fish and Other Wildlife

Seek expert advice from your Department of Fish and Wildlife agency when stocking fish or any other wildlife in your pond. Non-native species of reptiles, amphibians (especially bullfrogs), and fish create many serious difficulties for native populations if they leave your pond (see Chapter 7). They take over food supplies and may

introduce diseases to wild populations. In particular, wildlife purchased from pet stores are sometimes raised under poor conditions and frequently pass on disease.

Introducing fish will profoundly alter the pond's ecosystem. Fish eat amphibian eggs, tadpoles, and dragonfly larvae. Excess fish create stress, deplete oxygen, and add an unhealthy amount of ammonia to the pond from wastes.

Although small fish can provide a valuable service by eating mosquito larvae, predatory aquatic insects such as dragonfly larvae and water striders also eat mosquito larvae. Bats, birds, toads, and frogs eat the adults. Also, moving water (with a fountain or cascade) discourages mosquitoes from laying eggs. Fish are therefore unnecessary for mosquito control.

How to Improve an Existing Pond for Wildlife

No matter what size pond you have, there are ways to improve it for wildlife. For example, birds and other wildlife are attracted to the sounds, movement, and the flashing light of moving water. Falling water is also soothing to the human ear and masks noise. A small dribble or trickle over a log or rock is all you really need.

Here are some other ways to improve a pond for wildlife:

- Add a floating log that will remain anchored to the shore. In a small pond, fish tend to gather under such logs.
- Include a multi-forked stick that protrudes above the water's surface. (You can stick it in a sand-filled coffee can.) Songbirds and dragonflies will use the branch for a perch.

Making Your Pond Attractive to Dragonflies and Damselflies

Dragonflies and damselflies are closely related insects with bright, metallic bodies and large, gauzy wings, often seen darting over ponds, lakes, and streams in pursuit of prey. (See Plate 8.)

Dragonflies and damselflies regularly fly long distances in search of wet habitats, and if you have space to install a pond, you will probably attract them. However, they also feed and roost in fields and yards a mile or more away from ponds, for days at a time.

The nymph feeds on microscopic water organisms and small insects, including mosquito larvae. These nymphs themselves are important in an aquatic food chain, eaten by fish, turtles, frogs, and salamanders. Here are some things you can do to your pond to attract dragonflies and damselflies:

- Keep the pond filled year-round at a fairly constant level. Any potential breeding site for dragonflies

continued on next page

- Include a large rock that protrudes above the water's surface. Turtles, frogs, and butterflies will use it for basking.
- Submerge a small brush shelter in shallow water as a place for turtles, salamanders, frogs, toads, and aquatic insects to attach their eggs and to serve as a hiding place for fish or tadpoles. (See Figure 3.)
- Add a rock shelter next to or around part of the pond for animals such as salamanders. (See the example in Figure 2.) Add a group of large rocks in the pond as hiding places for fish and/or amphibians.
- Create a gentle slope and a better beach around a portion of your pond by adding sand, small rocks, or soil in steep areas.
- Install nest boxes nearby for cavity-nesting birds, such as violet green swallows and wood ducks. A bat house will help control a pond's mosquito population.

Maintaining Your Pond

The basic maintenance of a pond is a matter of common sense, tempered by the particular needs of different ponds. Possible tasks include dividing and repotting crowded container plants, tending or removing other plants in and around the pond, monitoring leaf drop, and maintaining the pump(s) for filters and waterfalls.

The best time to clean or repair a pond is in the fall. In the spring and summer, plants are growing, tadpoles may be developing, and other animals are raising their young. In the winter, many life forms are hibernating and shouldn't be disturbed.

Controlling Vegetation in and Around the Pond

Pulling a garden rake through the water can control the growth of most floating-leaf plants such as duckweed. In a large pond, a floating boom works better. This can be a long, thin piece of wood,

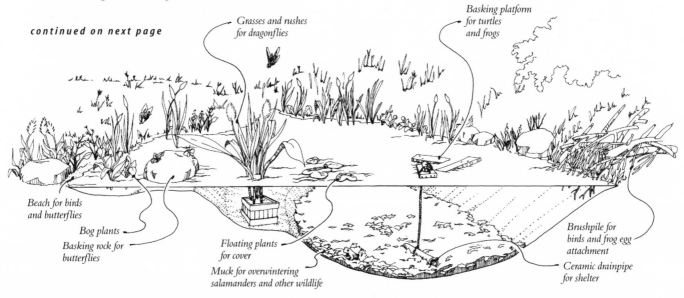

Grasses and rushes for dragonflies

Basking platform for turtles and frogs

Beach for birds and butterflies

Bog plants

Basking rock for butterflies

Floating plants for cover

Muck for overwintering salamanders and other wildlife

Brushpile for birds and frog egg attachment

Ceramic drainpipe for shelter

Figure 3. *Habitat features in and around a pond.*

or a heavy rope, pulled across the surface to collect floating plants in front of it. Neither technique will remove all of the plants, so be prepared to repeat the procedure.

When cleaning a pond, let the collected vegetation sit at the pond's edge overnight to allow excess water to drain and any aquatic wildlife to escape.

Don't use herbicides to manage aquatic and pond-side vegetation. They can contaminate the food source of your pond's wildlife. Herbicides may seem expedient at the time, but the long-term negative effects make them less attractive. If your pond is managed properly, any problems you encounter can be solved using environmentally friendly alternatives.

Cleaning and Emptying the Pond

After a storm and during the autumn, limit the amount of leaves that enter a small pond. Fallen leaves can make the water too acidic for many life forms, and decomposing leaves rob the water of dissolved oxygen. Fallen leaves and needles also create extra work for you if they clog a pump or drain. Don't worry about removing every leaf; a 3-inch layer of decaying vegetation on the pond's bottom is desirable if the pond is getting enough oxygen from live plants or from a fountain. Aquatic insects, turtles, and even some fish will burrow into the debris for the winter.

To keep the pond partially clear of leaves, place netting over its surface in the fall. After cleaning the netting, you can replace it tightly over the pond's surface to keep fish safe from predators throughout the winter.

A degree of turbidity is a sign of a productive pond, but if the water is too clouded, light will not be able to reach the deeper waters and the oxygen balance will be upset.

Unless the water is chemically polluted, to prevent the removal of important wildlife a balanced pond should not be completely emptied for cleaning. If you must empty a pond, relocate any fish and plants as you drain it. One way to do this is to create a makeshift pond above ground with a plastic tub or plastic sheeting. (Don't use a children's wading pool; they are often treated with chemicals.) Use pond water or a 1:3 mixture of new and old water. You can partially empty the pond with a sump pump or siphon. Use the nutrient-rich water to fertilize the plants around your pond. If your water supply is chlorinated, remember to add a dechlorinator to the water you use to refill the pond.

It is better to only partially change the water and keep the fish in the pond. In a small pond, you can do this by adding fresh water slowly: Let a hose trickle water into the pond, with the hose nozzle placed at the bottom of the pond. As the old water flows over and out of your pond, it will gradually be replaced by fresh water. The pond water can be replaced with up to 5 percent of chlorinated water and up to 25 percent of nonchlorinated tap water without killing the fish or plants. Never use irrigation water to refill a pond, however. It may contain chemicals that are deadly to both plants and animals.

must not dry out. Aquatic insects that dragonflies and damselflies eat are also more likely to be present in a permanent body of water.

- Include sunny spots at the edge of the pond, even if removal of some tall vegetation is required. Dragonflies are sensitive to prevailing temperatures and are most active on warm sunny days. Adults may also sunbathe for short periods on pond margins or on roost sites provided for them.

- Make sure there is mud or sand on the bottom of the pond. This provides egg-laying sites for adult dragonflies and appropriate burrowing habitat for the nymph stage.

- Design a shallow area in the pond to provide the most desirable growing conditions for aquatic plants, which are important to these insects. Plant stems provide places for adults to rest and to lay eggs. Also, nymphs of some species crawl up stems for their final molt. Aquatic plants also provide a food supply for smaller insects upon which dragonflies feed.

- Avoid the use of chemicals, including herbicides, insecticides, and fertilizers, in or near the pond. Situate the pond near natural vegetation such as grasses and shrubs; if not present, these should be added. Such areas provide adult dragonflies with terrestrial feeding, resting, and roosting sites. Roost sites can be added by placing bamboo stakes or other small branches in the ground around the pond at a slight angle.

BIRDBATHS, INCLUDING DIPPING POOLS

Providing wild birds with a reliable source of clean water for bathing and drinking is one of the most important contributions you can make to improving their habitat. All birds, from eagles to hummingbirds, bathe year-round. And, like all animals, they need water to survive.

In the wild, birds get water from moist food sources, snow, dew, rain puddles, ponds, lakes, and streams. But water can be scarce during summer dry spells and inaccessible during winter freezes. Species that are not otherwise likely to visit a yard may drop by a water source, especially during hot summer months and during spring and fall migrations. Since birds carry water to their young in their beaks, a water source also may encourage nest construction nearby. (For information on water developments for birds and other wildlife in arid areas, see Chapter 6.)

Birdbath Design

A birdbath can be almost anything that holds water—from a pie plate to a backyard pond. (Constructed below ground, a birdbath may be referred to as a "dipping pool.") Whatever form it takes, certain features are crucial:

Birds prefer a birdbath with margins that slope gradually, allowing them to wade in to a comfortable depth. A dry edge or beach gives birds a dry place to land before entering the water.

Many birdbaths go unused because they are too deep. Keep the water shallow—typically 1–3 inches at its deepest point—since most birds bathe in water that is no deeper than their legs are long. If you're building an in-ground pond, you can include deeper areas for plants and fish as long as you include a shallow area for the birds.

Few birds will try to bathe in a bath that has a slippery surface,

though they may perch on the edge to drink. A rough-textured bowl is much preferred.

In winter, some birdbaths can crack because, when water freezes, it expands about 10 percent. Because ice will exert pressure against an edge, this is another

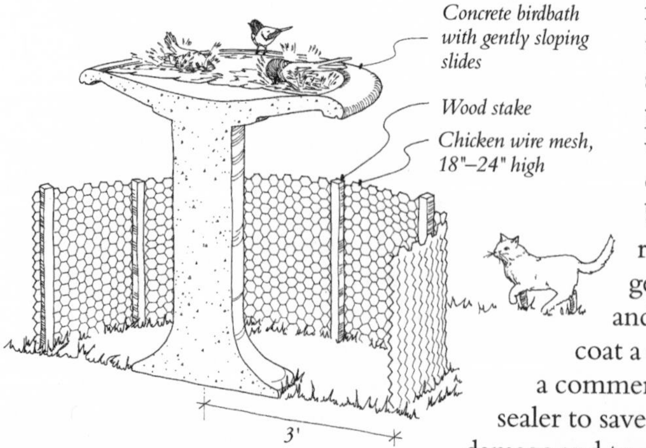

Figure 1. *Birds need a gradual slope that allows them to wade in and find a comfortable depth. An 18"-high wire fence can be added to keep cats from ambushing the bathing birds.*

Concrete birdbath with gently sloping slides

Wood stake

Chicken wire mesh, 18"–24" high

reason to avoid baths with edges that turn up sharply.

If you fall in love with a birdbath that is too deep or slippery for birds, you may be able fix the problem by adding flat rocks or gravel. Bathtub stickers or caulk sprinkled with sand also provide traction on slippery surfaces.

Any size birdbath may be used by birds. However, the bigger the birdbath the more birds will use it at the same time. For communal bathing, a birdbath should be at least 18 inches in diameter.

Commercial Birdbaths

A wide variety of ready-made birdbaths are available from local nurseries, garden centers, and specialty wildlife stores. All have advantages and disadvantages:

Metal baths are the most durable and won't break when the water in them freezes. However, they aren't favored by birds because they can get too hot in the summer and cold in the winter, and are slippery all year.

Concrete baths are widely available, have the feel of stone, fit into most landscape settings, and stay put even in strong winds. Birds like concrete baths because of their rough surface; wide, gently sloping bowls; and stability. You can coat a concrete bath with a commercially available sealer to save it from freeze damage and to mend any cracks. Concrete birdbaths can also be decorated with moisture-resistant paint. A disadvantage of concrete is that unsealed baths may crack if water is allowed to freeze in a bowl with edges that turn up sharply.

Plastic birdbath bowls placed in iron-rod or wood supports are lightweight and easily moved, hence easy to clean. However, the bowls often have steep sides and a slippery surface. Plastic may also crack if water is allowed to freeze in it.

While *ceramic birdbaths* can be quite ornamental, glazed ceramic is extremely slippery. Depending on how it is fired, ceramic may crack if water is allowed to freeze in it.

Hanging birdbaths are effective, with the advantage that they can be placed where it might not be convenient to locate a birdbath that rests on the ground. However, water spills easily from them, especially on windy days.

Ground-level ponds, or dipping pools, are an especially attractive feature in a garden and may be preferred by some birds. They can be made larger than a standard birdbath and may attract other wildlife, such as treefrogs, that might not visit a birdbath on a stand. However, they are more difficult to clean than are birdbaths that can be moved easily. Because they are at ground level, they may put birds at greater risk from local cats.

Making a Birdbath

You can build a birdbath at home with a little time and/or creativity. It can be as simple as an inverted garbage can lid set on rocks or on a tree stump. Some birds may even visit a pie plate or a frisbee filled with water. For a natural birdbath, try a large, flat rock with a depression, sometimes called a "dish rock." Dish rocks are available from rock quarries and can be delivered by truck. Concrete bowls for birdbaths are fun and easy to

make. (See Chapter 10 for information on how to make a dipping pool with a synthetic pond liner.)

Attracting Birds to Your Birdbath

To attract a wide variety of birds, place baths in different locations around your yard. A bath in a shady area with shrubs or small trees nearby can attract small, shy birds such as warblers and wrens. A bath at ground level can attract bigger, bolder birds such as juncos, as well as four-legged wildlife. (When placing a birdbath, also think about how you're going to keep it clean and full. For more on this see the discussion of birdbath maintenance.)

Proper landscaping around a birdbath can also help to attract birds. A small brush pile within 10 feet of a bath will attract birds that require nearby shelter. Birds also will use nearby open shrubs and trees with low branches for this purpose. However, be careful not to put so much shrubbery so close to the bath that you give local cats the opportunity to ambush birds while they are bathing. In areas with a lot of house cats, keep at least 10 feet of open space around your dust bath or birdbath to prevent birds from being ambushed. (For an example of protective fencing around a birdbath, see Figure 1.)

Birds are attracted to the sound and movement of water, especially the small drips and burbles of shallow areas and gentle streams. There are a variety of ways to bring moving water to your birdbath:

- Commercial birdbaths equipped with built-in fountains attract birds, but these fancy baths

How to Make a Concrete Birdbath Bowl

To build a roundish concrete bowl you can move onto a pedestal or elsewhere, or a ground-level dipping pool, you need the following: a rake, hoe, shovel, wheelbarrow, wire cutters, rubber gloves, a piece of thick-mil plastic about 40 inches by 40 inches, one or two 60-pound bags of quick-setting concrete (a 28-inch-diameter bath will take one bag of concrete, and a 36-inch-diameter bath will take one-plus bags), and a piece of chicken wire about 36 inches by 36 inches, depending on the pool size.

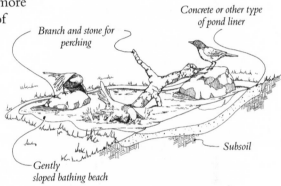

Figure 2. A homemade dipping pool can be made with a concrete, plastic, or rubber liner.

Follow these easy directions:
1. **Find a level area in the open with places nearby where birds can safely perch. Choose a spot that is near a hose bib, but not under a feeder.**
2. **Clear a circular area about 36 inches in diameter.**
3. **Dig a hole down about 6 inches deep in the center and extend the hole outward, getting shallower and shallower, until you reach the edges, where the hole should be about 3 inches deep.**

continued on next page

Table 1. Birds that will drink from or bathe in a well-designed and maintained birdbath

Band-tailed pigeons	Hummingbirds	Robins
Bushtits	Jays	Sparrows
Chickadees	Kinglets	Towhees
Crows	Mourning doves	Starlings
Finches	Nuthatches	Thrushes
Flickers	Pine siskins	Warblers
Grosbeaks	Quail	Western tanagers

4. Drape the plastic film in the hole, smooth out the folds, and hold it in place around the edges with some heavy rocks.

5. With wire cutters, cut out a circular piece of chicken wire 2 to 3 inches bigger around than your depression so you can roll the edges under, creating reinforcement around the lip of the bowl. Place the chicken wire over the depression, and mold it to fit.

6. Mix the concrete with water in the wheelbarrow. You should have a consistency that is a little looser than toothpaste.

7. Shovel the concrete into the form.

8. Wear rubber gloves to pat and shape the concrete into the pool form. Keep the concrete at least 2 inches thick (measure by placing a small stick in the concrete). Make sure your depression allows for 3 inches of water in the center. Keep the concrete moist while you work.

9. Cover the form with more plastic for two to five days to cure. After the cure, snip off any chicken wire that is sticking out. If you want to move the bowl after a minimum

continued on next page

rarely have the optimal bowl design. Modifications with stones or gravel may be needed to make bathing easier.

- Commercial drippers that hook onto outdoor faucets are available in some stores and mail-order catalogs. These devices, which require no electricity, release a slow stream of water droplets that are very attractive to some birds. One inexpensive way to create a small steady drip is to string "spaghetti tubing" (available in the irrigation section of stores), inconspicuously in a tree above the birdbath. A needle valve attached to the end of the spaghetti tubing makes it much easier to regulate the water flow.

- A small, recirculating waterfall can be constructed for a ground-level birdbath (see Chapter 10 on ponds for ideas and information).

- The same companies that make drippers often also make "misters." Installed just as the drippers are, misters send a fine spray of water into the air. They are particularly popular with hummingbirds, which like to fly back and forth in the spray.

Incidentally, don't worry if the ground around your bath becomes soggy. Barn swallows, robins, phoebes, and hummingbirds will use mud under the birdbath for nesting material, while butterflies will gather there to drink mineral-rich water from mud or wet sand.

Bathing Behavior

Small birds flutter in trees and shrubs after a rain for a leafy shower. Some birds use fog or dew condensed on grasses. Garden birds such as robins and towhees will bathe on the ground under the spray of a garden sprinkler. Many species will take advantage of a birdbath or dipping pool, wading in cautiously but then splashing around with abandon once they have found the right spot.

Bathing is part of an intricate array of grooming behaviors that birds use to keep their feathers clean, waterproof, and in good working condition. This includes an after-bath ritual in which most birds seek out a perch where they can sit to preen, running their bills through the ranks of feathers to spread oil from a gland located at the base of the tail. All told, birds spend many hours each day maintaining their feathers—a process essential for flight and for "all-weather" protection.

Maintaining Birdbaths

A well-maintained birdbath needs to be kept clean and full of fresh, thawed water.

Heating the Birdbath

Birds need to drink and bathe even on the coldest days. Although birds can eat snow and melting ice to get water in winter, a birdbath will be used and appreciated. To be sure it is a reliable source of water, keep it from freezing between dawn and nightfall, when birds are active. The water need only be kept just above freezing.

You can keep a birdbath free of ice by pouring warm water into the bowl, but this is tedious in extreme cold weather, as the water freezes rapidly. A stick of wood left in the water during cold snaps can help you pop out ice so you can add fresh water. (If the water does freeze, the stick will also help to prevent the birthbath from cracking.)

Birdbaths equipped with submersible, thermostatically controlled heaters will save your time and maybe your birdbath as well. Small heaters designed to operate at a depth of 1–3 inches are available at garden stores and hardware centers, and through mail-order catalogs.

You will need a source of electricity to run your birdbath heater. Exercise caution here. Outdoor outlets should be on a circuit or outlet protected by a ground fault circuit interrupter (GFCI), which will cut off the flow of electricity in the event of a short. Most outlets in newer homes are protected by GFCI. If yours isn't and you are comfortable with wiring and electricity, you can install your own. Otherwise, consult a qualified electrician. If you are not sure whether your outlets are protected, have them checked by a qualified electrician.

When using a heater, keep the birdbath full of clean water or you may ruin the heater and your birdbath.

Keeping the Bath Clean and Full

Diseases can spread quickly and easily in an untended birdbath. Change the water every few days in a small bath, and rinse a dipping pool every week with your hose, to get rid of regurgitated seeds and other debris. Change water more often if many birds are using the bath. (Locating your birdbath near a hose will make refilling and cleaning easier.)

Scrub small baths a few times each month with a plastic brush to remove algae and bacteria. Dipping pools can be swept out monthly. (See Chapter 10 for more on maintaining pond health.) Never add chemicals to kill algae or insects or to keep the water from freezing in your birdbath, because you may poison wildlife and neighborhood pets that drink from it.

If you irrigate your landscape with an overhead sprinkler, use it to keep the birdbath filled. You can also add a separate valve to the circuitry of an electronic irrigation system to fill the bowl automatically. You can even pipe the distilled waste water from your air conditioner to your birdbath to ensure a fresh supply of water during the hottest days of the year. A dipping pool can be built beneath a downspout of a house, especially if the overflow can be directed away from the outside wall.

Even a bath that is refilled automatically must be monitored at least weekly for cleanliness, however—more frequently during the summer months.

five-day cure, make sure you support it uniformly to prevent cracking.

Since concrete contains toxic chemicals, scrub the pool down with a mild vinegar solution, fill it with water, and empty it a few times before birds use it.

Dust Baths for Birds

Some birds "bathe" in dust, sifting and shaking it through their feathers. Experts believe that dust baths improve feather alignment and discourage fleas, lice, and other parasites. Birds that enjoy taking a vigorous dust bath now and then include Bewick's wrens, kinglets, sparrows, quail, grouse, pheasants, and hawks.

To build a dust bath, dig an area

Figure 3. A dust bath for birds.

4 inches deep, or build the sides up with brick or rock to a height of about 4 inches. A 3-foot-square area can serve many birds at once; a smaller area works well for fewer birds. Fill the bath with equal parts of sand, soil, and sifted wood ashes. Clean and refill it periodically.

CHAPTER 12

NEST BOXES AND OTHER NEST STRUCTURES FOR BIRDS

Many of the birds found in city parks, greenbelts, and urban settings in the Pacific Northwest build their nests in cavities—typically, holes in trees. These "cavity-nesting" birds include primary cavity nesters, such as woodpeckers, which excavate their own cavities in trees softened by death or decay, and secondary cavity nesters, such as violet-green swallows, which use cavities abandoned by woodpeckers or formed by tree decay, broken limbs, or other natural causes.

Unfortunately natural cavities are no longer abundant; the decaying or dead trees (snags) in which birds find or make cavities are often removed in the interest of safety and (presumed) aesthetics. Populations of some cavity nesters are in decline.

The best way to help cavity-nesting birds is to preserve and protect their habitat, including snags. The next best thing to do is to provide well-designed and maintained nest boxes that can serve as artificial cavities.

Nest boxes alone will not establish woodpecker and other wildlife populations where natural habitat is gone. Nor can nest boxes accommodate all the species that use snags. But nest boxes do have many advantages. They help to attract birds to areas where food is adequate but nest sites are limited. They can also draw species closer to you, to view and enjoy. A well-designed and well-constructed nest box can last five to ten years.

With simple woodworking tools, anyone with basic carpentry skills can build a basic nest box. Well-made nest boxes are increasingly available at many stores supplying the bird-feeding public; particularly when made by local woodworkers, these are likely to be very reasonably priced. However, bear in mind that many commercial nest boxes are designed primarily as decoration; some are not even meant to go outside. Examine a commercial box carefully to make sure it meets the specifications listed in this chapter. If you buy a purely decorative box, don't put it out for the birds to use.

For information on roost boxes for birds, see Chapter 6.

Nest Box Design

Before building or purchasing a nest box, know which species you are likely to attract. Different species of bird use different styles of boxes. If you put up a box for a species that cannot meet its other habitat needs in your area, you may be wasting your money and time. Conversely, if you put up a nest box that attracts aggressive, non-native species such as house sparrows and starlings, you may be doing a disservice to the native cavity-nesters in your area.

Any safe and sturdy nest box will have the following characteristics (requirements of specific species are listed in the section "Nesting Information for Common Species" below).

Large enough for the desired species. Nest boxes that are too small force nestlings dangerously close to the entry hole, where they can fall out or fall victim to predators such as raccoons, which eat both eggs and young birds. Larger boxes allow plenty of space for care and feeding of young birds. Dimensions appropriate for specific species of birds can be found in Table 1.

A roof that overhangs the front by at least an inch. This will protect the entrance from wet weather and keep predators from reaching in through the entry hole from above. A slanted roof works best.

Made of a material that has good insulating properties, is free of preservatives or other toxins, and blends into the landscape. Wood is excellent if it is thick enough and not treated with preservatives. (Thin wood will not last long and could be dangerous to birds because it doesn't offer enough protection from temperature changes.) Both cedar and redwood contain natural preservatives that keep them from rotting; these should be 1 inch thick. Exterior plywood ⅝-inch thick is also a good material. Small slabs of bark (from sawmill waste) can be added to the outside of a box to give it a natural look.

Made with durable deck screws, brass wood screws, or galvanized nails. Screws are preferable because nails may loosen as wood expands and contracts in extreme weather conditions. Screws also make it easier to fix mistakes.

At least two, ¼-inch holes near the top of the nest box. Young birds will suffocate if the nest box gets too hot; ventilation holes help to release heat from the box. A ¼-inch gap between the roof and one of the sides will also work as a vent.

Drainage holes at the bottom of the box to release any moisture that accumulates.

Area inside the box and under the entry hole grooved to help fledgling birds climb out. Hardware cloth (wire) is not recommended because a baby bird might get caught in it.

A smooth entry hole. Over a nesting season a mother bird's feathers can become very tattered if the entry hole contains rough edges. You can sand the entry hole to make it smoother.

Easy to check and clean. Nest boxes with a hinged side or bottom are the easiest to open and clean. Front-opening boxes with a predator block may be difficult to open fully. Top-opening boxes are the most difficult to clean because they can't be swept out.

Doors that stay closed and secure. Hinges should be made of galvanized nails, screws, or metal. Fastening devices that work well include a hook-and-eye latch with a locking device, two large-headed nails with wire wrapped around them, or a bent nail (see Appendix D). If a box is going to be installed in a public area, use a fastener that can't be opened without a tool; a screw is good because most people don't carry screwdrivers around.

Watertight. Seams meant to remain closed can be caulked. Apply a commercial water-seal to the outside of the backboard if the nest box is to be located in a moist place.

Easy to mount on a tree, post, or pole.

Equipped with nesting material as appropriate. Material added to the floor of a nest box mimics the typical interior of a natural cavity. Some birds will not use any other nesting material. Wood shavings or small wood chips work well. Livestock bedding can be purchased at animal feed stores. (For the needs of specific birds, see the section "Nesting Information for Common Species" below.) Avoid using fine sawdust as it tends to soak up water, then compact, making it difficult for birds to manipulate.

Finally, a well-designed nest box will deter use by house sparrows, starlings, and Eastern gray squirrels, all non-native species that compete with native birds for nest sites. See the following section for details on how to deter these competing species.

Dealing with Competitors

Nest boxes often attract species that are undesirable for one reason or another. Starlings and house sparrows, for example, compete with native cavity-nesters for nesting spots. They will peck holes in eggs, throw out nesting material, and kill young birds. Male house sparrows are aggressive defenders of their territory and will not tolerate other birds nesting nearby. They are known to peck at the necks of chickadees and swallows as they exit a nest box (the metal flashing in Figure 1 helps prevent this). Starlings will build nests on top of existing nests containing eggs and are so aggressive that they can even evict the larger wood duck from nest boxes.

Nesting Shelves for Robins, Barn Swallows, and Phoebes

Robins, barn swallows, and phoebes will nest on suitable platforms when faced with a scarcity of natural nesting sites. A favorite place for these birds to build a nest is on a flat supporting structure out of the reach of predators. If your property or the area around it contains a food source, a nesting platform (see Appendix D) may be all that is necessary to entice these birds to your property. To encourage these birds to nest on your home or outbuilding, construct a nesting shelf and attach it under a protective eave.

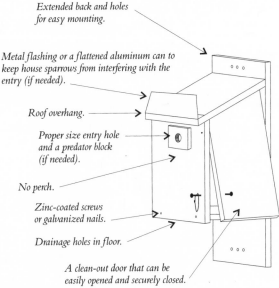

Extended back and holes for easy mounting.

Metal flashing or a flattened aluminum can to keep house sparrows from interfering with the entry (if needed).

Roof overhang.

Proper size entry hole and a predator block (if needed).

No perch.

Zinc-coated screws or galvanized nails.

Drainage holes in floor.

A clean-out door that can be easily opened and securely closed.

Figure 1. *A nest box designed to keep birds warm, dry, and safe from predators.*

Eastern gray squirrels will sometimes enlarge the smaller holes of nest boxes to take them over for themselves.

To help prevent house sparrows, starlings, and Eastern gray squirrels from using nest boxes, make sure that the entrance hole to your nest box is no bigger than it needs to be. Many desirable birds, such as chickadees and wrens, are able to enter nest boxes through holes too small to be used by house sparrows or starlings. To exclude house sparrows, make the entry holes no larger than 1⅛ inch in diameter. Starlings will be excluded by any hole smaller than 1½ inches in diameter.

To reduce the size of the entry hole on an existing nest box (or tree), drill the appropriate size hole (see Table 1) in a piece of wood or metal flashing and attach it over the existing hole. It is also possible to buy a pre-drilled metal plate that can be attached over the entry to a nest box. File down all sharp edges.

To discourage squirrels from chewing an entry to make it larger, surround the entry hole with metal. You can attach a piece of aluminum or sheet metal flashing to the box front (or tree) and drill the hole through the flashing.

Here are some other measures you can take to protect native cavity nesters from non-native competitors:

- Don't include perches on nest boxes.
- If your nest box is for a species that requires an entry hole larger than 1⅛ inches in diameter (for instance, flickers), don't put the box out until you see the desired species in your area. This may give a desirable bird a chance to occupy the nest box before undesirable species find it.
- Any time you see house sparrows or starlings entering a

box, open it to remove their nesting materials and eggs, or plug or tape over the hole for a few days. Because these birds aren't protected by state or federal laws, it is legal to remove their nests and destroy the eggs.

- Open a nest box in use by house sparrows or starlings and vigorously shake the eggs. If you leave the eggs in the nest, the eggs won't hatch, but the adults will continue to incubate the eggs and not attempt to build another nest.
- Attach a piece of metal roof flashing (or a flattened can) to the nest-box roof, extending this vertically above the box. This keeps house sparrows from perching on the roof and driving off desired species (see Figure 1).
- Don't attract house sparrows and starlings to your area. If you see these birds at your bird feeders, take measures described in Chapter 21.
- Mice, bees, paper wasps, fleas, earwigs, and sow bugs also may use bird boxes. To discourage mice and insects from using a nest box, leave it open or take it down after you are sure the birds' nesting activity is over. To discourage use by insects before birds arrive, coat the inside top of the box with bar soap; the smell may drive insects away.

How to Prevent Predators from Reaching into the Box

Racoons and some birds may be able to reach into a nest box entry and kill or prey on the occupants. You can make a predator block by drilling an appropriate-sized hole (see Table 1 for dimensions for specific bird species) in a piece of wood 1 inch thick. When this is mounted over the entrance hole to the nest box, it creates a short

tunnel into the nest (see Figure 1). Although a raccoon may still get its arm into the box, it won't be able to bend its arm down to reach the eggs or chicks. Similarly, a starling's head may be able to partially enter the hole, but it won't be able to bend and disturb the occupants. Examples of predator guards that prevent raccoons and other mammals from climbing a tree or pole that contains a nest box are shown in Chapter 21.

Nest Box Placement

When to Put Boxes Out

Most birds select nest sites from late March through May. This is the time when swallows and other migrating birds return to the Pacific Northwest from as far away as South America.

Mark the calendar when you first observe birds accessing the nest box and use this as a reference in following years.

Where to Place Boxes

Place nest boxes where they will be out of the reach of predators, vandals, and improper exposure to the elements. Other things to keep in mind when placing nest boxes in your landscape include:

- Place boxes so birds have a clear flight path to the entrance hole.
- Place boxes in a dry spot, preferably with early morning light and shade from hot sun.
- Place boxes so the entrance faces away from prevailing winds.
- Place boxes with the front tilted slightly downward so rain won't seep in at the entrance.
- Place boxes in an area convenient for cleaning, observing, and monitoring.
- Mounting the box at least 6 feet up will help to deter predators.

Table 1. Nest-box dimensions

Bird species	Floor of box (inches)	Depth of cavity (inches)	Entrance above floor (inches)	Diameter of entrance (inches)	
Black-capped chickadee	4 x 4	9	7	1 to 1⅛ ★	See Appendix D
Chestnut-backed chickadee	4 x 4	9	7	1 to 1⅛ ★	See Appendix D
Tree swallow	5 x 5	6–8	4–6	1¼	See Appendix D
Violet-green swallow	5 x 5	6–8	4–6	1¼ ★	See Appendix D
Purple martin	6 x 6	6–8	4–6	2¼	
House wren	4 x 4	6–8	4–6	1 ★	See Appendix D
Red-breasted nuthatch	4 x 4	9	7	1¼ ★	See Appendix D
White-breasted nuthatch	4 x 4	9	7	1¼ ★	See Appendix D
Western bluebird	5 x 5	8–12	6–10	1½	See Appendix D
Northern flicker	10 x 10	16–18	14–16	2½	See Appendix D
Hairy woodpecker	6 x 6	12–15	9–12	1⅝	See Appendix D
Downy woodpecker	6 x 6	9	7	1¼	See Appendix D
Wood duck	12 x 12	22–26	18	3 high, 4 wide	See Appendix D
American kestrel	8 x 8	12–15	9–12	3	See Appendix D
Western screech owl	8 x 8	12–15	9–12	3	See Appendix D
Northern saw-whet owl	8 x 8	12–15	9–12	3	See Appendix D
Barn owl	18 x 18	15–18	4–6	6	See Appendix D

★ This species will also use the diamond-shaped entry hole shown in Appendix D.

Nesting Information for Common Species

American Kestrel: Install the box 10–30 feet high in an open woodland or mature orchard, along a rural roadside, around a farm, or at the edge of a large field, also on the side of building close to perching trees. Place 3 inches of wood shavings or chips on the floor and install the box by early February. Monitor the box to prevent starling use. The female incubates the eggs, which hatch in 29–30 days.

Barn Owl: Install the box 12–18 feet high or higher in an open woodland, next to a large grassy area or a wetland scattered with trees. A nest box can also be placed in a large building that has a large opening for owls to enter. To prevent overheating don't place the box near the peak of a metal roof. Reduce human disturbance as much as possible, or owls will abandon the nest. Clean out all droppings and pellets and scatter a layer of wood shavings or dry hay inside the box in the fall. Barn owls mate for life and will return to the same nest every year. The eggs are incubated by the female and hatch in 32–34 days. The chicks fly at 8–10 weeks of age. When food is

Construction Tips and Useful Tools

Here are some things to remember when you construct nest boxes:

- Select straight boards with no large knots or split ends. Pieces of exterior plywood suitable for use in building nest boxes are often available free at construction sites.
- Self-drilling, self-countersinking deck screws tend not to split the wood. Screws also make it easier to fix mistakes and replace damaged parts.
- You can make a simple hinged side using two pivot screws or nails inserted directly across from each other (see Appendix D).
- To roughen the inside of the front panel of a nest box, cut parallel shallow (1/8-inch) grooves with a handsaw, or use the claw of the hammer. You can make a scratching tool by attaching six screws to a block of wood small enough to fit the palm of your hand. Rub the slightly protruding screw points across wood surfaces to quickly create numerous shallow cuts to which birds can cling.
- Drill entrance, ventilation, and drainage holes before assembling the box.
- The easiest way to drill an entrance hole is with an appropriately sized hole saw attached to a power drill. An entrance hole also can be created with a jigsaw after drilling a start hole, or you can drill multiple holes within the entrance section and complete it with a wood rasp or file.
- Pre-drill screw holes with a drill bit slightly smaller than the screw to prevent the wood from splitting when the screw is used.

continued on next page

scarce or after a hard winter barn owls will lay fewer eggs or not nest at all. Monitor the box to prevent starling use.

Black-capped Chickadee and **Chestnut-backed Chickadee**: Install the box 6–15 feet high. Locate it at the edge of a forest, in an open woodland, in a mature orchard, or in a residential area with mature trees. The box should receive about 50 percent sunlight throughout the day. Place an inch of wood chips or shavings in the bottom of the box. The eggs hatch in about 12 days and are incubated by the female. The box may also be used by house wrens and nuthatches.

Northern Flicker, Downy Woodpecker and **Hairy Woodpecker**: Install the nest box 5–25 feet high at edge of a forest or in an open woodland. If possible, attach the box to the side of a sturdy dead tree. Fill the box to the top with wood chips or shavings so birds can excavate a cavity to suit themselves. Tamp down the material so that it simulates soft heartwood in a dead tree. No additional nest material is brought into the box. The box may be used as a winter roost site and should also be filled in the fall. Incubation of the eggs is done by both sexes, and eggs hatch in about 12 days. Monitor flicker boxes to prevent use by starlings.

House Wren: Install the box 6–10 feet high near shrubbery, thick undergrowth, or under the eave of a building. Put the box out in early February. The male arrives first, establishes territory, and builds dummy nests of twigs in many available nest sites. House wrens are known to raise multiple broods, so clean out the nest box as soon as one family leaves the nest. Because assorted sticks are carried into the box, a diamond-shaped entry hole is recommended (see Appendix D). The eggs are incubated by the female and hatch in 12–15 days.

Northern Saw-whet Owl: Install the box 5–20 feet high in a deciduous or coniferous forest near water. Owls may use a box erected in a large shade tree near a large open space. Put an inch or so of wood chips or shavings in the box. An owl may appear at the entrance hole when the tree is tapped. When incubating, the female will not leave the box. No nest material is added to the nest except breast feathers. The eggs are incubated mostly by the female, and hatch in 21–28 days. Monitor the box to prevent starling use.

Purple Martin: Install the nest box 10–20 feet high (only 4 feet above water) on pilings or near a source of water, such as a stream or wetland. Boxes should be erected away from overhanging limbs or buildings as birds spend much time circling the nest site; 60 feet is preferable, but shorter distances may work. If more than one box is placed on a pole or piling, the heights should be staggered so that all boxes aren't on the same plane. Adults return to the same colony to nest each year, while their young return to the general area of the colony. Therefore, the probability of attracting martins will be highest if there is a colony in the area. Placing nest boxes over water will help prevent invasion from starlings. The three to five dull white eggs are incubated by the female and hatch in about two weeks.

Red-breasted Nuthatch: Install the nest box 5–15 feet high at the forest edge, in an open wooded area, or in a mature orchard.

Nuthatches will not nest near one another; each nesting pair of nuthatches needs several acres of territory around their nest that are free of other nuthatch nests. The birds heavily smear pitch around the entrance to the cavity. Pitch may prevent insects, small mammals, or other birds from entering. To avoid pitch, the female entering the cavity generally flies straight in. The five to six eggs are white or creamy, usually speckled, incubated by the female, and hatch in about 12 days.

White-breasted Nuthatch: Install the nest box 5–15 feet high at the forest edge, in an open wooded area, or in a mature orchard. Like red-breasted nuthatches, white-breasted nuthatches need several acres of territory around their nest. The nest consists of bark shreds, grass, twigs, and feathers. The female is fed by the male while she incubates the eggs, which hatch in about 12 days.

Tree Swallow: Install the box 6–15 feet high and near a wetland, creek, or open area near water. Tree swallows commonly nest alone but will nest in colonies where suitable tree cavities or nest boxes are available near water. The female tree swallow builds the nest, taking a few days to a couple of weeks. The eggs are incubated by the female and hatch in 13–14 days. If the box is cleaned out after the tree swallow family leaves, it may be used by bluebirds if they are in the area. Monitor the box to prevent house sparrow use.

Violet-green Swallow: Install the nest box 6–15 feet high at the forest edge or in a residential area with large trees and shrubs. This swallow commonly nests under the eve of a house. It has a strong

territorial sense, so place the box where the birds can't see the entrance holes of nests occupied by other swallows. The four to five oval white eggs are incubated by the female; they hatch in 13–14 days. Monitor the box to prevent house sparrow use.

Western Bluebird: Install the nest box 4–10 feet high near a rural roadside, pasture, or an open woodland that faces an open field. Tree swallows will sometimes compete for bluebird boxes. Bluebird territories can overlap a little if there is a major obstruction between them, like a building. Human activity is disruptive to them, so place their box out of areas frequented by people. If a house wren fills the box with sticks, remove the nest material and relocate the box away from a wooded area. The four to six eggs are blue, incubated by the female, and hatch in two weeks. Monitor the box to prevent house sparrow use.

Western Screech Owl: Install the box under the eave of an outbuilding or 10–30 feet up in a tree located in a forest, woodland, or mature orchard. Put an inch or so of wood chips or shavings in the box. The box may be used by owls for shelter in winter. The eggs hatch in 21–30 days and are incubated by the female, who will usually sit tight on the eggs when disturbed. Monitor the box to prevent starling use.

Wood Duck: Install the box 10–20 feet high (or at least 3 feet above the highest water level anticipated during nesting). Locate the box next to a marsh, slough, large ditch, pond, or creek with plenty of waterside trees and shrubs. Some still water is essential for brood rearing. Place the opening

An awl may also be used to create a screw hole. This step is especially important when youngsters or less-experienced woodworkers are helping. Using self-drilling, self-countersinking deck screws will reduce the need to pre-drill.

- Use an electric or cordless drill and screwdriver bit to make adjusting wood screws easier.
- Don't paint the interior of the nest box; paint can give off fumes and fill up the pores of the wood, lessening its insulating property.

Useful tools for constructing nest boxes include:

- Carpenter's square for measuring right-angles.
- Tape measure.
- Saw (hand or power).
- Drill (hand, electric, or cordless, with screwdriver and expansion bits) for assembly and holes.
- Wood rasp or sandpaper to smooth rough edges around entry holes.
- Hammer (if using nails instead of screws to assemble boxes).
- Phillips and/or flat-tip screwdriver (in place of electric drill and screwdriver bits).
- Awl or other small pointed tool for getting screws or nails started.

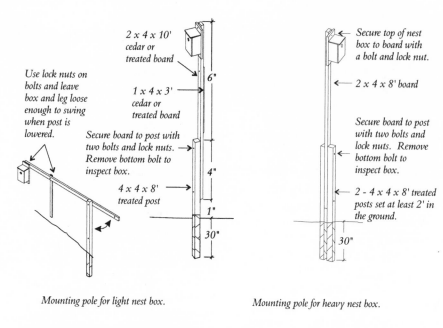

Use lock nuts on bolts and leave box and leg loose enough to swing when post is lowered.

2 x 4 x 10' cedar or treated board

1 x 4 x 3' cedar or treated board

Secure board to post with two bolts and lock nuts. Remove bottom bolt to inspect box.

4 x 4 x 8' treated post

6"

4"

1"

30"

Mounting pole for light nest box.

Secure top of nest box to board with a bolt and lock nut.

2 x 4 x 8' board

Secure board to post with two bolts and lock nuts. Remove bottom bolt to inspect box.

2 - 4 x 4 x 8' treated posts set at least 2' in the ground.

30"

Mounting pole for heavy nest box.

Figure 3. *Techniques for installing poles for nest boxes.*

A Method for Providing a Base for a Pole:

Step 1. Buy a length of pipe that's equal to the required height of the nest box from the ground plus 2 feet. Buy a 2-foot section of another piece of pipe of a slightly larger diameter than the first pipe. (This second piece will be placed in the ground, and the first pipe will slide into it.)

Step 2. Dig a hole about 2 feet deep and 12 inches in diameter.

Step 3. Place a 2-foot section of the short, larger-diameter pipe in the hole, standing upright and center on the bottom of the hole.

Step 4. Add some rocks to provide lateral stability and stiffness.

Step 5. Backfill the hole with dirt, rocks, or poured concrete.

Step 6. Affix the nest box to the top of the long, smaller-diameter pipe, and slide the bottom of the long pipe into the short pipe.

Step 7. To get at the nest box for inspection or cleaning, or to move it, lift the long pipe out of the short pipe.

toward the water. More boxes may be placed in heavily vegetated areas than in sparsely vegetated ones. Boxes may be placed 30 feet apart where dense foliage conceals them from one another. Boxes that are highly visible over water should be placed 100–150 feet apart. To attract wood ducks to regions where few are present or where production is low, place nest boxes in open areas. Once ducks occupy the conspicuous boxes, add new boxes to less conspicuous sites before the next nesting season.

The box should always be vertical or lean slightly to the front so the small birds can climb out. Also, the bottom of the box should have a few inches of wood shavings in it. Female wood ducks will often return to the same box year after year if it is maintained.

A fairly common occurrence among wood ducks is the "dump nest." These nests are used by a number of birds and may contain more than 30 eggs. Eventually one female may sit on the nest with varying hatching success. Incu-

bation period is about 30 days. Wood duck boxes are also used by kestrels, small owls, swallows, and other cavity-nesting ducks. Monitor the box to prevent starling use.

Mounting Nest Boxes

There are many options for mounting nest boxes (see Figures 3 and 4). They should be fastened securely and accessible for maintenance. The height at which a box is mounted is not critical if predation and vandalism are controlled. This means nest boxes for smaller cavity-nesters can be mounted using a small ladder.

TV antenna poles, galvanized metal pipe, and metal or wood fence posts make good mounting poles when set securely in the ground.

Mounting a large box or any box high in a tree is difficult and best done by two people. When installing a nest box on a tree it is a good idea to use a technique that will allow the tree to grow without putting pressure on the back of the box, causing it to split. One technique is to attach two 8-inch lengths of plumber's tape (perforated metal tape found in hardware stores) to the top back of the box, then nail the tape to the tree. Be sure to first bend back the ends of the tape to eliminate sharp edges.

Another approach is to use a lag screw and washer. These screws can be gradually loosened over the years to allow the tree to grow. Fasten large boxes with lag bolts where snow and heavy winds occur.

To temporarily secure the box while you screw or nail it to the tree, drill a small hole in the back of the box, pound a nail into the tree at the desired height, and then hang the box from the nail (see Figure 4). Note that this is not intended to hold the nest box in place any longer than it takes you to secure it properly.

Maintenance

All nest boxes attract insects. In small numbers, mites, lice, fleas, and flies generally are harmless. In larger numbers, however, most of these insects can cause injuries and fatalities to young birds.

Inspect all nest boxes each year in early spring to clean out insects. You may also need to tighten screws, loosen lag bolts, and unblock drainage holes. Used nesting material, dead nestlings, and infertile eggs also should be removed at this time. Removing nesting material keeps adults from building on top of an old nest, raising the nest contents dangerously close to the entrance hole. When cleaning out the nest material, spread it away from

the nest box so it doesn't create a smell that will attract predators.

If you are watching a nest box closely enough to know when the birds have finished raising a brood, clean the box between broods. Cleaning the nest box after each brood may encourage several pairs of birds to use the nest box throughout the nesting period, or the same pair to nest again.

Avoid bothering an occupied nest box. If for some reason young birds jump out of a box, don't panic. Pick up the birds, warm them in your hands, and then put them back in the nest. You need not worry that the adults will reject the nestlings if you handle them. Most birds don't have a well-developed sense of smell.

If an entire nest box falls, rehang it after making sure the nest and any young birds in it are all right. The adults generally will take over, but if they don't, call the Audubon Society chapter in

Platforms for Eagles and Osprey

While open-topped live or dead trees are preferred natural nest sites for bald eagles and osprey, a variety of artificial platform designs have been used for nests. Most of these consist of a frame or solid base that is mounted atop a tree or artificial support.

The benefits of artificial platforms include: provision of nests in areas that lack sufficient natural nest sites, replacement of insecure natural nests, relocation of nests away from excessive disturbance, and substitution for nests located on hazardous or conflicting man-made structures.

Artificial nest sites should be selected that provide maximum visibility of the surrounding terrain, unobstructed approaches to the nest

continued on next page

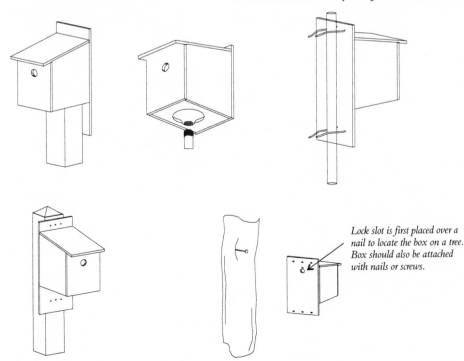

Lock slot is first placed over a nail to locate the box on a tree. Box should also be attached with nails or screws.

Figure 4. *Techniques for mounting nest boxes.*

so the birds can easily land, and a nearby resting perch with visibility requirements similar to those of the nest. Nests for osprey are frequently located next to beaver ponds, lakes, or rivers. Nest sites for bald eagles are often located in similar areas but may be up to 5 miles away from open water.

The support for the platform may be a snag, a live topped tree, a pole, or a tripod structure. A single support is generally sufficient, but tripods are more effective on lakes subject to heavy ice movement and in wetlands where the substrate is too soft to support a pole. Artificial structures often stand longer than snags.

When installed, the support should hold the platform at least 12–15 feet above the ground or surface of the water. Trees and snags should be topped to a level where the wood is solid. If predation is a problem at the site, a 4-foot-long strip of sheet metal can be attached around the middle section of the pole or tree; conical predator guards can be used on tripod supports (see Chapter 21).

Materials for the nest itself do not need to be supplied, although some sticks may be wired to the nest to help stimulate use by the birds. Osprey and eagles are very resourceful at finding branches and other material to construct their nests. Life preservers, rope, clothing, driftwood, and shingles have all been found in these nests. A typical nest for a bald eagle may be 4 to 6.5 feet in diameter and 1–2 feet deep; sticks up to approximately 7 feet long are used as nesting material.

For current design information and other help, contact your local Department of Fish and Wildlife office.

your area or a local rehabilitator for help. (See Chapter 23 for more information on orphaned birds.)

If a Nest Box Goes Unused

There is no certainty that birds will use the box you install, especially during the first year. Consider the following before becoming discouraged:

- Is the desired species found in the area?
- Is the nest box located in the appropriate habitat?
- Is the hole the right size?
- Is the nest box too near human traffic?
- Is there another nest of the same species close by?
- Did other birds, such as house sparrows or starlings, interfere with the nesting of the desired species?

- Was the nest box set up too late in the year?
- Has the nest box been cleaned of spider webs, wasps, and bees lately?

Public Nest Box Projects

You can volunteer to install and maintain nest boxes on property other than your own. "Adopt a Nest Box" projects also can involve groups such as a local Audubon chapter, school science classes, neighbors, a 4H group, or a scout troop. Farms, greenbelts, nature centers, parks, and schools make great places to install and maintain boxes. Attaching a small identification plate to each box with the name of an individual or organization may deter vandalism.

CHAPTER 13

BATS AND BAT HOUSES

Bats are mammals whose forelimbs are modified as wings (Figure 1). The fingers are elongated and serve as a framework for a thin layer of skin. The tail and hind limbs are connected by a similar membrane. Bats are not blind and do not become entangled in people's hair; if a flying bat gets close to your head, this is only because it is hunting mosquitoes and other insects that have been attracted by your body heat and emissions from your breath. The advantages of having bats around far outweigh any problems you might have with them. (For information on bats and rabies, see Chapter 5.)

Figure 1. The wing of a bat is actually a modified hand with greatly elongated fingers.

Bat Lives

A typical daily activity pattern for most Pacific Northwest bats begins at dusk when they leave their primary roosting sites, which they use during the day. These sites are generally secluded, crevice-like places, such as hollow trees, woodpecker holes, areas under shingles, attics, or bat houses. Some species characteristically fly in early evening, including twilight periods while others fly later. This difference in activity times, coupled with preferences for different types and sizes of insects, may reduce competition among species where their ranges overlap. Typical foods include caddisflies, mayflies, termites, moths, beetles, mosquitoes, crickets, and leafhoppers. Most insects are eaten in flight, but some bats catch them on the ground and in trees.

As primarily predators of night-flying insects, bats play a key role in preserving the natural balance of your property. Although many birds, including swallows, consume large numbers of flying insects, they are back in their nests by the time some insects, such as mosquitoes, are out in full force. But the little brown bat, flying at night, is capable of consuming one-half its weight in insects in one evening.

After an hour or two of intense foraging, the bats retreat to night roosts where they may remain for varying lengths of time. These can be caves, areas under the bark of large trees, or human-made structures such as bridges, barns, carports, or under eaves of houses. These sites are generally more exposed than are primarily daytime roost sites.

An additional foraging-activity period may follow the time of night roosting. After that, the bats return to the primary roost sides, where they spend the day. On average, mother bats give birth to only one young per year; babies are raised in the primary roosting site until they are grown.

In fall, usually October or November, some Pacific Northwest bat species migrate to Mexico and other warmer areas. Some species hibernate here using cave entrances, tree cavities, attics, or perhaps your bat house. They always return to the same hibernating places. Then, with the onset of warmer temperatures in March or April, the bats return to the places where they were born.

How they find their way back to these spots each year is still a mystery. They probably use mountain ranges and certain reference points along the way. It's likely that information is also passed along from generation to generation.

Considering their small body size, bats are extremely long-lived. Little brown bats, a species that commonly uses bat houses, often live ten years or more.

One way to attract bats to your landscape is to plant spearmint, phlox, stock, and flowering tobacco, to attract night-flying moths, which are eaten by bats. (For more on moths, see Chapter 18). Another is to provide open water. Bats drink by scooping up mouthfuls of water with their lower jaws as they fly over lakes, streams, water tanks, or ponds. Water also attracts insects that are important to bats' diet.

A well-designed, well-constructed, and properly located bat house may also attract bats if the appropriate species live in or pass through your general area. To tell if bats live in your area, watch for them at dusk or around street lights and other bright lights that attract insects at night. You might also check with local nature centers or pest-control companies to see if they have received calls about bats entering buildings; these are usually the same species most likely to use bat houses.

Bats that Use Bat Houses

Big brown bat, *Eptesicus fuscus*
California myotis, *Myotis californicus*
Little brown bat, *Myotis lucifugus*
Long-eared bat, *Myotis evotis*
Pallid bat, *Antrozous pallidus*
Yuma bat, *Myotis yumanensis*

Don't worry that adding a bat house to your property will encourage bats to move into your attic or wall space. If bats liked your attic or wall spaces, they probably would already be living there. For more information on bats in buildings, see Chapter 21.

Choosing and Constructing a Bat House

In designing or choosing a bat house, it is important to remember that bats need to conserve energy. In order to do this, they use day roosts and hibernation spots where air temperature is going to remain relatively stable; this allows them to divert precious energy to tasks other than regulating their own body temperature.

In summer, bats seek out warm places—similar to their own "operating body temperature" of a bit over 100 degrees Fahrenheit—to roost in during the day so they can convert all their food intake to growth and maintenance, fetus development, and/or fattening up for the coming winter months and hibernation.

In winter, bats actually seek out the cooler, humid, constant temperatures of places such as caves. Here they turn down their metabolism (respiratory and heart rates) and drop into a hibernation state in order to conserve energy.

Whether it is going to be used as a day roost in summer or for hibernation in winter, a well-designed bat house will provide its residents with relatively stable temperatures. If it is to be used as a day roost, it will also be designed and positioned so that the interior will maintain stable, warm temperatures during the day.

Like nest boxes for birds, bat houses can be purchased through the mail, from wildlife specialty shops, and from other commercial sources. While some of these bat houses include important design features described here, many do not and will go unused. You can design a custom bat house using the information in this section or you can use the drawings provided Appendix D. For additional bat house designs, contact Bat Conservation International (see Bat Houses in Appendix E).

Design

Bat houses don't have to be complex to be successful. Simple and inexpensive, yet large, single-chamber bat houses can be made from ½-inch-thick plywood and mounted on wood or masonry structures such as bridges, barns or other buildings. Bats often roost behind similar structures such as shutters and billboard signs.

The best bat houses are approximately 2 feet tall, at least 14 inches wide, and have a 3- to 6-inch landing area extending below the entrance (most species originally roosted where they could land on rough bark or rock). The house can contain one to four roosting chambers. Simple single-chambered houses are often successful when mounted against wood or masonry structures which help to buffer temperature fluctua-

tions. Houses with three or four chambers are more likely to provide appropriate temperature ranges; they also accommodate larger numbers of bats. Houses with more than four chambers may not absorb adequate solar heat in cool climates.

Roost partitions should be spaced ½ to ¾ of an inch apart. Some small bats prefer roosting crevices ½-inch wide, while large bats, such as big brown and pallid bats, may prefer ¾-inch widths.

The inside partitions and outside landing areas must be scratched or roughened to provide bats with traction. These surfaces can be scratched with a claw hammer or a scratching tool. To make a scratching tool, screw several screws through a block of wood small enough to fit into the palm of your hand. Then rub the slightly protruding screw points across wood surfaces to quickly create numerous shallow cuts that bats can easily cling to. It is important to remove any lose, sharp pieces of wood that might poke bats or give them splinters.

Materials

Half-inch exterior plywood is ideal for fronts, backs, and roofs of bat houses; 1-inch board lumber can be used for the sides. Using ¼-inch plywood for roosting partitions reduces weight and leaves more roosting space within a given house size. Never use chemically treated lumber for anything other than a mounting pole.

All seams that don't fit snugly after construction should be caulked, especially around the roof, prior to painting. Silicon caulk is easiest to use, but choose a brand that remains flexible over time and is paintable.

Ventilation

Providing sufficient warmth without overheating is a key element in attracting bats. Ventilated houses in hot areas allow bats to move up and down the inside of the box to find their preferred temperatures. Ventilation slots are critical in all houses to be used where average high temperatures in July are 85 degrees Fahrenheit or above; in other areas they are not needed. Ventilation slots can be located approximately one-third up from the bottom of the box. Vents should be ½-inch wide to reduce light and keep unwanted guests, such as house sparrows, from entering.

Wood Treatment

All outer surfaces of bat houses should be painted with at least two coats of exterior paint to protect against deterioration and air leaks, and to modify the box's interior temperature. Bat houses in cool areas need to absorb much more solar heat than those in hot areas and should be painted with a darker color.

If you're not sure about the temperature needs of local bats, you can mount two houses side by side so they receive similar sun, but paint one darker than the other. When bats move in, observe them during temperature extremes. Their choices will provide important clues to their needs, enabling you to enjoy improved success with future houses.

Bat House Location

Many bats prefer house locations in open areas away from branches and other potential perches for birds of prey. Because bats tend to fly straight down when exiting a bat house, there must also be a

Table 1. Bat house colors and sun exposure

Determine the color of your bat house according to the average high temperatures in July. Adjust to darker colors for less sun. For instance, if you put a bat house on your Puget Sound house, paint it dark brown or gray if you can locate it somewhere where it will get at least ten hours of sun. If the bat house will get less than ten hours of sun, compensate by painting it an even darker color.

Average high temperatures in July	Color of bat house	Recommended hours of exposure to daily summer sun
80–85 degrees F or less (Coastal areas and areas around Puget Sound)	Dark brown or dark gray	At least ten hours
85–95 degrees F (Areas west of the Cascade mountains and away from large bodies of water)	Medium-dark brown, gray, or green	Five to seven hours
95–100 degrees F (Areas east of the Cascade mountains)	White or other very light color	Five to seven hours

vertical clearance of at least a few feet under a bat house. Most bats also prefer to live within a few hundred yards of water, such as a stream, wetland, lake, or large pond.

Even if you don't have the ideal location available, it may be worth trying a house in a next-best site. Bat houses have been occupied that were a mile or more from water, for example.

Sun Exposure

When choosing a bat house location, sun exposure must be carefully considered. Bats need warm houses, predominantly between 80 and 100 degrees Fahrenheit, so their houses require solar heating in all but the hottest climates. Too little sun exposure is the most important known cause of bat house failure, even in relatively

hot climates. Overheating, also a possibility, can be greatly reduced by ventilation slots that allow heat to build up above the house but not below.

Refer to Table 1 for the recommended hours of direct sun.

Mounting

Most wooden bat houses can be screwed directly to a wooden post or building. The bat house should be mounted at least 12 feet above the ground.

A tree is not the best place to mount a bat house. Houses on trees tend to receive less sun, and branches and twigs make entry difficult. Tree-mounted houses are also more vulnerable to predators. (Because bats enter and exit bat houses rapidly, owls typically need to watch from a nearby perch to successfully catch them.)

Sheet metal painted black or heavy-duty roofing paper. Taper the roost site starting at the bottom with a 2" opening.

Figure 2. A quick and easy roost site for birds and bats. For bats, attach the "bat wrap" as high as you safely can and still in full sun. This will provide bats an option of moving around the tree to a warmer or cooler area as their needs change throughout the day. Bats also need to be able to fly up into the roost so keep the area below the roost free of side branches. Because bats need a high roost, full sun, and free access, large conifers are the best trees for bat wraps. For birds, the requirements can be more general.

Bat Watching

The bat's nocturnal lifestyle makes it one of the most elusive watchable wildlife species. You may not find bats on your first attempt, but follow these tips to improve your chances:

- **Dawn and dusk are the best times to spot bats as they begin and end their night flights.**
- **Pick an open spot where you can see bats silhouetted against the lighter sky.**
- **Look for areas where night-flying insects abound; areas near water are the best.**
- **Floodlights and streetlights that attract insects may also attract bats.**
- **When you find bats, listen to their clicks and squeaks so you can follow their flight paths after dark.**

DO NOT try to find day roosts or watch bats during the day. Bats may abandon roosts if they are disturbed in any way.

DO NOT disturb hibernating bats in the winter. One disturbance can cause a bat to use up to 60 days' worth of fat reserves needed over its winter hibernation.

Bats find houses mounted on poles or buildings faster than they find those on trees. Houses mounted under eaves where they are exposed to the sun are often used. Sun-warmed air rises up the building face and gets trapped by the eaves; stored heat is radiated back at night. Finally, don't mount a bat house where it will be brightly lit during the evening hours.

Monitoring Bat Houses

Brief observations with a flashlight are unlikely to frighten bats if you avoid touching the house, and are as quiet as possible. To observe houses high on poles, one person may need to hold a light while another looks through binoculars.

Check the bats' choice of roost chambers within a house on hot and cold days. It is especially important to observe where they are located at the hottest and coldest times. If they are mostly using only the hottest or coolest part of a house (hottest being front and top), you may want to provide a new house that is darker or lighter in color or exposed to more or less sun. Bats readily switch to houses that better meet their needs.

To determine actual numbers using a bat house, count the bats as they emerge at dusk. To discover whether they are rearing young, look inside about 45 minutes after sundown for the flightless, often pink, young.

Why do flightless pups fall out of the bat house? In nature, bat mothers sometimes abandon a pup if there is not enough food for both, since the healthy mother can reproduce again the following year.

Why Bats Leave or Don't Use a Bat House

Bats may not be attracted to your bat house for a variety of reasons. The house may be poorly built and too drafty, or access may be impaired by branches. Most often bats reject houses that are not located appropriately to meet their need for solar heating. Try moving your bat house or painting it a different color if it has not attracted bats by the end of the second season.

Also, bats may not be able to live in your area due to heavy insecticide use, an inadequate food supply, or the presence of predators including house cats and owls. They may also already have sufficient local roosts.

There are several other reasons for bats abandoning a house. It may have developed cracks that cause water leaks or drafts; branches may have grown up to prevent access into the house. It may be occupied by wasps, or predators such as cats, raccoons, hawks, or owls may be attacking the bats. Sometimes bats simply find a better roost nearby, but typically they leave for a reason. Loss of an essential hibernation cave or mine, or foraging site, may also be the cause.

For information on how to use predator baffles to prevent predators from gaining access to a bat house, see Chapter 21.

CHAPTER 14

BIRD FEEDERS

Washington state bird-feeding surveys reveal that a surprising number of bird species visit feeders. While feeding wild birds doesn't make up for lost habitat, a well-placed feeder can allow you to see and enjoy birds close up and help foster an appreciation for nature.

In feeding birds, an important consideration is the potential for spreading disease. Diseases such as avian pox (the "canker" seen on pigeons), salmonella (food poisoning), and respiratory ailments are generally spread when too many birds crowd too close together. Disease is spread through direct contact with open cuts and saliva, and by birds sneezing and excreting around each other and in the food. Birds can also become ill from eating seed that has become moldy; this disease is called aspergillosis.

In order to prevent the spread of disease, you will have to make sure that the food offered to birds is always in good condition, and clean your feeders and the areas around them regularly; it's also helpful to prevent large numbers of birds from congregating in one area. These points are discussed at greater length throughout this chapter.

You'll also need to decide what kinds of birds you want to feed. Different birds have different food preferences; they may also have different preferences in the style of feeder they will use. Chickadees, nuthatches, and house finches, for example, will use a feeder placed well above the ground; other birds, such as juncos and sparrows, prefer to eat close to the ground. Information on types of foods and feeders found in this chapter will help you attract the birds you want and discourage those you would rather not see. For information on hummingbird feeders, see Chapter 17.

Supplemental or "Feeder" Foods

The type of food you offer is probably the single most important factor in determining whether birds will come to your feeder. It also plays a significant role in determining what kind of birds you will attract.

Seeds

Seeds are the food most often provided in feeders. They're rich in carbohydrates and supply valuable energy. You can purchase seed in bulk at stores that cater to the bird-feeding public.

Since different birds prefer different seeds, choosing the right seed is important. It will help to control which birds visit your feeder and to minimize waste. See Table 1 for details on the types of seed commonly offered at feeders and the types of birds they attract.

It's best not to offer seed mixes in your feeders. Not only can this encourage overcrowding at your feeder, but it also can cause birds to waste seed. This problem is likely to be especially severe if

Why Feed Some Birds but Not Others?

Many people who feed birds decide that they wish to feed some birds but not others. There are a number of reasons why you might make such a choice.

One is concern about "mess" around feeders. Some species, notably house sparrows, finches, and starlings, tend to remain at feeders for long periods of time, making a great deal of mess in the process. In contrast, chickadees, jays, and nuthatches come to the feeder, get what they want, and leave. If you're concerned about mess, you'll want to cater to the latter group.

Another concern is economy. Large birds and birds that travel in large numbers—finches, house sparrows, starlings, non-native pigeons (rock doves)—naturally tend to eat much more seed than do small birds that don't travel in huge flocks, such as chickadees and nuthatches.

People may object to jays and crows because they are noisy. Others enjoy watching these intelligent birds and appreciate the beauty of their plumage.

Many people prefer not to feed starlings and house sparrows because these non-native species compete with native birds for nesting sites. Many people prefer to discourage not only these birds but also other aggressive birds, such as crows and rock doves, which may discourage other bird species from visiting a feeder.

Homegrown Sunflower Seed Is for the Birds

Many feeder foods (corn, wheat, milo, millet, barley, sunflower) can be grown in a garden. One way to grow these is to sprinkle and lightly rake a small bag of mixed birdseed into a designated bird garden in spring. But what comes up may not be what you expected. For instance, sunflower plants may reach only three feet in height and have heads only six inches in diameter. These are the varieties preferred by the large-scale growers because the yields are higher and mechanical harvesting is easier.

For the flower border, sunflowers are available in a wide range of sizes, from a few feet tall to giants. They are easy to grow from seed, but short ornamental varieties are often available as plants at nurseries. Plant sunflowers in full sun in

continued on next page

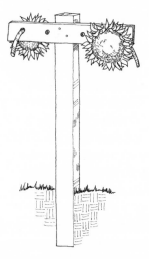

Figure 1. A sunflower-head feeder can easily be made by drilling several 1.5-inch holes in a 1 x 4-inch board mounted on a post or pipe. Drop the short piece of stem left attached to a sunflower head through one of the holes.

your seed mix contains mostly small seeds, such as millet and milo, that are not preferred by the perching birds that visit feeders placed high above the ground. These birds, which generally prefer sunflower seed, are likely to spill many small seeds on the ground as they hunt for the larger seeds they want. Unless ground-feeding birds (which do like millet and milo) eat the scattered seeds, the result will be wasted seed, higher feeding costs, and a food source for mice and rats near your home.

No matter what type of seed you buy, remember to give birds only fresh, dry seed that is free of mold or other contamination. Store seed in airtight containers to avoid moisture build-up, and keep it out of reach of raccoons, rodents, and grain-eating insects. Because squirrels and rats can chew through even the heaviest plastic container, an aluminum garbage can or other metal container with a tight-fitting lid should be used for outside storage.

Keep bulk seed in a cool, dry place, or buy smaller amounts to be able to store it safely. Always buy seed from sources that offer the proper varieties of fresh seed. Most seed deteriorates after a season or two; if it gets moldy or smells musty, it should be thrown away.

Suet

Suet is a type of food especially popular with insect-eating birds such as woodpeckers, bushtits, and chickadees. Other birds that are particularly attracted to a suet feeder are juncos, nuthatches, and red-winged blackbirds. The word "suet" originally referred to the hard fat that surrounds the kidneys and loins in cattle and sheep.

Suet Recipes You Can Make at Home:

Basic suet formula
Heat to boiling 1 part suet and 6 parts water
Add 2 parts cornmeal, ½ part flour
Cool, pour into cupcake molds, and allow to harden
Serve in appropriate feeder

Suet/peanut butter mix
(The relative proportions of these ingredients may vary)
Melt 2 parts suet, let it harden, and then melt it again
Blend in 1 part yellow cornmeal and 1 part peanut butter
Allow to thicken, pour into molds, and allow to harden
Serve in appropriate feeder

Oatmeal/lard mixture
Melt 1 pound of lard and add 5–6 cups oatmeal
Let cool
Store in container

However, it is now used to refer to any fat used to feed birds. Suet can be obtained from a grocery store or meat market, or neatly packaged from stores that cater to the bird-feeding public.

Suet doesn't freeze solid, so birds can feed on it in very low temperatures. However, it will melt in warm weather and can present a problem for the birds if the melted fat coats the feathers around their beaks. Birds can also get "rancid fat disease," which is akin to food poisoning, in warm weather. For this reason, all fat products should be provided only on days when temperatures will be low enough to keep them from melting. If you provide suet during warm weather, use a commercially available type with a high melting point or put out only a small portion and monitor it daily.

Other Feeder Foods

Fruits, either fresh or dried, and nuts may attract birds otherwise not interested in your feeders. (See Figure 2 for a drawing of one type of fruit feeder.) Nuts are rich in protein, fat, and minerals. Peanuts, chestnuts, walnuts, acorns, and other nuts are most attractive to birds when the shells are opened and the meats are crushed or chopped.

In winter, when protein and fat are scarce, peanut butter can be a valuable addition to a bird's diet. It can be offered alone—smeared on a branch, pine cone, or other surface—or be included in a suet mix. It can be mixed with oats to make it less gooey. (Contrary to a widespread rumor, peanut butter does not cause birds to choke.)

a place where you will be able to watch chickadees, finches, grosbeaks, and jays eat the seeds that will ripen in late summer or fall.

When the plants are mature, some of the heads can be cut and hung to dry in a shed or attic for later use in a special feeder (Figure 1). Leave 18 inches of stalk attached and hang the heads in bundles of five or six with the heads facing out. Make sure there is adequate air flow or the seeds will mold.

Table 1. Types of seeds used in feeders

Seed Type	Information
Black-striped sunflower seed	Large size is awkward for smaller birds, especially when seeds become wet and leathery. Easily grown in the garden.
Black oil sunflower seed	Very popular with birds. Small, thin hull is easy for birds to manage. Easily grown in the garden. Hulls can accumulate under feeder and inhibit plant growth.
Hulled sunflower seed	Manageable size is very attractive to many small birds, including pine siskins and goldfinches. The least messy way to offer sunflower seeds. Oils in hulled seed can turn rancid in warm weather. Can get gummy if wet.
Millet	White or red proso millet is the main ingredient in less expensive commercial bird seeds. Both are popular with ground-feeding birds such as sparrows, juncos, and mourning doves. Easily grown in the garden.
Peanut kernels and peanuts in the shell	Highly attractive to jays, chickadees, and nuthatches; also squirrels. Use roasted peanuts only. Raw peanuts contain toxins that can cause diseases in birds and other animals.
Thistle seed	Not related to North American thistle and will not escape into gardens and wild areas because it is steamed to prevent germination. Seed is usually distributed from a specially designed "thistle" feeder. Especially attractive to pine siskins and American goldfinches.
Mixed seed	May contain mostly millet and milo with a small amount of other seeds in varying ratios. May be wasted when birds flip through the feed in search of preferred seeds.
Grains	Include oats, rye, wheat, barley, and corn. Whole grains are attractive to ground-feeding birds such as pheasants, quail, grouse, pigeons, juncos, mourning doves, towhees, and thrushes.
Chicken scratch	Contains cracked or crushed grains and is easily consumed by native sparrows, finches, and ducks. It can cake up and spoil quickly in a damp climate and should be replaced frequently.

Table 2. Feeders, feeder foods, and natural foods for common feeder birds

Feeder Types:
1: Tube feeder; 2: Thistle feeder; 3: Hopper feeder; 4: Platform feeder; 5: Suet feeder; 6: Ground feeder

Birds	Feeder Types	Feeder Foods	Natural Foods
Black-capped chickadee and Chestnut-backed chickadee	1, 3, 4, 5	Black oil and black-striped sunflower seed (hulled and unhulled), whole peanuts and peanut kernels, safflower seed, peanut butter, suet and oatmeal/lard mixtures. Commonly stashes feeder foods for later use.	Tent caterpillars, flies, wasps, aphids and other insects and spiders in all stages. Seeds from conifers, birch, and various weeds. Fruits of poison-oak, blueberry, serviceberry, thimbleberry, and other brambles. Also a mild scavenger.
Bushtit	4, 5	Peanut butter, suet and oatmeal/lard mixtures.	Insect eggs, larvae, pupae, and adults. Some fruit and seed.
Red-breasted nuthatch and White-breasted nuthatch	1, 3, 4, 5	Sunflower seeds, peanuts, suet and oatmeal/lard mixes.	Searches tree trunks for insects in all stages. Seed from oaks, conifers, wheat, and oats. Also some fruits, and a mild scavenger.
Varied thrush	4, 6	Cracked corn and grains, millet, raisins, apples, and suet mixes.	Beetles, millipedes, centipedes, ants, crickets, and snails. Fruits of madrone, snowberry, raspberry, apple, prune, honeysuckle, and poison-oak. Also seeds and nuts, including acorns.
Spotted towhee	4, 6	Sunflower seeds, whole peanuts and peanut kernels, millet.	Caterpillars, beetles, grasshoppers, ants, crickets, and flies. Fruits of crabapple, mulberry, dogwood, serviceberry, elderberry, blueberry, gooseberry, holly, raspberry, and Oregon-grape. Seeds and acorns.
Black-headed grosbeak and Evening grosbeak	1, 3, 4, 5	Black oil, black-striped, and hulled sunflower seeds, whole peanuts, and safflower seed. Seed salt from livestock licks.	Wasps, bees, and spiders when feeding nestlings. Seeds and fruits of maples, pine, spruce, fir, cherry, juniper, cascara, manzanita, dogwood, serviceberry, chokecherry, and mountain ash. Will crack the pits of the chokecherry for the inner meat.
American goldfinch	1, 2, 3, 4	Black oil and hulled sunflower seed, millet, and thistle seed.	Heavily reliant on thistle and dandelion seed. Also eats seed of ragweed, goldenrod, and chickweed; buds of fruit trees, and occasionally caterpillars, aphids, and other insects.

Birds	Feeder Types	Feeder Foods	Natural Foods
House finch and Purple finch	1, 3, 4, 6	Black oil, black-striped, and hulled sunflower seed, millet, thistle, and safflower seed.	Weed seeds including filaree, mustard, pigweed, chickweed, and miner's lettuce. Occasionally berries, nuts, and buds of fruit trees. Eats very few insects.
Song sparrow	3, 4, 5, 6	All sunflower seeds; also cracked corn and grains, peanut kernels, millet, thistle seed, and suet mixes.	Weed seeds including pigweed, knotweed, miner's lettuce, thistle, filaree, and chickweed. Insects added when feeding nestlings.
White-crowned sparrow	4, 6	All sunflower seeds; also cracked corn and grains, peanut kernels, and millet.	Weed seeds, flower buds from trees and shrubs, and berries.
Pine siskin	All	Small sunflower seeds, thistle seed, millet, and suet mixes. Especially fond of thistle seed.	Eats mostly seeds during the winter including alder, birch, elm, maple and conifer; also dandelion, thistle, and other weed seeds.
Dark-eyed junco	3, 4, 6	Black oil and hulled sunflower seed, peanut kernels, millet, thistle seed, and occasionally suet mixes.	Eats mostly insects during the nesting season. In colder months eats weed seeds, vegetable/flower garden seeds; also seeds from birch, hemlock, pine, juniper, and sumac.
Northern flicker	1, 4, 5	Sunflower seeds and suet mixes.	Ants, mostly from the ground, and other insects. Seasonal fruits from the wild, garden, and orchard. Also nuts and weed seeds including grasses, clover, pigweed, ragweed, and mullein.
Downy woodpecker and Hairy woodpecker	1, 4, 5	Sunflower seeds and suet mixes.	Primarily insects and other invertebrates excavated or gleaned from trees. Also some nuts, berries, and seeds.
Steller's jay and Scrub jay	3, 4, 5, 6	Black oil, hulled, and black striped sunflower seed; also cracked corn and grains, whole peanuts, peanut kernels, safflower seeds and suet mixes.	Insects, scavenged road kills, acorns, seeds, and fruits. Stashes feeder offerings.
Red-winged blackbird	1, 3, 4, 6	Black oil, hulled, and black-striped sunflower seed; also cracked grains, millet, and suet mixes.	The seed heads of field crops including small grains, corn, and sunflowers; also weed seeds, grubs, and various insects.
Band-tailed pigeon	3, 4, 6	Black oil and black-striped sunflower seed, cracked corn, and grains.	Nuts, especially acorns; also fruit of manzanita, madrone, holly, and elderberry.
Mourning dove	3, 4, 6	Black oil and black-striped sunflower seed, peanut kernels, safflower seed, and millet.	Seeds, some nuts, berries and fruits, and corn left in fields.
California quail	4, 6	Cracked corn and grains, and millet.	Weed seeds, especially clover; also grain, leaves, and grass shoots. Crickets, grasshoppers, ants, and beetles.

Peanut butter isn't recommended as a summer food because it melts and can become rancid in warm temperatures.

Bread, crackers, and dough-nuts often attract starlings and house sparrows. Furthermore, bakery goods attract rats and mice to your home and lack nutrition-al value. They are not recom-mended as feeder foods.

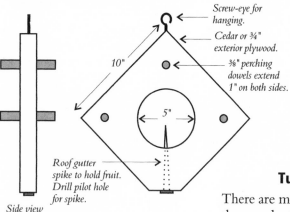

Side view

Figure 2. *A fruit feeder.*

Types of Feeders

A wide variety of feeders are avail-able from retail stores and mail-order catalogs that specialize in bird feeding. A visit to your local bird-feeding specialty store or a look at a mail-order catalog (see list in Appendix E) will help you familiarize yourself with the different feeders.

For those interested in build-ing feeders, there are woodwork-ing ideas and plans in the books listed in Appendix E. Some uni-que suet feeder designs are shown in Appendix D; and see Figure 5.

In choosing a commercially available feeder, you'll want to consider a number of factors. Among these is the type of bird you want to attract. Larger birds require feeders with longer or wider perching areas, while small,

agile birds happily use feeders with perches as short as half an inch. Some birds, such as chicka-dees, nuthatches, and woodpeck-ers, can even cling to wire mesh or an opening in the side of a feeder without needing a perch at all.

You'll also want to be sure that your feeder minimizes the likelihood of spreading disease. Such a feeder will:

- Limit usage by many birds at one time.
- Be easy to clean and maintain.
- Protect the seed from wet weather.
- Keep spillage and waste to a mini-mum.

Tube Feeders

There are many variations on the popular tube-shaped feeder. They're usually made of clear plastic and have two or more feeding ports (holes from which seed is dispensed), with metal or wood perches (Figure 6). (Because birds do not have sweat glands in their feet, they will not stick to metal perches in cold weather.) The clear plastic makes it easy to see when the feeder needs a refill. Tube feeders can accommodate only as many birds as there are perches and openings. This design ensures that seeds are removed with minimal waste. A feeder made of good-quality plastic will also be long-lasting and easy to clean. Short perches discourage larger birds.

Most tube feeders have large feeding ports to accommodate sunflower and other large seeds. The tube feeder for thistle seed has very small, slit-like openings that prevent birds with larger bills from removing the feed. Goldfinches, pine siskins, and redpolls will often use thistle feeders.

Platform Feeders

Platform feeders provide a spa-cious feeding area where food can be spread out for several birds at a time. They can be placed high up or set slightly above ground, where they are ideal for catering to ground-feeding birds such as juncos. Platform feeders provide the extra room needed by larger birds such as jays, but can be covered with chicken-wire if you want to use the feeder for smaller birds (Figures 3 and 7). A roof pro-vides protection from the elements and helps keep the seed dry. A rim around the edge reduces spilling due to wind or scattering by birds; however, seed hulls may build up in the tray unless they are discarded daily. Most commercially available platform feeders now are made with a wire-mesh tray that holds seeds while allowing water to drain

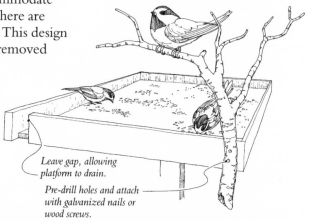

Figure 3. *A platform feeder with clean-out gap and a perching branch.*

away quickly. If your platform feeder has a wooden floor, it should have drainage holes (which you can drill). Screening can be used to prevent seeds from falling out.

Because of their size, these feeders allow many birds to gather at one time. Hence they require more careful management than other types of feeders.

Hopper Feeders

A hopper feeder has a container (a hopper) that releases seed to a bottom tray as birds eat it. These feeders range in size from that of a small loaf of bread to that of a 55-gallon barrel. A design for a large feeder for quail and other large ground-feeding birds can be provided by your local Department of Fish and Wildlife office.

Some of the larger hopper feeders, particularly those with wide perching areas, will accommodate larger birds than will tube feeders with short perches. However, the design of hopper feeders often makes them more susceptible to accumulation of hulls, seeds, and bird droppings in the feeding area, so they may have to be monitored more closely and cleaned more often than tube feeders are.

Although it is possible to find hopper feeders made of other materials—including tough, long-lasting recycled-plastic lumber—many hopper feeders are made of wood. Wood can be difficult to clean, can be chewed by squirrels, and may take a long time to dry out if it becomes damp.

Ball-shaped Feeders

These round plastic seed feeders come in a variety of sizes (Figure 4).

They are designed without perches, in such a way that they

Figure 4. A ball-shaped feeder allows one bird to feed at a time.

can be used only by birds capable of clinging to the seed port; this design tends to eliminate use by house sparrows and starlings and to discourage use by house finches, so this is a great feeder for those who want to keep mess to a minimum.

Suet Feeders

You can offer suet and suet mixes to birds in a variety of different feeders. An inexpensive feeder made of rubber-coated wire, designed to hold commercially available suet blocks, is easy to come by. Suet can also be softened by heating and then pressed into the crevices of small hanging pine cones, or placed in a special feeder for woodpeckers and smaller birds such as nuthatches. Jays, crows, starlings, and magpies work hard at getting suet. Your local bird specialty store can give you information on new designs to deter these birds if they're a problem. (See Appendix D for examples of suet feeders you can build.)

Managing Feeders

The best feeding arrangement gives you the opportunity to observe a variety of different birds who can feed in safe, comfortable condi-

tions. Problems occur when feeders containing the same food are hung close together and where too much of one type of food is placed in an area where many birds have access to it. Therefore, moderation is the key to a successful supplemental feeding program. Too much kindness can be harmful.

The preferred way to attract a variety of birds is to place single types of food in separate feeders designed for that food. This way, different bird species can select their preferred food with less crowding, waste, and mess. For instance, to observe a variety of birds, you might have a suet feeder, a ground-level platform feeder for millet, and a tube feeder for black

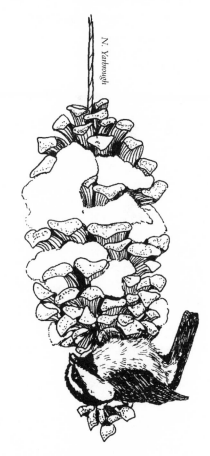

Figure 5. Suet can be placed into a pine cone and hung for birds.

oil sunflower seed. Because these feeders will attract different species of birds, they can all be located in the same general area.

Placement and Mounting

When you place your feeders, choose a location where they can be easily watched, enjoyed, and maintained. Many birds will come right up to a house in order to reach a feeder; placing your feeder next to a window allows you to see them a few feet away. Young children are especially thrilled to see the birds so close, as are the aged or infirm, who may not be able to see birds at a distant feeder.

Feeders should be placed in an area that is as quiet and undisturbed as possible—away from noisy traffic, pets, and house entrances, and out of strong winds. Placement of your feeder will affect maintenance: don't put a feeder where you will have to get out a ladder to reach it. Placement of your feeder will also have a pronounced impact on whether it is raided by squirrels.

If possible, place feeders where they can take advantage of winter sun. In rainy areas, try to place the feeder where it will stay as dry as possible, especially if the feeder is not watertight. Under your eaves is a good location; if this isn't possible, consider putting your feeder under a cover of some sort. Bird-feeding stores sell clear plastic domes, often called "squirrel baffles," that work well for this purpose. You can also make a bonnet for your feeder out of a piece of cardboard with foil or plastic folded over it for waterproofing.

All feeders should be placed with predators (especially domestic cats) in mind. In an area where

Figure 6. A tube feeder on an iron pole that can be placed on a lawn or elsewhere. A chicken-wire fence can be placed around a feeding area to give ground-feeding birds protection from domestic cats.

birds gather under a feeder, you should try to leave 10–15 feet between the feeding area and any hiding place a cat could use to prepare for an ambush. This will allow at least one pair of the many eyes in a group of feeding birds to spot a lurking cat and warn the others. You can also place chicken wire around a feeding area or between a feeder and the areas where cats lurk (see Figure 6).

Birds will feed much more readily if they have a place close by where they can perch and escape from other predators, including hawks. Birds also need a place to perch before making a visit to the feeder. A large open shrub, a thicket, a brush shelter, or a tree with low branches are all ideal forms of cover. If a natural perching area doesn't exist within 5 feet of your feeder, you can add

a perching branch. See Chapter 6 for information on perches and an example of a perch you can build.

Most styles of feeders can be mounted in any of several ways: hanging, on a pole, or on a window. A feeder may be hung from a tree, or from a bracket or overhanging eave in front of a window; it may also be hung from the arms of a pole. Feeders may be mounted at the top of poles made of metal, milled lumber, PVC, a slender tree trunk, or anything else that will support them.

Poles are handy in that they can be placed almost anywhere; there are some styles on the market that can be placed without the necessity of pouring a concrete footing. If you wish to prevent squirrels from gaining access to your feeder, this is often easier to do if your feeder is mountedon a pole. (See "Mounting Nest Boxes" in Chapter 12 for more information.)

Many styles of feeder are now available in designs that allow them to be attached to windows, usually using suction cups. An advantage of a window feeder is that is may help prevent birds from colliding with a window by making them aware that a barrier is there. If suction cups that attach a feeder to a window pop off, heat them in warm water, then dry and coat them on the inside with a thin layer of light oil. Place the cups on squeaky-clean glass while they are still warm.

Note: You may find the feeder on the ground if it's located within a squirrel's reach.

When to Feed

Some people choose to feed birds throughout the year. For birds that use them, however, feeders may be particularly important during the winter. While birds can find most of their own food during the summer, they may need supplemental food when their natural food has been depleted or covered with ice and snow. During very cold weather, a well-stocked feeder may be a lifesaver. Other birds may also benefit from a feeding station while migrating during harsh conditions in early fall or early spring.

Depending on the harshness of the weather, start your winter feeding activity around mid-October. It's best to continue feeding through mid-April (or later, depending on the weather), when food supplies are lowest. By spring, the natural foods needed to raise babies are in good supply.

Often, people worry that once they start feeding birds they

Figure 7. *Wire mesh placed over a platform feeder prevents larger birds like jays from dominating it.*

will need to continue without interruption—otherwise birds that have become dependent on the feeder will starve. It's more likely that birds visit a number of different feeders and utilize natural food sources in their daily search for food. Recent research has shown that the black-capped chickadee, a bird commonly found at feeders, takes only 25 percent of its food from feeders.

So if your feeder is temporarily empty, they will just go to the next stop. However, if you know you will stop feeding for a couple of weeks during the winter, you should try to wean the birds from the feeder slowly or get a friend to continue feeding while you're away. It is also a good idea to keep your feeder full during exceptionally cold weather.

People also wonder if feeding birds will change their migration pattern. Migration is triggered by photoperiod (length of day) and complex interactions of hormone levels and environmental factors.

By itself feeding cannot override the urge to migrate.

How Much to Feed

Never put out so much food that it will spoil before being eaten. Ideally, put out only enough for one day. This is particulary important with platform feeders and when you are putting seed on the ground because these situations can attract a large number of birds and rodents. Start with a small quantity and add more, if needed. If you are consistent, the birds will learn what time of day food will be placed out for them and will eat it by nightfall.

Feeder Maintenance

Maintenance of your feeders is important for the health of the birds you attract. Refill feeders frequently rather than putting large amounts of seed out for long periods of time. Keep feeders clean by washing them with a solution of vinegar and warm water at least once a month, weekly during heavy feeding times. Rinse and dry them thoroughly before refilling.

It's also important to keep areas around and under feeders free of contaminated, moldy, or otherwise spoiled seed. A variety of strategies will help you do this:

- Mount feeders above a surface such as concrete, which you can sweep regularly.
- Move feeders periodically.
- Cater to birds (jays, chickadees, nuthatches) that create less mess around feeders.

Tools for Feeder Maintenance

When assembling items for your feeding program, a few things to have on hand include:

1. A bottle brush for cleaning tube feeders and hummingbird feeders.
2. White vinegar for use as a disinfectant.
3. A good, stiff-bristled hand brush for sweeping off a platform feeder.
4. A putty knife for chipping away ice, droppings, and clumped-up seed.
5. A small ladder to make reaching high feeders easier.
6. A container to use in filling seed feeders from bins or bags.

• Offer shelled peanuts, hulled sunflower seed, or suet to eliminate hulls that would otherwise be dropped on the ground.
• Avoid mixed seed, especially mixes containing mostly milo or millet.
• Use small feeders that allow only one or two birds to feed at a time. You can also plug all feeding holes in a tube feeder, removing extra perches as needed.
• Place a tray under your feeder to catch spilled seed. These trays are available commercially. However, since birds will gather on a tray to eat spilled seed, trays must be cleaned frequently.

Food Poisoning

In winter, people who feed birds may notice that some appear lethargic and perch with their feathers puffed out and their eyes closed. Often these birds will sit at a bird feeder with their heads tucked into their feathers.

These birds likely have become infected with salmonella bacteria, the bacteria associated with a common form of food poisoning. Birds that feed in large groups—such as pine siskins, goldfinches, house finches, and juncos—are particularly susceptible. Infected birds become weak, dehydrated, and emaciated.

Messy platform feeders and ground-feeding areas are particularly conducive to the spread of the infection. To prevent spreading or catching salmonella, wear rubber gloves when cleaning feeders or handling dead birds. Cats should be kept from scavenging birds that you suspect are dying from the disease.

Coexisting with Wildlife Around the Feeder

Dealing with Aggressive Birds

A few aggressive, non-native birds can dominate feeders or exhaust the supply for other birds. Three of these species are the starling, the house sparrow, and the common pigeon (not to be confused with the native band-tailed pigeon). For additional information on controls for these species, see Chapter 21.

Because of their size, starlings and pigeons may be kept from seeds by using feeders with small (or no) perches and small openings. Some commercially available feeders designed to frustrate squirrels may also prevent large birds from reaching seed. (See the next section for more on this.) The cage design in Figure 7 will prevent starlings, pigeons, and squirrels from entering a feeding area for small birds.

Starlings are often attracted by suet. However, because starlings have trouble clinging upside down, a suet feeder that requires the birds to clasp the feeder in this position will help discourage starlings from using it.

Starlings are also reported not to care for safflower seed, hulled sunflower seeds, or peanuts in the shell.

Both starlings and house sparrows will be deterred by small feeders that swing and twirl whenever a bird lands on the perch. To discourage all three species, don't place large amounts of seed on platform feeders or on the ground.

Squirrels

For many people, squirrels are a delight to watch. However, when a squirrel invades a feeder, devours the expensive seed, and sends the empty feeder crashing to the ground, it may no longer seem so cute.

If squirrels are a problem, you may be able to keep them away from your feeder; an added advantage is that any setup that prevents squirrels from reaching your feeder probably will stop rats from getting to it, as well.

Because squirrels are world-class jumpers, you will need the right location for your feeder: at least 5–6 feet off the ground, and at least 6–8 feet from the nearest tree, building, or overhanging branch. To protect against Eastern gray squirrels (a non-native species and the most common squirrel in urban areas), these distances should be the maximum of those ranges.

Having found a location squirrels can't jump to, you now must prevent squirrels from reaching it by climbing either up or down to it. If the feeder is reachable from the top, a commercially available dome- or cone-shaped "squirrel baffle" can be placed over it. (This will also protect the feeder from rain; see Figure 8.) You can also use a series of smooth metal discs, such as pie pans, metal lids, or phonograph records held in place by short sections of garden hose or plastic tubing to prevent squirrels from climbing down to a feeder.

Squirrels can climb almost anything, including a narrow-diameter pipe. However, they may be unable to climb a smooth pole made of metal or PVC if its diameter is so large that they cannot wrap their feet around the pole. Five inches is the minimum diameter for Eastern gray squirrels. Other squirrels may be deterred by a smaller diameter pole.

You can also install a squirrel baffle to prevent squirrels from climbing a pole. These may be either flat or cone-shaped or they may be shaped like cans. They too are available commercially; see Chapter 21 for examples of homemade baffles.

Be aware that a tray placed under a feeder to catch spilled seed is an invitation to squirrels if they can reach it by jumping to it. Another invitation to squirrels is a feeder hung with rope, string, and sometimes even plastic or thin wire: squirrels will chew through lightweight materials in order to knock a feeder (and its delicious contents) to the ground.

Another way to discourage squirrels is by modifying your choice of food. Safflower seed is generally unpopular with squirrels, though it will be eaten by birds.

Old World Rats and Mice

These non-native rodents may be attracted to seed that is left on the ground or in platform feeders. Rats can also climb up and down wire to get to feeders. To prevent this, use the same methods that deter squirrels.

Old World rats and mice are generally most active at night, so provide only as much seed to ground-feeding and platform-feeding birds as they can consume during the day. For more information on rat and mouse control, see Chapter 21.

Figure 9. Hawks are a natural part of a wildlife landscape.

A Note About Hawks

A large concentration of birds around your feeder may attract a hawk or two, especially during the winter. During these lean months, birds of prey may venture into urban habitats in search of a meal. An occasional foray of a sharp-shinned or Cooper's hawk into a backyard wildlife sanctuary should be welcomed rather than treated as a problem.

Predation is a natural part of a well-functioning ecosystem. Hawks weed out the unfit and help maintain the health of the prey population. Healthy songbirds can usually protect themselves from hawks by taking to cover quickly. Trees, shrubs, thickets, and brush piles that you provide can serve them well.

Jays are quite often the target for predatory birds, so they make their loud alarm call when a hawk is spotted in the area. Encouraging jays on your property gives all birds a warning system. If you have a persistent hawk hunting around your feeding station on a regular basis, remove the feeders for a day or so and the hawk may move on.

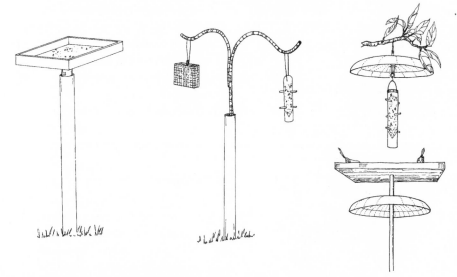

Figure 8. Feeders that are designed to keep squirrels from reaching the seed.

CHAPTER 15

BRUSH AND ROCK SHELTERS

If your yard lacks mature shrubs, thickets and tangles, or other forms of natural shelter, you can attract wildlife, particularly songbirds, almost immediately by building a shelter of brush or rocks. Even if your landscape has natural shelter, a brush or rock shelter in the right location can enhance your wildlife-viewing opportunities.

Even a single log or large stone will harbor some wildlife; logs or stones can be thoughtfully placed almost anywhere. But more elaborate brush and rock piles are fun to build, and the necessary materials are often found on site. A small pile can be located in any size yard. A large pile can be constructed in a secluded place on a larger property or a school ground, cemetery, or golf course.

Once your pile is established, you may never see some of the animals that use it. But tracks, well-worn paths, droppings, songs of birds, the interest of a hawk above, or sounds of scampering are all signs the pile is being used.

Brush Piles

Importance to Wildlife

Wildlife will use parts of a typical brush pile in different ways (Table 1). The inside will attract insects and other wildlife that are food for other animals; it will also protect wildlife from sun, rain, and predators. During strong winds, birds that would ordinarily use an evergreen tree for evening shelter may instead use a brush pile located on the ground out of the wind. Far into a pile, mammals and some birds find nesting cover in the tight network of strong twigs. The outside, where sticks protrude from the pile, provides places for birds to perch and sing, preen, and catch insects. If the base of the pile contains large limbs or logs, salamanders, snakes, and lizards may hibernate there. Ants, worms, beetles, and other insects

Figure 1. "Half-cuts" can create quick cover for wildlife.

Making a Quick Shelter in the Woods

In an area with plenty of trees, quick shelter and a food source can be created several ways. One method is to cut down a tree, or a few small trees without completely cutting through the trunk. Good choices are trees in need of thinning or those destined for removal. To do this, make a single cut a little over halfway through the trunk wherever it's convenient, then push the tree over (see Figure 1). This creates a "hinge" that may keep the tree partially alive for a year or two. This provides food for deer, rabbits, and a variety of birds. Maple, aspen, Oregon ash, alder, willow, and chokecherry trees provide excellent shelter.

These "half-cuts" are especially effective where shrub-level cover is sparse or absent. They are also helpful along the edges of woods or in wooded areas where light can later penetrate after the tree fully dies. Be careful not to damage existing habitat (such as your house!) when felling trees, and check for nests before selecting trees to be cut.

Another way to create quick cover is to cut partway through the lower limbs of a tree. This works best on conifers, such as Douglas-fir, hemlock, pine, and red-cedar, with limbs low to the ground. Deciduous trees can also be used. Partially cut the first two or three whorls of branches to form a "teepee" around the trunk.

Figure 2. A living teepee-style brush pile.

Table 1. Wildlife that use an average-size brush pile

Birds that will use the inside of the brush pile:	Birds that will use the outside of the brush pile:	Mammals that will use the inside of the brush pile:	Reptiles and amphibians that will use the base of the brush pile:
Bushtits	Grouse	Chipmunks	Alligator lizards
Chickadees	Hummingbirds	Cottontail rabbits	Salamanders
Dark-eyed juncos	Jays	Ground squirrels	Snakes
Flycatchers	Pheasants	Fox	Toads
Golden-crowned sparrows	Robins	Mice	Turtles
Grouse	Song sparrows	Rabbits	
Pheasants	Towhees	Shrews	
Quail	Warblers	Skunks	
Song sparrows	White-crowned sparrows	Voles	
Thrushes	Woodpeckers	Weasels	
Towhees		Woodrats	
White-crowned sparrows			
Wrens			

will live and feed in the rich soil beneath a pile.

When snow covers a brush pile, a complex array of snow-free spaces and runways provides important habitat for protection and foraging by small mammals.

Where to Start the Brush Pile

If you're mostly interested in attracting a variety of songbirds, locate your pile where you can easily see it through a window or from an outdoor seating area. Other good locations include the following:

- In an unused corner of your backyard or any wild portion of the landscape, particularly near a hedgerow, thicket, or group of mature trees or shrubs.
- Near a wildlife food plot planted in your vegetable garden.
- Near a pond, irrigation canal, or birdbath.
- Near a constructed blind used for photography or wildlife

viewing. (See Chapter 20 for information on how to construct a viewing blind.)
- Next to another rock pile or brush pile, or next to a snag or large stump.
- In an area that was recently cleared of blackberries or other tangles.

If a lot of brush-pile material and places to put it is available, make several piles in areas with different sun exposures or vegetation types. In areas with little rain and hot summers, wildlife will benefit from a brush pile mostly in shade. In cooler regions, deep shade may be too cold and damp for most wildlife species. An ideal spot for a brush pile in cooler regions is at the edge of a clearing or anywhere the pile can be in some sun.

Brush piles need not be permanent. A knee-high brush pile you build in fall can be added to in winter, then moved in early spring if necessary. A pile can also be the

slow-compost type, containing limbs, leaves, and lawn clippings. It can serve as habitat for a few years and then be spread out in an area to be planted with new trees and shrubs.

Don't start a large brush pile too near a heavily traveled road, potentially putting both wildlife and vehicles in jeopardy. Also, unproductive low areas are not the best spots for habitat piles because they tend to hold and collect cold air and excess water, which may limit use by burrowing and overwintering species.

Do not place a brush pile where it might become a fire hazard. Natural resource agencies can provide information on adequate fire breaks and other ways to reduce fire hazards associated with brush piles on your property.

Size of the Brush Pile

The larger the brush pile the larger the number of wildlife species that will use it. However, a loose heap of limbs and branches 3 feet high and 5 feet wide is adequate for most songbirds, especially if it is near shrub and tree cover. Even a smaller pile constructed from an armful of twigs placed over a small pile of old firewood will interest small wrens and sparrows.

Materials for the Brush Pile

The ideal materials for building brush piles are handy, abundant, and asking to be reused or recycled: old fence posts, wood pallets, and prunings from trees and shrubs. Avoid using chemically treated lumber, including creosote logs.

Build brush piles when the materials are readily available—for instance, while pruning shrubs or trees in your yard. Recycled Christmas trees—tied to the trunk of a large tree or along a fence—make good temporary shelters when there is a shortage of other materials.

How to Build a Brush Pile

There are many ways to build a brush pile, and all work well as long as the pile isn't too tightly packed. One way is to create a base of one or more layers using old wooden fence posts, wood pallets, large limbs, or logs, each layer at right angles to the next. You can also start a pile over an old stump.

Place the logs or other sturdy material 6–12 inches apart horizontally within each layer. Then add small limbs and branches in a semi-random arrangement over the base.

Be sure to leave some openings at the bottom so animals can get inside. To create some permanent access points into the pile, use a few 6-inch-diameter (or smaller for smaller creatures) plastic, concrete, or ceramic pipes, about 18 inches long, in the bottom layer (see Figure 3). To provide moist, safe shelter for insects and perhaps an overwintering salamander, pack some organic material like rotting twigs or leaves into some but not all of the lower portions of the pile.

Fallen leaves, straw, or black plastic placed over part of a brush shelter will change the quality of the shelter the pile provides. It may then serve as a hibernation site for wildlife that might not otherwise use it.

If you are at all concerned about attracting unwanted wildlife by providing a brush shelter, keep the pile open and loose as shown in the top portion of Figure 3.

Maintaining a Brush Pile

A large, well-constructed brush pile will last as long as five years if maintained. As time passes, top branches will collapse, brush will settle, and the base will be exposed. When this happens, add branches to the top or sides, making sure small animals can still get to the center of the pile. When you add branches to the top, expand the base to keep the pile from toppling. If possible, avoid disturbing a large brush pile in spring as this is nesting time.

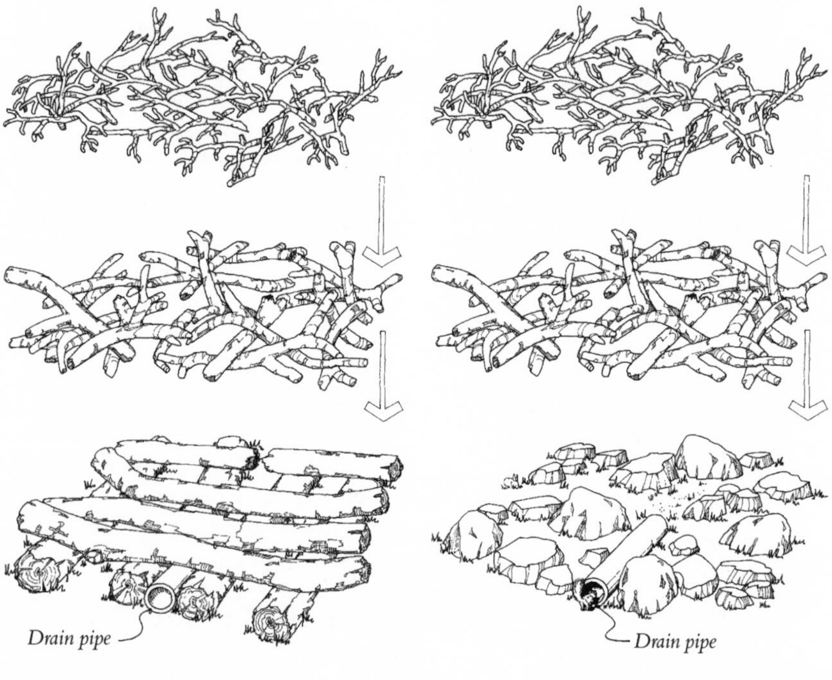

Rock Shelters

Depending on a rock shelter's location, size, and proximity to undisturbed wildlife areas, these mini-habitats attract a variety of interesting wildlife species (see Table 2).

If their other habitat requirements are in place, amphibians, reptiles, small mammals, and insects will seek out a rock pile, a heap of old concrete, or a rock wall to escape from predators and weather, and to raise their young.

Drain pipe

Drain pipe

Figure 3. Examples of how to construct a brush pile. The top portion alone is large enough to attract songbirds to a small yard.

Table 2. Wildlife that use rock piles and large rock walls

Birds	Mammals	Reptiles/Amphibians	Other
Chukars	Chipmunks	Frogs	Butterflies
Sparrows	Ground squirrels	Lizards	Centipedes
Towhees	Mice	Salamanders	Ground beetles
Wrens	Rabbits	Snakes	Slugs
	Shrews	Toads	Snails
	Skunks		
	Voles		
	Weasels		
	Yellow-bellied marmots		

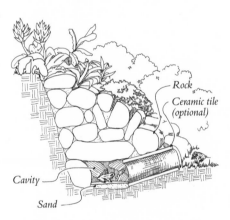

Figure 5. *Rocks assembled in any form in the landscape offer crevices, crannies, and cavities that provide shelter to some wildlife. Rocks in a small garden wall or large retaining wall provide a variety of wildlife shelters. Here, toads find a home in a rock wall.*

Long-toed and other salamanders make homes under moist rocks and decayed vegetation. Toads hide in cool nooks and crannies in garden walls during the day, waiting until dark to come out and eat flying insects, spiders, and other invertebrates living in and around the wall (Figure 5).

Reptiles and amphibians regulate their body temperature by absorbing the heat that rocks give off at night. On cold days, many will use stored and reflected heat to raise their body temperature. During a hot day, snakes will find a cool refuge in rocky crevices. To escape from the cold, some amphibians and reptiles spend winter hibernating in or below a rock pile. For examples of hibernation mounds you can construct for reptiles and amphibians, see Chapter 7.

Shrews are often found around moist, rocky areas where slugs and bugs are common. Shrews, in turn, are eaten by snakes, hawks, and larger mammals like fox and coyote. The Townsend chipmunk nests under stumps, logs, and rock piles and hibernates in these places for short periods of time.

Raised Brush Piles for Quail and Other Birds

You can create permanent night roost habitat for California quail in a variety of ways. The best habitat is established by planting thickly foliaged trees and shrubs such as native juniper and other conifers. If there aren't any evergreen trees around, you can collect clumps of brush and pile them in the crotches of any large trees. In the absence of large trees, or if they are too difficult to establish, you can create artificial cover with an elevated brush pile. This kind of brush pile gives quail a nighttime roost area and protection from predators, including domestic cats. The roost will also be used by other birds during the day and may be a valuable source of shade for mammals, especially in hot, dry regions.

A roost for quail can be any type of open frame and support structure elevated 6 feet above the ground and 6–16 feet long and wide (see Figure 4). Brush is piled on top of the structure to provide elevated night roosting.

Artificial roosts should be located within 50 feet of some other cover source, out of high winds, and not in a deep, narrow gully. Providing two roosts approximately 150 feet apart is helpful, so that if the birds are flushed they can fly from one roost to another.

Figure 4. *The quail roost design consists of a rectangular wooden frame elevated approximately 6 feet above the ground on four posts. A series of 2-by-4-inch boards span the width of the interior of the roost, and wire mesh serves as the roost floor. Brush piled into the roost is supported by the boards and the wire floor.*

Where to Start a Rock Shelter

In areas where the summers are hot and dry, a collection of rocks that receives some sun and some shade during the day is an ideal location for a rock shelter of any kind. Where the summers are cool the wildlife will benefit from a rock shelter in full sun most of the day.

Small mammals favor a rocky area near plants. This provides some additional cover from predators like hawks and owls and a place to find food. A rock pile built next to a pond or creek of any size will attract creatures associated with water. A loosely built rock wall, a small rock pile, or even a flat boulder at ground level in a sunny vegetable garden may attract a garter snake looking for a basking site and a supply of slugs.

A rock pile built next to a brick chimney or other masonry in the sun can hold stored heat into the evening. Additional rock-shelter locations that benefit wildlife include the areas listed under "Where to Start the Brush Pile."

Remember to locate rock piles away from driveways or heavily traveled roads to avoid vehicle/wildlife unpleasantries.

Size and Materials for a Rock Shelter

As with a brush pile, any size of rock shelter will attract some wildlife species. A good rule of thumb for choosing rocks is that the base-layer rocks, and the space between them, should be as large as the largest animal for which you are creating the shelter. Rock piles created for rabbit-sized animals need openings at least 5 inches in diameter.

A small rock pile for reptiles, amphibians, and small mammals can include stones dug up in the garden, broken brick, and used concrete. As with a brush pile, pieces of plastic, ceramic, or concrete pipe can be laid at the base of the pile or wall to create access into an interior cavity.

Rocks for a pile or a wall can be collected or purchased. When buying rock you have the option of picking it up from the quarry or having it delivered. If you collect rocks, pay attention to habitat you may alter and possibly harm in the process. Preserving existing habitat is far more crucial to the survival of local wildlife species than creating new habitat.

Constructing a Rock Pile

Small rock piles are easily assembled with hand tools. The largest rocks should form the bottom layer, and be spaced far enough apart to leave openings and a central cavity.

To stabilize the interior temperature and humidity of a rock pile, use large rocks throughout as much of the pile as possible. In interior areas of the Pacific Northwest where the winter temperatures are colder, sections of the pile can be placed underground (below freezing level).

Like a brush pile, a rock pile may be enhanced by addition of a partial cover of fallen leaves, straw, or black plastic.

CHAPTER 16

DEAD TREES AND DOWN WOOD

Snags, or dead and dying trees, occur naturally as a result of damage from disease, lightning, fire or animals, or from shade, drought, or root competition. Living trees may also contain decayed wood and function as snags; hollow trunks and dead branches are a normal part of a tree's development and aging process.

Snags, along with other forms of rotting wood, have tremendous value to wildlife: birds, flying squirrels, bats, and other wildlife use snags for homes, nurseries, hunting territories, and perching sites.

"Down wood" such as rotten logs and fallen branches also provides shelter for many animals. In fact, all rotting wood can host insects that provide food for many wildlife species.

Unfortunately, many old trees are thought of as "senile" or "dying" and so are cut down. In an effort to tidy their landscape, people often remove wood on the ground. But it is often possible to preserve snags without creating any danger to life or property, and down wood can be an enhancement to a naturalistic landscape. Both can be added to landscapes to enhance their wildlife value.

Types of Snags

All trees are potential snags. Because of their size and stability, the conifers—cedar, fir, larch, and pine—tend to rot more slowly than do deciduous trees such as alder, aspen, and wild cherry. However, large cottonwoods, big-leaf maples, and oaks can last many years as snags. And, as they age, they tend to develop cavities in large live and dead branches and in their trunks, providing homes for small mammals and birds.

All sizes of snags are important to wildlife. Small trees rot rapidly, quickly creating wildlife habitat. Red-breasted nuthatches and black-capped chickadees nest in snags as small as 8 feet tall and 8 inches in diameter.

Hard and Soft Snags

A partially dead tree, or one that has recently died, is called a hard snag. Hard snags tend to still have their bark intact and their heartwood (the inner core of a woody stem, wholly composed of non-living cells, generally darker than sapwood) and sapwood (the younger, softer, growing wood between the bark and heartwood that conducts water and minerals to the crown) are still firm.

A snag with considerable decay in its heart and sapwood is called a soft snag. Fungus fibers infiltrate the heartwood, and the tree becomes soft or hollow in the center. A soft snag rarely has limbs, and its top may be missing. Over the years, a soft snag gets shorter as weather and animal activity weaken sections. Eventually the soft snag falls over. But even on the ground, it continues to provide food and shelter for many kinds of wildlife.

How Wildlife Use Snags

Snags have been essential to wildlife in the Pacific Northwest for thousands of years. Nearly 75 species of birds and mammals in the Pacific Northwest nest or den in snags. Nearly 45 species of birds and mammals may forage for food in snags. In winter, when insects are scarce and snow covers the ground and down wood, woodpeckers and other birds rely even more heavily on snags for food.

Large snags—ones more than 12 inches in diameter and 15 feet tall—offer ideal hunting perches for hawks, eagles, and owls because they provide unobstructed views. They also are resting perches for swallows, band-tailed pigeons, and mourning doves, and food storage areas for mice, squirrels, woodpeckers, and blue jays. Woodpeckers also use large dead tree trunks as a way to announce their presence during courtship, hammering their bills against the resonating surface to make loud drumming sounds.

Small snags may be used as song posts by bluebirds, hummingbirds, and flycatchers attracting mates and proclaiming nesting territory boundaries.

This high use underscores the importance of preserving snags and including them in your landscape.

Snags that Make the Best Nest Cavities

The best snags for cavity-nesting wildlife are those with sound sapwood and decayed heartwood, particularly near the upper third

N. Yarbrough

Figure 1. A saw-whet owl is just one of the bird species that nests in tree cavities.

Dead Tree/Wildlife Condo

A snag may harbor beetles, ants, and many other insects that are food for wildlife. You can see where wildlife find food and shelter if you look carefully at a snag:

- **The outer surface of the bark is where bark beetles, spiders, and ants are eaten by birds such as brown creepers, nuthatches, woodpeckers, and sapsuckers.**
- **The inner bark is where larvae and pupae of insects are eaten by woodpeckers and sapsuckers.**
- **The heartwood is where carpenter ants and termites are preyed upon by strong excavators such as the pileated woodpecker.**
- **The space between partially detached bark and the tree trunk is where nuthatches, winter wrens, and brown creepers roost or search for food. Treefrogs and several species of bats and butterflies also find shelter there.**

of the snag. The sound sapwood provides protection from predators and insulation against weather, while the softened heartwood allows easy excavation deep into the snag. Many birds avoid very soft snags because extremely soft wood can be wet or crumbly, leaving nest sites vulnerable to predators.

Some woodpeckers select living trees with decayed heartwood because they can penetrate through the sound layer of sapwood and excavate the nest cavity in the soft heartwood. Generally, the sapwood remains fairly intact and forms a shell surrounding the decaying heartwood. The excavated interior may remain a desirable shape for many years and may be used by several different wildlife species over the years.

Snags in the Landscape

Incorporate snags into your landscape plan. This means leaving old, damaged, unhealthy trees whenever you can. Sap runs, splits in the trunk, dead main limbs, and fungi on the bark offer clues to a future snag. Other potential snags may show evidence of animal use, such as woodpecker excavations.

Also note the trees you may want to make into a snag. These may include:

- A tree creating a hazard, for example one with a forked top, weak wood, or disease.
- A tree shading an area where you want sun.
- A tree with invasive roots threatening a drainage or septic system.
- A tree in a group that needs thinning out.
- A tree in an area where there aren't any snags.

When clearing an area for development, retain trees and tall shrubs near any snag or potential snag to protect the snag from wind and to provide a more hospitable environment around the snag for wildlife.

Snag Location and Distribution

The location of a snag often determines which wildlife species will use it. Many wildlife species use snags along streams, ravines, marshes, lakes, and the edges of greenbelts. Snags occurring along streams eventually will fall into the water, adding important woody debris to aquatic habitat.

In many hot, arid areas of the Pacific Northwest, trees are rare. They're usually isolated along streams, in wetlands, or in draws, or they have been planted in rural and urban areas. Consequently, the few snags that do form are vital to wildlife survival, and their protection is crucial.

In urban areas, the best snags are located away from high-activity areas, where they won't pose a hazard if they fall. On larger properties, snags should be left in as many different habitats as possible.

How to Manage Snags in Public Areas

Retain all dead trees, including fruit and ornamental trees, where they do not pose a safety hazard. Attach a sign to the tree that lets people know it's a wildlife tree (see Figure 3).

If possible, remove the top/upper portion of unsafe snags to a safe height rather than removing the entire snag.

If you must remove an existing snag for safety reasons, consider

Table 1. Snag locations and some of the native birds that use them

(This assumes that the size of the snag, its condition, and the surrounding area are favorable to the species.)

Streams, wetlands, or open water: Osprey, bald eagle, purple martin, kingfisher, wood duck, hooded merganser, bufflehead, goldeneye

Grasslands and shrublands with scattered trees: Tree swallow, violet-green swallow, house wren, bluebird, black-capped chickadee, American kestrel

Forest edges or open-canopy forest conditions: Hairy woodpecker, downy woodpecker, flicker, Lewis' woodpecker, red-breasted nuthatch, black-capped chickadee, chestnut-backed chickadee, white-breasted nuthatch, saw-whet owl

Dense or mature forest stands with abundant large-diameter snags: Pileated woodpecker, Williamson's sapsucker, three-toed woodpecker, red-breasted sapsucker, red-breasted nuthatch, Vaux's swift

Clearcuts: House wren, violet-green swallow, tree swallow, flicker, Western bluebird, mountain bluebird, hairy woodpecker, American kestrel

Suburban yards with mature trees: Downy woodpecker, flicker, red-breasted nuthatch, black-capped chickadee

Figure 2. The pileated woodpecker is a primary excavator.

The Woodpecker: Cavity Creator

Woodpeckers create most new cavities in snags. They have a thick-walled skull supported by powerful neck muscles, with a beveled, chisel-like bill. A wood-pecker's strong, grasping feet with sharp, curved nails form a triangular base, with the specially-adapted ridged tail feathers offering support in the vertical position. The woodpecker's tongue, with barbs on the tip, and sticky saliva help the woodpecker obtain insects from deep crevices.

Woodpeckers excavate several cavities a year and rarely nest in the same cavity for two successive years. In the winter, however, they may roost in a recently excavated hole. This behavior creates many old cavities. Secondary cavity users, such as bluebirds, tree swallows, nuthatches, flying squirrels, and some small owls, cannot excavate a cavity but use existing ones for nesting, denning, or shelter.

Unlike other cavity-nesting birds, woodpeckers rarely use nest boxes because they are biologically adapted to dig their own cavities: the physical motions of cavity ex-cavation stimulate reproduction.

relocating the snag to a safe location by using one of the techniques described later in this chapter, or leave it nearby as a log.

To prevent aggressive, non-native European starlings and house sparrows from nesting in a snag, reduce the size of existing cavity holes to 1⅛ inches. Wood or metal covers can be used to make the existing entry hole smaller.

Hazard Tree and Snag Removal

If not managed properly, a snag can be a hazard to vehicles, buildings, power lines, fences, and people. However, trees that lean away or are downhill from structures and other areas of potential human activity present little or no risk.

In public locations such as parks, greenbelts, cemeteries, and

commercial open spaces, it may be necessary to keep the public away from a large preserved snag and warn them of the possible danger. But it is also important to let people know how valuable the snag is and why it is being saved. This can become an excellent educational opportunity through the use of an interpretive sign.

If a dead or dying tree's imminent collapse threatens something that can be moved, such as a swing set or lawn furniture, consider moving those items before getting rid of such a beneficial and hard-to-come-by part of your wildlife landscape.

As an alternative to removing the entire tree, you may be able to remove only the dangerous section(s) and leave the rest standing (see Figures 3 and 4). Professional tree services can give you an estimate on partial remo-

Figure 3. *A small snag such as this can provide safe wildlife viewing. Look at every hazard tree as an opportunity to create habitat. An alternative to removing an entire tree is to keep a portion of it for wildlife; even a snag 15 feet tall and 8 inches in diameter can be a home or a source of food for wildlife.*

val, which may save you money over having the whole tree cut down. The remaining parts can be cut off over time if they pose further problems. Often, once the unsafe limbs or portions of the trunk have been removed, the tree will be safe.

Even a small dead tree can be incorporated into the wildlife landscape. Even though it may not be large enough to be used by a cavity-nesting bird, it will be used as a perch on which birds will preen, rest, and sing.

When a tree has to be cut down, maximize its habitat value by leaving as much of the tree as possible near the area where it was cut down. In hot, dry areas, move the material into the shade of nearby trees or large shrubs. Bringing branches in contact with the ground will cause them to rot faster.

When hiring a tree-service company, try to work with someone who is familiar with the needs of local wildlife and who considers these needs in the process of his or her work.

Creating and Moving Snags

If your property doesn't have a snag or snags on it, you can create one from a live tree. It is also possible to transplant a small snag from somewhere else.

Creating Snags from Live Trees

If you want to create a wildlife tree where you can watch wildlife using it, or if you want to add to the habitat diversity of a property that contains many live trees, consider creating a snag from a living tree. Ways to do this include (see Figure 4):

- Remove the top third of the tree and half the remaining side-branches.
- Leave the top the way it is and remove a majority of the tree's side-branches.
- Leave the top and sides as they are and girdle the trunk.

An expert in tree work will be required to remove the top or side branches of large trees. Make sure that whoever does the work understands your intention to make a wildlife tree.

By removing the top third of the tree along with about half of the remaining side branches (drawing 1 in Figure 4) you will ensure that the tree begins the preferred inside-out decay process. Leave some shortened branches at the top of the tree for perch spots and to make the snag look natural.

You can also make the top look natural by creating a jagged top with a chain saw. This can create the appearance of a wind blow-out or a lightning strike. A jagged top also provides an avenue for infection by wood-decaying fungi and other rot-causing

(1)   *(2)*   *(3)*

Figure 4. *Ways to create a snag from a live tree. A jagged top and shortened branches at the top help the snag look more natural.*

organisms. Water and bird feces will collect and speed decay. Sowbugs, earwigs, and other invertebrates will also find their way to the top and assist in the decay process. The decay of the tree will soften the heartwood enough to allow a bird to excavate an entry hole a few feet down from the top. These sites are premium spots for cavity-nesting birds.

Another way to encourage rot to form slowly from the inside out is to cut off roughly three-quarters of the live side-branches of a tree (drawing 2 in Figure 4). Douglas-fir, hemlock, and pine respond well to this technique. Western red-cedar is a tough conifer to kill in this way, but it makes an excellent snag because it is extremely wind-resistant.

Avoid neatly sawed ends and flush cuts. Jagged ends are more susceptible to invasion by microorganisms and fungi, are

more likely to form rot holes, and look more natural.

All the branches you remove from the tree can be added to or used to create new shelter for songbirds and other wildlife. (See information on brush shelters in Chapter 15.)

To girdle a tree, remove a 4-inch belt of inner and outer bark around its trunk (drawing 3 in Figure 4). This will stop water and nutrients from moving up and down the tree; the tree will die, and the decay process will begin. Big-leaf maple, aspen, and poplar may send up sprouts, which can be removed or left to grow around the tree as temporary cover. Later these can be cut down and added to a brush shelter, or girdled and left as perches. A tree girdled in winter may not show signs of decline until well into spring, after it has utilized its stored energy. Some tree species, alder for example, are difficult to kill even when carefully girdled.

Girdling creates a dead but intact top, providing a taller snag, but it leaves the snag more susceptible to breaking in the wind at the wound site. If girdling is done at breast height and the tree falls, this leaves very little remaining vertical area. Therefore, try to make the girdling cut as high up into the tree as you can to preserve the trunk as potential habitat. Similarly, if you are cutting down a straight-trunked tree for firewood, you can leave a large portion of the trunk as habitat. To do this, drop the tree from a securely positioned extension ladder at whatever height is safe (Note: Unless you are an extremely experienced tree cutter, this is a job for a professional.)

Remember, a tree can provide habitat for wildlife even when just part of it dies. For instance, if a large conifer has a fork in it, you can girdle one of the forks. This can create an excellent perch site, and if the girdled trunk is large enough in diameter, a future

The Decay Stages of a Large Log

(Adapted from: Maser, Chris, and James M. Trappe. *The Seen and Unseen World of the Fallen Tree.* Pacific Northwest Forest and Range Experimental Station, USDA Forest Service, General Technical Report PNW164, 1984.)

Logs, like snags, go through recognizable stages of decay. When the stage changes, the wildlife that use the log also change. Logs in the early stages of decay provide lookout sites for ground squirrels, chipmunks, and yellow-bellied marmots. Ruffed grouse use them to drum on, and Western fence lizards use them to sunbathe. Raccoons, skunks, and foxes use hollow logs as dens. Woodpeckers get insects from logs in all stages.

As the log begins to settle into the ground, small animals, such as deer mice, search for food and nest in it. When the bark loosens, spaces between the log and the bark provide shelter from the cold for more wildlife, such as beetles and salamanders. The log's contact with the ground creates a moist microclimate, which animals such as treefrogs and voles prefer. Small animals hunt for insects on the ground alongside the log and inside the loose bark.

When the log is soft enough for small animals to burrow under and inside, shrews, deer mice, and voles dig burrows and tunnels in it. Amphibians and reptiles, such as toads, skinks, and gopher snakes, may also use the log in this stage.

Finally the log becomes soft and powdery and mostly buried in soil and duff, which provides a moist

continued on next page

cavity site. In addition, if the tree is not dying after the side branches and top have been removed, some individual side branches can be girdled to help the tree decline.

Unfortunately, girdling tends to cause a tree to rot from the outside in, instead of by the preferred inside-out method. As a result, by the time the rot has progressed far enough for woodpeckers to excavate a cavity, the tree has become fragile and may easily fall in a windstorm. Furthermore, a cavity in a girdled tree may not be safe because the hole is likely to be shallow, which exposes the young to weather and predators.

Roosting slits for bats and some songbirds, including brown creepers, may be added to created snags that are tall enough and wide enough in diameter to accommodate them. The slits should be at least 8 inches deep, 1 inch wide, and angled sharply upward. Bats need to have an open flight up into the slits so the slits should be located in an area free of branches. The higher up the snag they are, the more likely these roosting slits will be used.

Moving a Snag

In some instances, a snag (or a portion of it) may be salvaged from a construction site or a logging site where it would otherwise be removed or burned. (Be sure to get permission from the landowner before removing anything.) Snag relocation is not easy and will require professional help and special equipment. A dead tree is generally heavier and more fragile than it looks; a 20-foot snag may weigh several hundred pounds. An old snag, too rotten to support its own weight, is best used as a log.

After you have successfully removed the snag you will need to relocate it to a place where it will remain upright and secure. If you are moving a snag from one place to another on your property, try to install it as close as possible to its original location to minimize disturbance to any wildlife that have been using it. If possible, locate the snag in a wind-protected area near live trees and shrubs.

The best procedure for installing a large snag is to place it in a deep hole and firmly tamp soil around it. The hole depth should be approximately one-third of the height of the snag. Before setting the snag, cut its base flat, or excavate some dirt so the snag will stand straight.

To "plant" a small snag, you can do any of the following:
- Place the snag in a hole and secure it with tamped soil, gravel, or a concrete footing.
- Lower a firm, hollow snag over a metal or wooden post that's been securely placed in the ground.
- Wire the snag to a sturdy post.

Creating a Cavity in a Live Tree or a Stump

There are several ways to create a cavity in a live tree without killing it (see Figure 5). A long-term approach is to drill a 1-inch-wide hole at a 10-degree angle downwards into the heartwood of the tree (1). A good place to drill is below a crotch of a branch or anywhere water normally accumulates. This starts the cavity-making process and, depending on the rate of decay, will eventually allow woodpeckers and other cavity-using birds and mammals to use that area of the tree.

Another long-term approach is to remove a large (4 inches or larger) limb and leave the jagged, broken stub (2). When exposed to moisture this stub can develop conditions suitable for invasion by bacteria, fungi, and insects. Eventually decay will extend through the hole into the heartwood, and a cavity will form. Because most diseases attack the dead heartwood, the outer layer can continue its growth around the rotten core, and the rest of the tree can continue to grow for many years (3).

A short-term technique to create a cavity in a tall (4 feet or more) stump is to saw, drill, or chisel a cavity out in the top of the stump. Cover the top flush with a board to keep water out, and then drill an appropriately sized entrance hole through the side of the stump into the new cavity (4).

Another short-term technique is to gouge out a cavity in a live or dead tree using a chainsaw or chisel. Cover the cavity with a wood or metal plate and drill the appropriate size entry hole into the face of the plate (5).

Down Wood

Wood on the ground, or down wood, is an important element of any landscape for wildlife because it provides many small animals with food and shelter and it slowly returns nutrients to the soil. Dead wood can occur in a wide range of forms and places on your property and can vary in size from a log, old fence post, or rootwad to a small branch or a group of twigs. It can be in sun or shade or in a wet or dry place. Like a snag, sometimes a log or rootwad can be salvaged from a construction or a logging site where it would otherwise be burned, buried, or removed.

Down Wood and Wildlife

Over a hundred species of birds, mammals, amphibians, and reptiles use large logs for nests, dens, food, cover for rest, and to sun, drum, preen, and dust. A log crossing a creek will provide a bridge for a squirrel or a mouse. A log that's mostly submerged in a pond may be used by a duck as a preening spot or by a frog or turtle as a sunning platform. A more exposed log may be used as a resting and feeding place by a muskrat or river otter.

In a small backyard, wherever some type of wood is lying in con-

microclimate most of the year. Insect populations increase in this last stage, attracting insect-eating animals. Tree squirrels use the log to stash cones, and voles eat fungi growing inside the log. Some mammals will bed down where the rotted wood makes a soft resting place. Grouse and other forest birds use the decomposed wood to take dry dust baths.

Eventually, the entire log is incorporated into the soil, enhancing its organic content and productivity.

(1) (2) (3) (4) (5)

Figure 5. Long-term ways to create a cavity in a tree or large stump.

Figure 6. A log will look more natural, be more stable, and provide more opportunities for wildlife if it is located in partial shade and buried slightly in the ground.

and other small mammals can gnaw on them.

Although any down wood is valuable, the richest variety of wildlife tends to be in wood that is in partial shade. In sunny, arid regions, down wood in full sun may become too hot for many species to survive. In rainy regions, down wood in deep shade may be too cold and damp for some wildlife.

A log placed parallel to a slope will encourage more use by small animals than will one located vertical to a slope. Logs parallel to contours help reduce soil erosion and surface-water runoff, trap sediments and nutrients that are suspended in runoff, and provide excellent nurseries for tree and shrub seedlings. To make logs look more natural in a landscape and to create a better environment for wildlife, excavate the area around the log so that about a quarter of the diameter of log is buried in the ground (Figure 6).

tact with the soil it will support a distinctive group of animals, especially insects and other invertebrates. Many of these are food for birds that can be seen scratching in wooded areas where the ground layer is left undisturbed.

Down Wood in the Landscape

The simplest rule about down wood on your property is to leave it where it can decay naturally and serve as habitat for some wildlife. However, this is not always possible because of access, safety, or aesthetic reasons. If you must move a log or other large piece of wood, try to shift it as little as possible, and try to keep it as intact as possible (just as when you move a snag). Smaller pieces of down wood can be used to create or augment an existing brush shelter. They can also be placed out of the way and in an area where rabbits

CHAPTER 17

HUMMINGBIRDS AND HOW TO ATTRACT THEM

The name hummingbird comes not from the bird's voice but from the whirring sound of its wings whipping the air 70 to 80 times a second.

Hummingbirds are like living helicopters. They can hover and fly straight up and down, sideways, backwards, and even upside down. This is possible because their wings rotate from the shoulder, so they get power from both the downbeat and the upbeat. While their average flight speed is 25 miles per hour, they can travel up to 50 miles per hour, with their wings beating 200 times per second. While 320 different species of hummingbirds live in North, Central, and South America, four grace the Pacific Northwest: the black-chinned, calliope, Anna's, and rufous hummingbirds.

Pacific Northwest Hummingbirds

Black-chinned hummingbirds are found east of the Cascade mountains. The adult males are dark green above and whitish below. They have a black chin and violet and blue fringe on the throat. Females are bronzygreen above and grayish below. They're common in canyons and waterways, but are occasionally seen in suburban areas during summer. They winter in Mexico.

Calliope hummingbirds are also found east of the mountains and also winter in Mexico. They are the smallest of the Northwest hummers. Males have metallic-green upper parts and rose neck-feathers; females have bronzy-green upper parts. They are common in brushlands and open woodlands.

Anna's hummingbirds stay year-round in some areas west of the Cascade mountains. Male Anna's hummingbirds have greenish sides and an iridescent dark-rose cap and throat. Females have a small rose throat-patch. The male performs a steep diving display during late winter and spring, and it defends its food source from other hummingbirds. The call is a sharp *tsik*, in series. Anna's hummingbirds are probably restricted to urban areas, especially in the winter months. In areas where few flowers are blooming, Anna's subsist largely on insects, though they may get help from feeders, especially in cold weather.

Rufous hummingbirds can be seen throughout the Pacific Northwest. Male rufous hummingbirds have fire-red feathers on their throats and reddish-brown feathers on their backs. Females have green upper parts and whitish under parts, and may have red or golden-green spots on their throats. They are found in residential areas, parks, flowering meadows, and woodlands. They return to Northwest lowlands from Mexico in early spring and head to higher elevations following the bloom of wildflowers. They start to migrate south in late summer along the spine of the Cascades.

The hummingbirds that live in the Pacific Northwest are only 3–4 inches long from head to tail, and they weigh no more than a nickel. Yet they expend more energy for their weight than any other animal in the world. This energy is used mainly for flying and for keeping their tiny, heat-radiating bodies warm. "Hummers," as they're sometimes called, meet their high energy demand by eating more than half their weight in food and by drinking up to eight times their body weight in water every day.

Most hummingbirds eat nectar from flowers for instant energy, and in so doing they pollinate the plants. They also eat aphids, beetles, flies, gnats, mosquitoes, and spiders for protein, fats, vitamins, and minerals. These are picked from flowers or caught in midair.

Except for the non-migratory Anna's hummingbirds, hummingbirds generally arrive in early spring when the native plants whose nectar they feed upon begin to bloom. Male scouts arrive around the first of March, two to three weeks earlier than females. Nesting and rearing of young generally occur in May and June. Hummers usually depart this area by October for warmer, flower-producing weather in the southern United States and Mexico.

Hummingbird Lives

Although hummers often nest in lower tree branches and shrubs, people rarely notice their golf-ball sized nests. The female assumes all nesting duties. She sculpts a cup of plant parts, mosses, and lichen held together with spiderweb. In this tiny nest she lays two pea-sized, white eggs and incubates them for 14–21 days. Once they are hatched, she feeds the young ones a rich diet of regurgitated nectar and small insects. After about 25 days, the youngsters leave the nest to survive on their own.

Figure 1. *Most people don't notice a hummingbird's golf-ball sized nest.*

During the day, hummingbirds eat roughly every 15–20 minutes, but spend about 60 percent of their time perching. Part of this time is spent digesting the contents of their tiny crops (throat pouches). They also like to watch a food source and may attack another hummingbird if they see it trying to use a food source they consider especially good. Hummingbirds fill their crops before dark, then slowly digest this stored food throughout the night. Some hummers enter a brief state of torpor at night, reducing their metabolic rate and body temperature by as much as 50 degrees below its normal 104 degrees Fahrenheit level, thus conserving energy. This state of torpor may continue for several days in inclement weather. (Hummers face increased danger from predators at this time because they cannot respond quickly.) As soon as the sun appears, they warm up their bodies with shivering movements and fly away in search of food.

Hummers are in jeopardy while feeding because their heads are usually buried in a blossom. Natural predators include kestrels, jays, crows, cats, and bullfrogs. Storms, insecticides, and other pesticides are also responsible for deaths. Increasing loss of habitat in their wintering range is a continuing concern.

Hummingbird Feeders

There are two ways to attract hummingbirds: the first is with a special feeder filled with a nectar-like sugar solution that resembles what they would naturally find in flowers.

Feeders

There are many types of nectar feeders for hummingbirds. All should have red on them somewhere to attract the passing hummer, and should come apart easily so they can be cleaned thoroughly. A feeder with a larger opening will make filling and cleaning easier. Glass is more resistant to the elements than plastic is. A perch is not necessary but allows hummers to conserve precious energy. Because hummingbirds can be territorial about their food sources, most people find it's better to have multiple small feeders if they want to feed more birds, rather than one large one.

Nectar Solutions

To fill your feeder, you can use either a commercially produced solution or you can make your own hummingbird nectar at home. Any solutions with dye, food coloring, or flavoring are considered unsafe. Red coloring isn't necessary because most feeders have red parts to attract hummingbirds.

A simple, safe nectar solution can be made at home by mixing one part cane (white table) sugar with four parts water. Boil the solution for 30 seconds to retard mold growth, and let the solution cool before filling your feeder. (Make a lot of solution at one time and freeze the rest; it will ferment in the refrigerator in two weeks.) Use a funnel or a container with a spout to fill the feeder without spilling.

Don't use honey, brown sugar, or artificial sweeteners in your feeder solution. Honey encourages molds that contain botulism toxins and will kill hummingbirds. Hummers may quickly starve to death eating only artificial sweeteners because they contain no calories.

Location

Place your hummingbird feeder where it will be easy to clean and refill. To retard mold growth, place your feeder in a shady spot; if necessary, you can make a bonnet to place over the feeder to keep it shaded. It's easy to make one with a piece of cardboard with foil or plastic folded over it for waterproofing.

Try to place the feeder in the vicinity of nectar-producing plants; the birds will use these plants and the insects they attract and will derive a more complete nutritional balance from the insects and nectar. If you illuminate the feeder area, hummers may visit it after dark.

Since hummers are territorial, place individual feeders far apart or out of sight of each other. Some hummers may also shy away from a busy seed or suet feeder, so don't place them right next to your hummingbird feeder.

When to Use a Feeder

If you live in an area where Anna's are found year-round, you may want to keep your feeder up throughout the year. In some areas where alternative food sources are not available during winter months, Anna's may rely on feeders in order to survive the winter. Feeders may also help birds get through cold spells.

If you don't get overwintering Anna's in your area, you'll want to put your feeder up around March 1st to greet migratory birds returning to the area from their winter homes. Hummers remember from year to year where food sources were found; the same birds will probably return to your feeder if you put it up consistently every spring.

Many people find that hummers make heavy use of feeders in the spring but cease visiting feeders in the summer. This may be because hummers have abundant alternative food sources during the summer. However, feeders may be valuable to hummers again in the fall when the birds are getting ready to migrate or are passing through our area on their way from the more northern end of their range. Many authorities recommend that you keep your feeders up until at least two weeks after you've ceased seeing hummers in your area.

Don't worry that keeping feeders up in fall will prevent birds from migrating. Authorities generally agree that migration is an instinct and is not affected by availability of food. Banding studies indicate that migratory birds that don't migrate are often underweight or have an injury that prevents them from making the trip. In the event that you do find a yourself hosting a bird that should have migrated but didn't, you may be able to help it survive the winter by providing food.

Problems

Hummers remember from year to year where food sources were found and can rely on the feeders you provide for them. So, if you wish to phase out a hummingbird feeder, do so gradually. A sudden change could mean trouble to part of the hummingbird population if there aren't other feeders or suitable flowering plants in the area to sustain them. Make arrangements with a neighbor before you take a vacation if many birds are using your feeder.

A Caution on Red Insulators: If you have an electric fence around your property, be warned that the red insulators look like tubular, nectar-holding blossoms to hummingbirds. In some designs it is possible for birds to insert their long tongues into the tempting tubes and immediately get zapped with lethal electricity. Please take the time to paint your insulators with white or black paint, and the hummers will leave them alone.

Maintenance

The bacteria and molds that form in nectar feeders in warm weather can be deadly to hummers. A sugar-water solution is also prone to fermentation caused by wild yeasts, which will make your solution unappetizing to hummingbirds.

Clean and change the solution in your feeders about every four to five days (more often when the temperature is over 80 degrees), when the solution begins to look cloudy, or if you notice the development of wild yeasts, which will appear as flecks on the surface of the liquid around the edge. Clean the feeders thoroughly with a bottle brush, hot water, and a little vinegar.

If your sugar solution attracts ants, bees, wasps, or yellow jackets, apply petroleum jelly around the openings of the feeder and on the wire from which it hangs. Or try moving the feeder to another spot. Don't use insect sprays or repellents to control insects on or around the feeder.

Nectar feeders can attract other birds—including woodpeckers, finches, orioles, tanagers, and warblers. These birds can be a threat to hummingbirds and should be given their own feeder. A special feeder, designed for these other birds and often sold as an "oriole feeder," is available commercially. These require the same nectar solution and maintenance as a hummingbird feeder.

If you are feeding overwintering hummingbirds, the challenge is to keep the liquid in your feeders from freezing during cold weather, when birds are likely to need it most. Sometimes a hummingbird feeder attached to a window with suction cups will receive enough heat through the glass to prevent freezing. Sometimes hanging the feeder near a light bulb (turned on, of course) can do the trick. A very time-consuming method is to alternate two feeders: One is inside warming up while the other one is outside getting cold.

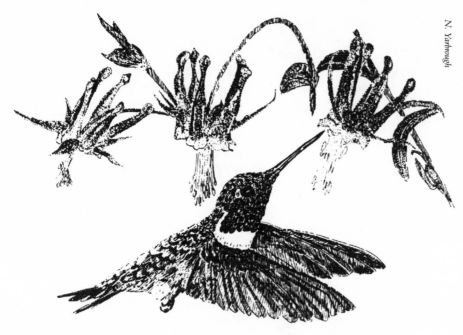

N. Yarbrough

Figure 2. A hummingbird obtaining nectar from columbine.

Grow a Bean Tepee

From late July to September, children can play in the cool, shady confines of a scarlet runner bean tepee that also attracts hummingbirds. Scarlet runner bean plants are showy and ornamental, with bright scarlet flowers that are popular with hummingbirds. The flowers are followed by flat, dark-green pods that are delicious when young.

To construct the frame for the teepee, place a vertical, 10-foot center pole 18 inches deep in the soil. Pack it in well. Place bases of at least four 10-foot-long bamboo poles (or other skinny wooden sticks) 3 feet out from the base of the center pole and tie their tops together 6 to 12 inches below the top of the center pole. In a circle just outside the pole bases, plant the beans 1 inch deep and 1 to 3 inches apart. Train the vines up the outside of the tepee, keeping main stems out of the interior.

Hummingbird Gardens

Another way to attract hummingbirds is by creating a special hummingbird garden, or adding known hummingbird plants to an existing garden. By planting the appropriate trees, shrubs, vines, and perennial flowers, you can have a hummingbird garden that produces blossoms each year and over a long period of time.

Hummingbirds apparently are most attracted to nectar-rich plants with bright red, orange, or red-orange tubular-shaped blossoms. Hummers will visit other flowers, but the brightest red flowers are perhaps the most effective. Hummers prefer old-fashioned, single-flowered blossoms because these usually have more nectar or the nectar is more accessible to them.

Where house cats roam freely in and around your flower beds, select plants that grow to be at least 2 feet tall. Birds will also safely visit hanging pots full of fuchsias, red impatiens, and geraniums.

If there are no large twiggy trees or shrubs around your yard, try attaching perches to flower pots and boxes to see if you can provide them with a place to rest and digest.

Hummingbirds use birdbaths and are attracted to the sound of running water. They especially enjoy devices that create a spray or fine mist of water. See Chapter 11 for information on birdbaths and Chapter 10 for information on ponds.

For more ideas on hummingbird gardens, see Chapter 18 on "Butterfly Gardens." For ideas on how to plan a hummingbird garden, see Chapters 2 and 3.

For lists of plants for hummingbirds, see Appendix B.

When planning your hummingbird garden, include native plants. These plants have evolved in association with hummingbirds in the area and are adapted to the local climate.

Be patient while hummingbirds discover your garden. It may take a year. You may coax them in with a feeder wrapped in red or by hanging bright red or orange ribbons or windsocks nearby. Meanwhile, enjoy the other wonderful wildlife that will visit the landscape.

CHAPTER 18

ATTRACTING BUTTERFLIES, MOTHS, AND CATERPILLARS

Butterfly watching ranks high among outdoor pleasures, right alongside viewing birds and wildflowers. Fortunately, all areas of the Pacific Northwest are home to some butterfly species. (Some of the common butterflies are described at the end of the chapter.)

The best way to watch butterflies is to invite them to an outdoor area by offering plants that they and their larvae (otherwise known as caterpillars) use as food sources. You can begin to meet the needs of butterflies by adding flowers and herbs to an existing flower bed or container garden. However, a colorful grouping of butterfly-attracting plants will help butterflies locate your garden when they are flying through the neighborhood. No site is too small to create a butterfly garden.

An added bonus of creating a butterfly garden is that it will probably attract not only butterflies but also moths and hummingbirds.

Where to Locate a Butterfly Garden

Because butterflies rely on the heat of the sun to warm them enough to fly well, your butterfly garden should be located in an area where it receives sunlight throughout most of the day. (Plants for hummingbirds and moths do not have to be in sun, however.) Because butterflies use up more energy flying in windy areas, they prefer places where they don't have to fight the wind. So choose a sunny site out of the wind; in a windy area create a windbreak by building a fence or a planting a hedgerow.

For your own enjoyment, determine where the butterfly garden would be most visible, enjoyable, and easy to maintain. Bear in mind that some of the best butterfly plants require fertile, well-drained soil. Good locations include:

- Outside a kitchen or other frequently used window.
- Next to a walkway, patio, or other seating area.
- Near a frequently used entry.
- Near a neighbor's flower garden.
- In a vegetable garden.

Wherever you locate the garden, add a bench so you can observe butterflies drinking nectar, laying eggs, basking, chasing mates, and defending their territory.

Plants for Adult Butterflies

Most butterflies eat by sucking nectar from flowers. (A tube-like mouthpart called a "proboscis" is

Watching Butterflies and Conducting a Butterfly Survey

Few other insects can be as pleasing to watch as butterflies, not only for their fascinating flight patterns but also for sheer beauty of color and pattern. Butterfly watching can also give you a new awareness of the plants and habitats around your property.

Butterflies are best found in open, sunny areas that have flowers. Your own yard is a good place to start. Any rural roadside will also do—provided it hasn't been sprayed with herbicides. Powerline cuts, irrigation ditches, sunny streamsides, and a city bed of marigolds are other good sites. You can sometimes find butterflies, including anglewings, tortoiseshells, and wood nymphs, sipping at willow sap or fallen fruit. Swallowtails, blues, and crescents sip moisture and salts at mud-puddle margins. If you spread manure in your yard, you may find that this attracts butterflies.

Take notes on what plants butterflies visit. You can use these notes later to decide which plants to include in your butterfly garden.

Butterfly activity tends to be restricted to warm daylight hours; indeed, many species disappear from sight when the sun goes behind a cloud. Furthermore, butterflies are best observed when feeding or basking in the sun. They sometimes become so involved in drinking that you can approach to within inches. When approaching butterflies, move slowly and fluidly. Try offering them some

continued on next page

sweat from your brow by way of a finger.

You can survey what types of butterflies appear in your neighborhood during the warm times of the year. Use the colored photographs in this book to assist you with identification. If you take a butterfly field guide with you, mark the pages containing the common species for quick reference. Binoculars are almost as helpful to the butterfly-watcher as to the birder. They enable you to survey a large field for butterflies, or to sit on your porch to view your butterfly garden. Lower-powered binoculars that focus closer than most are best. Eventually you'll be able to identify certain butterflies "on the wing." Finally, when looking for butterflies think small; many common species have a wingspan of an inch or less.

used for this purpose.) A wide variety of flowers, including many popular garden plants, can provide nectar for butterflies. Brightly colored, fragrant nectar plants are especially attractive. (For a list of the best plants for butterflies, see Appendix B.)

Some ornamental flowering plants have been hybridized to produce showy flowers with many petals. Unfortunately, these recently developed plants may not be good sources of nectar. When selecting plants for nectar avoid showy flowers, such as those described as "double," and choose simple, old-fashioned varieties instead.

To keep your butterfly garden from looking bleak during winter, include some butterfly plants with evergreen foliage such as lavender and hyssop.

Choose plants so that you will have something blooming from early spring to late fall. To extend the blooming season still further, add annual flowers and plan to remove dead flower-heads to extend blooming periods (see Chapter 4).

Good plants for containers include fuchsias, sweet alyssum, garden sage, dianthus, and lavender. For containers, avoid tall annuals such as tall marigolds, tall zinnias, and cosmos.

Some of the plants on the butterfly list—dandelions, for instance—are commonly considered weeds. If you don't feel comfortable letting such plants occur in your butterfly garden, consider the advantages of letting a few weeds thrive elsewhere. Leave some dandelions in the lawn. Let a weed patch thrive in a hidden sunny corner of the yard. Not only will you decrease

the amount of time you have to spend weeding, but you may be surprised at some of the insects and other animals that will be attracted to that area of the landscape.

Plants for Caterpillars

A butterfly begins life as an egg, which hatches into a tiny caterpillar, which is basically a small eating machine. Typically, caterpillars eat plants. The caterpillars of each butterfly species have preferred larval food plants. Specificity is so strong that most caterpillars will starve to death if they can't find their larval food plants soon after emerging from the egg. Adult female butterflies probably won't venture great distances from the plants that will be needed as larval food plants, especially if there is an ample supply of nectar nearby.

So if you really want to attract butterflies to your garden it's important to provide not only nectar plants but also plants eaten by the caterpillars of the butterflies you most want to attract.

Fortunately, many larval food plants are common. Your yard probably already has some. However, if you know what butterflies to expect in your area, you can make a point of planting larval plants listed for those species at the end of this chapter. (See Appendix B for additional plant suggestions.)

It's generally a good idea to group larval plants just as you would nectar plants. This will help females locate future nursery sites and provide caterpillars with ample nourishment.

Enhancement Features for Butterflies

Water Sources

Butterflies don't feed from nectar alone. They may also take water and trace minerals from a patch of wet sand or soil. Mud around the edge of a pond, or under a hose bib or birdbath may already be a popular spot. To create a small damp area, dig out a couple of inches of soil about 24 inches wide in a frequently watered area. Another way to provide a drinking place is to sink a small bucket in the ground and fill it almost to the top with wet sand. Remember to place these water sources in a sunny area out of the wind and near nectar plants. If cats are a concern, put wet sand in a birdbath or other elevated container.

Basking Sites

On cool days, in the morning, and periodically throughout the day, butterflies warm their blood and flight muscles by basking with their wings open and their bodies perpendicular to the sun. Place a few large stones or rocks facing south to serve as basking sites. Again, if cats are a concern, put the rocks in a birdbath or other elevated container.

Wild Patches

Many butterfly species seek shelter among weeds and tall grasses at night and during bad weather. If you can, leave or add wild patches in out-of-the-way portions of your yard, or leave a patch of lawn unmowed. (A bonus is that you'll probably be growing larval plants, too.) To avoid complaints, mow a strip around the unmowed area and let neighbors and local officials know what you are trying to accomplish.

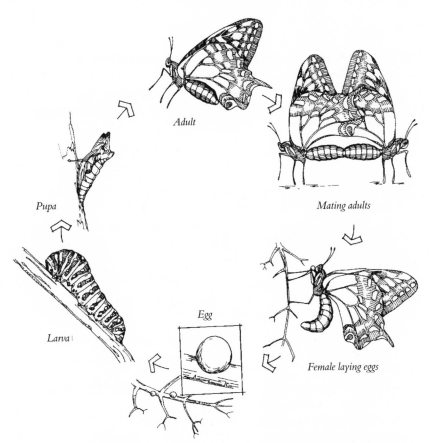

Adult

Mating adults

Pupa

Female laying eggs

Larva

Egg

Moths

In addition to being important pollinators, moths serve as an important food source for breeding birds, bats, and spiders. Of the 6,000 species of moths in North America, only two species have caterpillars that favor woolen garments and carpets.

There are at least ten times as many moth species as butterflies in the Pacific Northwest. A particularly useful plant for moths is the alder (*Alnus* spp.), which may be the preferred food plant for more than 50 species.

Moths are fascinating visitors to the evening garden. Adult sphinx moths extract nectar from deep-throated, fragrant flowers that open at night. Like hummingbirds, they hover in flight while feeding. Instead of the long beak of the bird, though, they have a long (3–6 inch) tongue like a drinking straw. The larvae feed on a variety of plants, including apples, azaleas,

How a Caterpillar Becomes a Butterfly

The change a caterpillar undergoes when it becomes a butterfly has long fascinated human beings. When the caterpillar reaches its full size, it finds a stable, sheltered location and sheds its skin, forming a hard shell called a chrysalis. Inside the shell the caterpillar dissolves into a "soup" of undifferentiated cells. This "soup" then is "reconstituted" into the adult butterfly.

When the adult emerges, its wings are tiny and shriveled. Immediately it hangs upside down and begins to pump fluid into the wings, which expands them to full size. The wings then dry and harden into structures capable of supporting flight. At this point, the butterfly flies off to find food, water, and a mate. After mating, eggs are deposited on or near a host plant and the cycle begins anew.

Table 1. Differences between moths and butterflies

Butterflies	Moths
Day fliers	Mostly night fliers
Often brightly colored	Generally less colorful (with some dramatic exceptions)
Antennae knobbed at the ends	Antennae may be feathery, are not knobbed
Pupa has no silky cocoon around it	Pupa often in a silk cocoon, may be encased in leaves
When they first land, wings are generally closed	When they first land, wings are generally open

fuchsias, grapes, cottonwood, poplar, willow, snowberry, and cherry.

Moths and butterflies take nectar from many of the same plants. Flowers that attract night-flying moths include:

Sweet William, *Dianthus barbatus*
Fireweed, *Epilobium angustifolium*
Jasmine, *Jasmine* spp.
Honeysuckle, *Lonicera* spp.
Four o'clock, *Mirabilis jalapa*
Bee balm, *Monarda didyma*
Catmint, *Nepeta* spp.
Evening primrose, *Oenothera* spp.
Petunia, *Petunia* x hybrida
Toadflax, *Linaria purpurea*
Mock-orange, *Philadelphus lewisii*
Tall garden phlox, *Phlox* spp.
Lilac, *Syringa* spp.
Yucca, *Yucca filamentosa*

If you include some of these plants in your landscape, be sure to go "moth watching" at dusk on some summer evening. Use a flashlight after dark. Try covering the flashlight with red cellophane so as not to distract moths from feeding. (See Plate 7.)

Helping Butterflies Through the Winter

Different species of butterflies may spend the winter in varying life stages, and in separate areas of your yard. Leaf litter, garden mulch, and dead and dry stalks of summer perennials may be homes for butterflies that pupate on or near the ground. Other species spin up a protective shelter and survive winter as larvae or chrysalids among the leaves on their host plants. (If you find one while pruning, clothes-pin the twig and leaf to a lower branch, where you can watch the awakening larva in spring.) Other places a butterfly may spend the winter as a chrysalis include in a protected area under an eave or trellis, or in a brush or wood pile. Adult mourning cloaks, Milbert's tortoiseshells, and satyr anglewings hibernate in tree hollows, rock crevices, and open buildings.

Recent research indicates that so-called "butterfly hibernation boxes," which you may have seen in gardening catalogs, have not been effective at attracting overwintering butterflies. The best way to help butterflies survive the winter is to adopt a maintenance plan that meets your aesthetic requirements and the needs of butterflies without disturbing all the butterfly habitat. Don't be too concerned about tidiness in all areas of your property. Overzealous fall cleaning of yards and gardens can remove the very stuff that many butterflies depend on to get through the winter.

Pesticides Kill Butterflies

If you want butterflies, you must accept that caterpillars may eat some plant parts. In most cases, the plant will not be seriously harmed, and munched leaves or limbs can be removed if objectionable or unsightly. Insecticides—which kill butterflies and other beneficial insects along with whatever pest species you are trying to target—cannot be used in the butterfly garden. Even Bt (*Bacillis thuringiensis*), often promoted as an alternative to traditional pesticides, is actually a biological agent that kills caterpillars.

Should you have too many caterpillars, share them with a friend who also grows their host plant. Caterpillars, like their later forms, are fascinating to watch. Observing the life cycle of butterflies will easily captivate the minds of children.

Some Common Pacific Northwest Butterflies

The following list includes some of the common butterflies found in different areas of the Pacific Northwest. For identification use the colored plates provided in this book. After you've identified the species found in your area, you can use the plants listed to attract them to your yard. See the Appendices for additional information on individual plants. Common names of plants are used here. For a list of common and scientific plant names, see Appendix B.

Key

Adult description = Describes the adult (winged stage) of the butterfly.

Food plants = Plants eaten by butterfly larvae (caterpillars); also called host plants.

Nectar sources = Nectar-producing flowers and other nectar sources, such as manure and rotting fruit that are used by adult butterflies.

Comments = Information on broods (generations of butterflies hatched from the egg laid by the females), flight times, plus other information.

Swallowtails and Parnassians

Swallowtails are large, brightly colored butterflies. They usually have a well-developed "tail" extending from the rear edge of each hind wing. Swallowtails may have been given their common name because their "tails" reminded people of the long, pointed tails of barn swallows. The caterpillars are the only

group with a Y-shaped, orangish, retractable organ behind their heads for protection from predators. When disturbed the caterpillar extends this organ, which not only emits a foul-smelling chemical, but also may scare off predators because of its appearance.

The very different-looking parnassian butterflies have black and yellow or black and orange caterpillars that pupate in primitive cocoons.

Anise swallowtail
(See Plate 6)

Adult description: Smaller than other swallowtails; yellow bands across black wings; abdomen mostly black. *Food plants*: Desert-parsley, fennel, dill, carrot, garden parsley, cowparsnip, seaside angelica. *Nectar sources*: Butterfly bush, desert-parsley, penstemon, garden mint, zinnia, lantana, coltsfoot. *Comments*: One or two broods per year; overwinters as a chrysalis. Easy to rear. Found from sea level to mountain tops. Males commonly gather on hilltops and at mud puddles.

Western tiger swallowtail
(See Plate 6)

Adult description: Large; yellow with black "tiger" stripes. *Food plants*: Big-leaf maple, willow, aspen, poplar, cottonwood, sycamore; perhaps cherry, plum, alder, apple, serviceberry, hawthorn. *Nectar plants*: Common lilac, butterfly bush, mock-orange, rhododendron, blackberry, thistle, phlox, milkweed, garden mint, lily, lavender, verbena, wallflower, honeysuckle, sweet William, Barrett's penstemon. *Comments*: Two broods in some lowland areas. Overwinters as a chrysalis. Males commonly visit streamsides and mud puddles.

Very common and conspicuous in urban areas from May through early September.

Pale swallowtail
(See Plate 6)

Adult description: Large size; broad black stripes; pale or sometimes almost white. *Food plants*: Buckbrush, cherry, plum, hawthorn, cascara, alder, hardhack spirea, oceanspray, currant, coffeeberry. *Nectar sources*: Oceanspray, sweet William, penstemon, lily, columbine, garden mint, thistle, blackberry, and those listed for Western tiger swallowtail. *Comments*: One brood per year. Flight times vary with altitude, generally May through mid–August. Found from sea level to timberline. Overwinters as a chrysalis.

Clodius parnassian
(See Plate 6)

Adult description: Larger than the common cabbage white butterfly. Milk-white with black checks, gray patches, and red spots on the wing (males). The female often has transparent areas on outer wings. The caterpillar is usually black with rows of yellow or reddish spots; overwinters in decayed leaf litter and pupates in thin silken cocoon in spring. *Food plants*: Bleeding heart. *Nectar sources*: Blackberry. *Comments*: Found in forest edges at sea level; moist, cool mountains and shaded canyons and ridges in drier parts of their range. One brood; June–July in Washington.

Whites and Sulphurs

These butterflies are principally white, orangish, or yellow, with blackish markings and borders on their wings. The word "butterfly" was used originally to describe the yellow color of European sulphurs.

Pine white
(See Plate 6)

Adult description: Medium size; about the same size as the common cabbage white. Male is white with black markings; female cream, heavily veined, with outlines on underside of wing. *Food plants*: Pine (especially western white and ponderosa pine), Douglas-fir, fir, hemlock, red-cedar, Deodar cedar. *Nectar plants*: Butterfly bush, dusty miller, daisies, coreopsis, lobelia, goldenrod, strawflower. *Comments*: Most abundant in late summer. Often seen flying up in conifers; comes down to ground only to feed on nectar. Considered a pest by some foresters, although damage is modest even in an outbreak. Overwinters as egg.

Orange sulphur
(See Plate 6)

Adult description: Medium size; bright orange, males have solid black borders. *Food plants*: Alfalfa, clover, and other legumes. *Nectar plants*: Alfalfa and other legumes, mustard, thistle, aster, red-twig dogwood. *Comments*: Multiple broods from mid-April through October. Common in alfalfa fields and open, weedy wasteplaces in residential areas. Difficult to identify due to their quick, evading flight. Overwinters as a chrysalis.

Cabbage white
(See Plate 6)

Adult description: Medium size; female has two black spots on wings, male has one. *Food plants*: Mostly plants in the cabbage family, including cabbage, collard, broccoli, radish, winter cress, mustard; nasturtium and spiderflower. *Nectar plants*: Butterfly bush, money plant,

blackberry, coreopsis, dandelion, thistle, wild pea, sweet pea. *Comments*: Overwinters as a chrysalis; three or more broods per year, from last to first hard frost. Successful enough to be considered a pest in some gardens, although it is often unjustly blamed for the damage done by the cabbage looper, whose green caterpillar is similar, but whose moth stage goes unnoticed. Often valuable as the only butterfly around in urban areas. Unintentionally introduced in Quebec, Canada, in 1860. Occurs throughout the Pacific Northwest except where extreme climatic conditions exist.

Sara orangetip
(See Plate 6)

Adult description: Small to medium; female yellowish, male white; both have brilliant orange wing tips. **Food plants**: Mostly plants in the cabbage family including hedge mustard and winter cress; perhaps nasturtium, moneyplant, rockcress, fringepod. **Nectar plants**: Cherry, plum, strawberry, monkey flower, dandelion, violet, rock cress, fringepod, coltsfoot, tansymustard. **Comments**: Most common in spring, and easily seen due to its low-flying, avid nectaring behavior. Never moves very fast but seldom stops to rest. Overwinters as a chrysalis.

Hairstreaks, Elfins, Coppers, and Blues

When at rest, these small butterflies characteristically hold their wings folded over their backs. The color patches or tails of these species resemble butterfly heads. Some caterpillars of this family produce a sugary substance, called honeydew, that is "milked" from the caterpillars by ants.

Brown elfin
(See Plate 6)

Adult description: Small; dark brown to orangish-brown. *Food plants*: The flower parts, buds and seed pods of apple, salal, buckbrush, bitterbrush, manzanita, rhododendron, azalea, bog-laurel, Labrador tea, oceanspray, blueberry, sedum, kinnikinnik. *Nectar plants*: Cherry, plum, willow, osoberry, bitterbrush, winter cress, blueberry, wild-buckwheat, kinnikinnik. *Comments*: One brood per year in April or May. Overwinters as a chrysalis. Males perch and dart out at females and other passing objects. Difficult to observe because of their small size and rapid flight. Found in open woodlands, bogs, along forest edges, and in urban parks.

Purplish copper
(See Plate 6)

Adult description: Small; female has an orange and dark brown pattern with orange zigzag on the hind wing; male is brown with a brilliant purple iridescence when struck directly by sun. *Food plants*: Knotweed, cinquefoil, dock, sorrel. *Nectar plants*: Fennel, mint, heather, clover, and many composites. *Comments*: Multiple broods. A hardy butterfly that persists late into the fall. Found in urban weed fields and wet areas from sea level to over 10,000 feet.

Spring azure
(See Plate 6)

Adult description: Small; female, dull with dark borders along outer wing edges and blue highlights; male, violet blue. *Food plants*: Flower parts and seeds of dogwood, oak, buckthorn, apple, madrone, viburnum, cherry, plum, sumac, blueberry, escallonia, cotoneaster, hardhack, manzanita, oceanspray, cinque-foil, salal. *Nectar plants*: Cherry, plum, willow, mountain-lilac, holly, privet, rock cress, winter cress, escallonia, blackberry, cotoneaster, milkweed, forget-me-not, dogbane, coltsfoot, dandelion, violet, miner's let-tuce, many plants in the mustard family. *Comments*: One or two broods per year. Overwinters as a chrysalis. Appears early in the spring in lowlands and later in higher elevations. An avid puddler.

Silvery blue
(See Plate 6)

Adult description: Small; female brownish with blue highlights; underside has row of black spots, each ringed with white. Male is brilliant silvery blue. *Food plants*: Mostly lupine; also wild pea, vetch, clover and other legumes. *Nectar plants*: Cherry, plum, coneflower, desert-parsley, lupine. *Comments*: One brood per year. Overwinters as a chrysalis. A slow flier that is among the first butterflies to appear in spring in lower eleva-tions, even in chilly, windy weather.

Brush-foots

All members of this large, di-verse family have front legs that are relatively short compared with the other two pairs. The stunted front pair of legs is useless for walking, and somewhat hairy or brush-like in appearance, hence the family name. The caterpillars typically are spined and dark-colored. Many feed only at night. The adults commonly have an orange coloration, are active fliers, and feed on a wide array of food sources. These include flowers, tree sap, rotting fruit, and, for males, animal wastes and carrion.

Lorquin's admiral
(See Plate 7)

Adult description: Medium size; black with broad white bands across wings, and red wing tips. *Food plants*: Willow, chokecherry, aspen, oceanspray, cottonwood, hardhack spirea, cherry, apple. *Nectar plants*: Thistle, dogbane, mustard, blackberry, privet, giant-hyssop, Barrett's penstemon; also rotting fruit, animal droppings, carrion. *Comments*: One or two broods per year with flight times from early June until October. Very bold; males will inspect or attack large birds in their territo-ries. Found in ditches and along forest edges; common in parks and gardens in June and July; rare along coastlines or in high moun-tain areas. Overwinters as a small caterpillar in a leaf which it curls around itself.

Red admiral
(See Plate 7)

Adult description: Medium size; black with broad white bands across wings and rusty-orange wing tips. *Food plants*: Mostly stinging nettle. *Nectar plants*: Butterfly bush, daisy, aster, thistle, dandelion, gumweed, goldenrod, gayfeather, dahlia, ageratum, milk-weed, candytuft, alfalfa, sedum, dogbane, wallflower, fireweed, red clover, mallow, sea-holly, garden mint, red-valerian, giant-hyssop, Barrett's penstemon, spirea, germander. *Comments*: Two broods from spring to fall. Mi-grates north from as far as Mex-ico, but adults and chrysalises may overwinter in milder areas. Fluctu-ates in numbers from year to year. The black caterpillar has yellow branched spines with wicked barbs designed to cause major digestive upset to predators! Originally called the red "admirable."

Painted lady
(See Plate 7)

Adult description: Medium size; red-orange and black; spots are black, blue, and white. *Food plants*: Mostly thistle; also knap-weed, burdock, sunflower, pearly everlasting, stinging nettle, borage, hollyhock, checker-mallow, cheeseweed, legumes. *Nectar plants*: Oregon-grape, rabbitbush, butterfly bush, zinnia, dandelion, thistle, gayfeather, aster, Shasta and Michaelmas daisy, cosmos, dahlia, bee balm, garden mint, sweet William, red-valerian, red clover, milkweed, pincushion flower, mallow, wallflower, candytuft, purple coneflower, aster. *Comments*: Two or more broods per year. Infiltrate into the Pacific Northwest in spring from California. Numbers fluctuate widely from year to year. Adults often bask on bare ground and visit moist sites, vacant lots, and flower gardens.

Mourning cloak
(See Plate 7)

Adult description: Large; dark brown, with yellow to whitish wing margins. *Food plants*: Elm, cottonwood, poplar, willow, birch, hackberry, hawthorn, wild rose. *Nectar plants*: Willow, butterfly bush, milkweed, clove pink, rockcress, dogbane, Shasta daisy, daphne; also tree sap and rotting fruit. *Comments*: The number of broods varies with altitude. Over-winters as an adult and may appear on sunny midwinter days. Rarely seen nectaring. Found in vacant lots, flower gardens, woodland clearings, and paths.

Milbert's tortoiseshell
(See Plate 7)

Adult description: Medium to small; dark, with bright orange and yellow band down wing. **Food plants**: Stinging nettle. **Nectar plants**: Willow, butterfly bush, garden lilac, thistle, sneezeweed, daisy, goldenrod, marigold, ageratum, stonecrop, wallflower, aster, dandelion, calendula. **Comments**: Two or three broods per year. In flight from spring to fall. Adults overwinter and appear in early spring; adults occasionally seen midwinter on sunny days. Larvae feed communally and may immediately drop off host plant if disturbed.

Mylitta crescent
(See Plate 7)

Adult description: Small; mottled orange and brown. **Food plants**: Thistle. **Nectar plants**: Pearly everlasting, hawkbit, goldenrod, aster. **Comments**: Several overlapping broods from early spring to late fall; most abundant in August. Tend to occupy same nectaring areas for days at a time. Males will patrol back and forth. Popular spots are open woods, meadows, water edges, roadsides, vacant lots, and other weedy areas.

Satyr comma
(See Plate 7)

Adult description: Medium size, about the same size as the common cabbage butterfly; underside golden brown; top golden-orange, resembles tree bark. **Food plants**: Stinging nettle. **Nectar plants**: Dandelion, aster, blackberry; also rotting fruit, tree sap. **Comments**: Two or more broods from early spring to late fall; adult overwinters and may be seen flying very early in the year. Often seen sunning on a board, a rock, or on the ground. The larva makes a nettle leaf tent.

Common ringlet
(See Plate 7)

Adult description: Small; varying shades of orange-yellow (ochre above, olive below). **Food plants**: Grasses. **Nectar plants**: Dandelion, sweet clover, buttercup, and others. **Comments**: Two overlapping broods; caterpillars of second brood overwinter at the base of grass stems, even below the snow. Flight time is from April to September.

Common wood nymph

Adult description: Medium to large; cocoa brown to nearly black above. **Food plants**: Grasses. **Nectar plants**: Coneflower, garden mint, sunflower, fleabane, penstemon, spirea, mock-orange, alfalfa, clematis; also rotting fruit, tree sap. **Comments**: One brood per year. Can often be seen perching on tree branches or boughs. Since they often choose backgrounds the color of their wings, they become nearly invisible. In the air, the eye-spots near the wing edges help to distract birds from vital body parts. Absent in outer coastal areas. Overwinters as larva.

Skippers

This group of small butterflies is easily identified by their swift, bouncing, erratic flight. This "skipping" flight pattern gives the group its name. The adults are usually brown, orange, or black. If you get close to a skipper, you will see it has a rather stout, hairy body, with a large head and triangular wings. Although all butterflies have clubbed antennae, you will also see under close observation that skippers have distinctive hooks at the ends of their antennae.

Woodland skipper
(See Plate 7)

Adult description: Small; upper side orange and brown, underside duller. Often holds forewing and hindwing at different angles. **Food plants**: Grasses; caterpillars feed at night. **Nectar plants**: Bluebeard, lavender, butterfly bush, oxeye daisy, garden sage, oregano, coreopsis, pearly everlasting, statice, black-eyed Susan, thistle, dandelion, cat's ear, marigold, fall sedum, lobelia, aster, and others. **Comments**: One staggered brood July to October; overwinters as young larva. Larvae are pale green or yellow with lateral lines. Vary adaptable. Found in most sunny habitats from sea level to mid-elevation mountains; common in lawns and gardens. Likes mud at the edge of ponds and streams.

Moths and Larvae

For photos of Anise swallowtail larva and Western tiger swallowtail larva, see Plate 8.

For photos of the clear-winged sphinx moth, polyphemus moth, and ceanothus silk moth, see Plate 7.

CHAPTER 19

HEDGEROWS

A hedge is a closely-planted row of tightly pruned trees or sheared shrubs. Although a hedge may meet the needs of those who maintain it, there may be better ways to meet those needs and to provide for the needs of wildlife.

An alternative to a hedge is the hedgerow. It functions as a hedge and also provides cover for nesting, shelter in winter, food sources, and a safe travel lane for a variety of wildlife. Hedgerows are planted in a more natural arrangement and require less intensive management than hedges.

A hedgerow composed of a variety of trees, shrubs, and smaller plants can form a tapestry of flowers, fruits, seeds, and foliage that serves the needs of wildlife throughout the year (Figure 1).

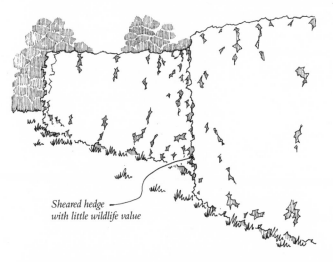 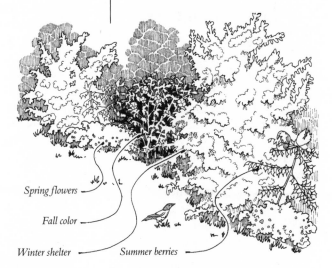

Sheared hedge with little wildlife value

Spring flowers

Fall color

Winter shelter   Summer berries

Figure 1. *A sheared hedge has little wildlife value. It offers few places where birds can enter and lacks the openness that allows flowers, fruits, and nuts to develop and be used by wildlife. When a hedge is composed of a single plant species, it will attract a limited number of wildlife species. However, a hedgerow that contains several plant species of different sizes will be used by a variety of wildlife.*

Functions of Hedgerows

A homeowner may plant a hedgerow along a property line to screen a view of the neighbors' house and to create more privacy. However, if the hedgerow includes a mixture of plants such as disease-resistant, semi-dwarf plum trees and native hawthorns, huckleberries, and elderberries, it will also help wildlife. As an added benefit, the homeowner may provide seasonal food for the family.

Hedgerows may provide a variety of other benefits, as well, including:
- Buffering a sensitive wildlife area such as a wetland or streamside.
- Hiding an unsightly building, parking area, or glaring light.

Thickets

Hedgerows are in some ways similar to the wild thickets that develop along fences and ditches. These tangles are created from seeds dispersed by wildlife or the wind, or through vegetative means such as suckering and layering.

The large number of bird-dispersed plants found in many thickets is a result of "cooperation" between plants and birds. Plants such as wild rose, elderberry, and currant produce fruit that is eaten by birds, which in turn disperse the seeds through droppings to produce more plants. In fact, thickets often develop along fences and other places birds like to perch.

Unfortunately, as a result of urban sprawl and the development of large corporate-farm fields, these wildlife areas are being removed. They are missed by wildlife and by wildlife enthusiasts.

- Buffering a home from noise and glaring lights.
- Providing winter wind protection for people, pets, wildlife, and buildings.
- Directing cool summer breezes to create a micro-climate for both wildlife and people.
- Beautifying the landscape with displays of bloom, fruit, twigs, textures, colors, and forms.

Figure 2. Waxwings are one of the many bird species that will visit a hedgerow that features berries.

How Hedgerows Support Wildlife

Most wildlife will use a newly planted hedgerow as soon as plants begin to flower, fruit, or provide a perch. As the plants mature, more and more wildlife will use the area to meet their needs. A hedgerow should there-fore be considered a long-term investment that increases in value the older it becomes.

Hedgerows meet many wildlife-habitat requirements, concentrating food, shelter, and space in a small area. A well-designed hedgerow will provide food and shelter throughout the year. During winter, the shelter of a hedgerow can be critical for many species, particularly in areas where little other cover exists.

Not only the hedgerow it-self but also the areas immediately adjacent may be used by wildlife. Quail and other birds, for example, may roost and nest in a hedgerow, and feed in adja-cent areas.

A hedgerow can provide a protected corridor along which animals can travel without being exposed to predators or human interference. This may be espe-cially important in areas where there are many homes; in such areas, wildlife habitat may be bro-ken up into many small parcels. In order to find enough food and other resources to survive, wildlife species such as raccoons, rabbits, squirrels, foxes, and skunks may need to move about from one fragment of habitat to another. Properly sited hedgerows make this possible. (See Figure 1 in Chapter 1 for an example of how yards can be landscaped to create a corridor.)

Different species have differ-ent requirements, however: a fox may need a 30-foot-wide travel corridor, while a rabbit or skunk may require half that width.

A hedgerow also may offer an isolated sanctuary in a large expanse of cropland or grassland. These "vegetated islands" often function as stopover points for migrating songbirds. They also provide elevated song perches for breeding grassland and wood-land birds.

Many insects and related creatures are found in hedge-rows, including species that are beneficial to gardeners. Insects are also an important food for birds, some small mammals, and other insects. A hedgerow contain-ing a small number of pest insects can even serve as a "larder" for predators: the small reservoir of pest species in the hedge allows predators to survive in numbers sufficient to check pest outbreaks when they occur.

Designing a Hedgerow

A hedgerow can be located any-where you'd plant a traditional large hedge, including along a fence, a property line, and/or the side of a house. Use of a hedge-row increases the closer it is to an undisturbed area or a water source. Whenever possible, use the planting to connect a larger wild area, such as a stream corri-dor, woodland, or wetland to a habitat planting on your property. However, don't create a driving hazard by placing plants where they may block drivers' lines of sight—for instance, at the corner of a road or at the entrance to a driveway.

The goal in designing a hedgerow is to create a concen-trated group of plants that will establish themselves quickly and soon become nearly maintenance-free. As with any successful plant-ing, after the planning and design, pay careful attention to site prepa-ration, planting, and maintenance during the establishment period. Since time, labor, land, and money are all at stake, make every effort to choose the proper plants. (See Chapter 3 for information on obtaining plants and planting them.)

Choosing Plants

Choose plants adapted to the site conditions, such as wet or dry soil and sun or shade (see Appendixes B and C). Native plants that occur, or that historically occurred in your area are good choices. Remember that many native plants, such as wild rose, elderberry, and tall

Oregon-grape, grow quickly and spread widely when watered and fertilized; use caution when planting these in a small space. Non-native species that don't spread aggressively, and disease-resistant fruit trees that grow well in your area could also be used. Look around your neighborhood to see what grows in conditions similar to those of your future hedgerow.

When selecting plants:

- Include some thorny plants and some suckering plants.
- Include fast-growing plants and slower-growing plants. Fast-growing plants create quick food and shelter and can be removed at any time.
- Include both evergreen and deciduous plants. If you have only a small area to plant, include at least a couple of evergreen shrubs that offer a food source for wildlife. Include plants that provide flowers, fruits, and seeds at different times of the year.
- Try to include some plants that hold their fruits into the winter.

- Include sun-tolerant species on the sunny side and shade-tolerant species on the shady side of the planting. Include butterfly plants, including perennials, on the sunny south and west sides.
- If the planting is to be viewed from only one side, choose taller species for the back and smaller plants for the front.

(For more information on choosing plants, see Chapter 3.)

Installing a Hedgerow

Although this style of planting is loosely modeled after the traditional hedgerows grown in northern Europe, it relies more heavily on a variety of plant types for its structure, rather than on closely spaced plants and intensive maintenance.

The optimum spacing for the plants in a hedgerow depends upon the growth rates of the plants. Fast-growing shrubs need fairly wide planting distances. Slow growers that get large can have fast-growing plants between them. In time, the

Ways to Improve an Existing Hedge (or Windbreak) for Wildlife

You can improve an existing hedge or windbreak for wildlife by taking any of the following steps:

- Add plants to provide additional food for wildlife.
- Add plants to create a mixture of evergreens and deciduous plants, to increase options for shelter.
- Leave unharvested vegetable-garden plants near the area to provide food plots for wildlife.
- Add brush and rock shelters nearby for additional shelter.
- Retain all snags and dead limbs within the hedge or windbreak where they don't pose a danger. These may be used as nesting, perching, and foraging sites.

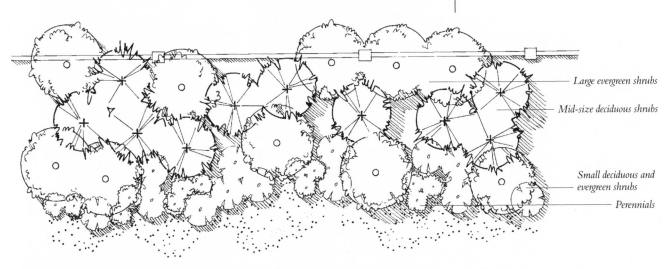

Figure 3. A sample of a planting plan for a wide ornamental hedgerow along a fence. Trees can be included every 30 feet to add vertical structure.

Modifying Temperatures for Butterflies and Other Insects

Reflected light and radiation from hedgerow shrubs, as well as protection from cooler winds, exert a considerable warming effect on the south and southwest sides of a planting. Temperatures are highest on the south side of an east–west and on the west side of a north–south hedgerow. Because butterflies and other insects require a warm, relatively wind-free area, butterfly nectar plants are excellent candidates for the south or west side of the planting. See Chapter 18 and Appendix B for a list of plant possibilities.

For a three-row hedgerow, a good distribution pattern is as follows:

D A B C D A B

A D A B C D

B C B C D A C

Key

A = tall permanent species
B = shorter permanent species
C = shrub–understory species
D = fast-growing species

fast-growing species may have to be removed to make room for the permanent species.

Your choice of plant sizes will be partly determined by how long you are willing to wait for the plants to grow to the desired height. Other considerations are cost, availability, ease of handling, and the speed at which the plants grow.

When the proper plants are chosen for the site, they should be able to fend for themselves after an establishment period of two or three years. In hot, dry areas, however, the establishment period may be longer. In general, pruning will be limited to keeping more aggressive plants in scale with the planting, but a hedgerow located in a more traditionally maintained landscape may require additional tending. In the long term, graceful spreading and blending of plants should be encouraged to maximize the food and shelter provided for wildlife.

Grasses, especially sod-forming ones, compete with young plants and should be kept 2–3 feet away from them during the first three years after planting.

One or more types of mulches should be used to control weeds and to conserve moisture in new plantings. Mat-type weed barriers made of synthetic materials are cost-effective and are especially useful at remote sites. Drip irrigation systems are the most efficient way to get water to large linear areas; they also conserve water and help prevent blanket weed growth.

If the planting is next to a road that is sprayed by the department of transportation or your county roads division, contact them to stop roadside spraying. It may be possible for you to form an agreement making you responsible for managing plant growth along the road next to your property.

Windbreaks and Living Fences

Windbreaks are barriers used to reduce and redirect wind. They are planted to shelter homes and farmsteads or are incorporated as part of a crop or livestock operation to enhance production, protect livestock, and control soil erosion. Windbreaks usually consist of trees and shrubs, but may also include perennial or annual crops and grasses, fences, or other materials.

Living fences are small-scale windbreaks designed to hold livestock, divide farm fields, and, unlike traditional fencing, when properly maintained they rarely need replacing. Living fences work best when bordered only on one side by a livestock field.

Both windbreaks and living fences can create an attractive visual boundary in the landscape. When plant species and arrangements that give wildlife the basic essentials of food and cover (and water, if possible) are provided in their design, windbreaks and living fences will also support and protect a wide array of wildlife species. Wildlife activity will increase if windbreaks and living fences are connected to other planted or natural sources of cover, streams, or ponds. If the windbreak or living fence cannot be designed to connect, plant travel lanes to connect to other food, cover, or water sources.

Important considerations when planning a windbreak include:

- Plant food plots alongside the windbreak or leave a few rows of standing crops. Cultivating a strip to let annual plants grow can be a good source of food and cover.
- Generally speaking, wider plantings are better. A single-row windbreak is less valuable to wildlife than multiple rows are. The ultimate windbreak might be 10 rows of trees and shrubs up to 150 feet wide. However, few people are willing to give up this much land or maintain this large of a planting.
- Try to mix different yet compatible plants in the rows to give natural "feel" to the windbreak.

Important considerations when planning a living fence include:

- The most common planting plan is the double line. Offset the two rows to give the necessary even distribution of plants (see "Installing a Hedgerow").
- Select a good "backbone" plant which will constitute 60 to 70 percent of the hedgerow. Next, add four to six additional shrubs or small trees to the design to add value as wildlife habitat, and allow for minimal gapping if a particular species dies out. Choose plants such as wild rose

and Oregon-grape that produce shoots close to the ground, thus allowing for both small and larger animals.

- To improve the stock-proofing properties, include 70–95 percent thorny species (see Appendix B). Plants should have a good growth rate, be strong enough to resist the efforts of animals to escape, not be edible or overly attractive to the animals within the field, and be able to withstand severe pruning.
- Trees in a hedgerow should be spaced a minimum of 20 feet apart. They are beneficial for shade and as a wind barrier for livestock. They also provide great cover for wildlife. Hedgerow trees can be particularly important to hawks and owls.

To create a good stock-proof living fence that can be trimmed to create density, the inner row (field side) in the double line should be planted at 18–24 inch spacing. The outer (second) line of the hedge should be spaced 7 feet away and planted with a variety of other species allowed to grow with less formal training to provide a better habitat for wildlife.

Fencing must be erected at a distance great enough to prevent stock from browsing the tops off the hedge plants. To insure strength and vigor in a hedgerow, protective fencing from livestock must remain in place for a minimum of five years.

A mower or weed whacker can be used to cut competing vegetation as close to the ground as possible. To further discourage weed competition, a weed mat can be installed at planting time; a thick mulch can be applied at planting time and then reapplied as needed.

How a Windbreak Works

As wind blows against a windbreak, air pressure builds up on the windward side (the side toward the wind), and large quantities of air move up and over the top or around the ends of the windbreak. Windbreak structure—height, width, density, plant species composition, length, and orientation—determines the effectiveness of the windbreak in reducing wind speed and altering the microclimate. Windbreak height is the most important factor determining the downwind area protected by a windbreak.

On the windward side of a windbreak, wind speed reduction is measurable upwind for a distance of two to five times the height of the windbreak. On the leeward side (the side away from the wind), wind speed reductions occur up to thirty times the height downwind of the barrier. For example, in a windbreak where the tallest trees are 30 feet, lower wind speeds are measurable for 60 feet to a150 feet on the windward side, and up to 900 feet on the leeward side.

Windbreak density is the ratio of the solid portion of the barrier to the total area of the barrier. Wind flows through the open portions of a windbreak; thus, the more solid a windbreak, the less wind passes through. In designing a windbreak, density should be adjusted to meet your objectives. A windbreak density of 40–60 percent provides the greatest downwind area of protection. Windbreaks designed to catch and store snow in a confined area usually have densities in the range

continued on next page

of 60–80 percent. If density is above 80 percent, excessive leeward turbulence may reduce windbreak effectiveness beyond eight times the distance of the windbreak.

Windbreaks are most effective when oriented at right angles to prevailing winds.

Drip Irrigation Basics

(Adapted with permission from Harmony Farm Supply and Nursery, (707) 823-9125, www.harmonyfarm.com)

If you have ever put something together with Tinkertoys, you have the basic ability to put together a drip system. Small, backyard systems can be purchased in most hardware stores. For a drip system that waters over a hundred plants at one time, you should get the components you need from an irrigation supply center. Look under "Irrigation Supplies" in your telephone directory.

The main question you need to answer concerns the quality of your water. The dirtier your water is, the more careful you will have to be with the components you choose. The worst water is from ponds or lakes that have algae or suspended sediment. The next worst are water containing iron slime bacteria and water that contains calcium or magnesium that precipitates out when exposed to the air. For these situations, large-orifice emitters such as 1 or 2 gallons per hour are recommended because they won't clog as easily. Soaker-type tubing clogs the easiest.

What the System Needs

All drip systems should have filters, regardless of the water source. They are a very cheap way to ensure that contaminants in the water will only clog the filter, not the emitters.

Most drip systems should also have a pressure regulator — especially if you are on city water or have a well set to run between 40 and 60 psi (pounds of pressure per square inch). If you are on a spring box or have a gravity-fed tank with low pressure, you probably won't need a pressure regulator.

Most home landscape systems will require half-inch drip hose as a mainline. Large systems should use three-quarter inch hose, or break the system into segments, or use more than one faucet hookup.

Individual emitters are punched into the main line wherever the plants are spaced. On hilly ground (greater than 20 feet vertical change up or down a hill), you should definitely choose pressure compensating emitters so that all of your plants get the same amount of water.

System Design

Start by making a sketch or scale drawing of the areas you want to drip. Note the location and size of plants to be irrigated, whether they are drought tolerant or require more water, and the location of your water source. Note distances needed for main-line tubing.

Next, check the capacity of the water source. To do this, simply time how long the valve takes to fill a measured container. (Example: If it takes 30 seconds to fill a 5-gallon bucket, then the maximum flow available is 600 gallons per hour. The formula to use is: GPM x 60 = GPH.)

Standard garden valves will almost always provide more than enough water for your landscape needs. To determine how much water the drip system will require, simply add up the total number of emitters and their flow rates. (Example: 20 x 1-gph emitters = 20 gph; 20 x 2-gph emitters = 40 gph; 20 + 40 = 60 gph total.)

Installing Tubing

Allowing the tubing to sit in the sun before installation will make it easier to work with. When assembling fittings, cut the tubing with pruning shears or a sharp knife and be careful to keep dirt out of tubing and fittings. Be sure not to bury the ends of the tubing. You will need to have access to them for periodic line-flushing.

Placing Emitters

Punch holes in tubing with a punch. Using an ice pick or a nail is not recommended because the hole is not evenly punched and often results in stress cracks and leaks around the hole. Also, you can go all the way through to the other side!

It is easiest to attach emitters when the weather is warm and water is filling the drip hose. This will help make the tubing firm and easier to punch into, as well as make the hole the right size for the emitters (no leaking around barbs).

Insert emitters carefully. For small plants, emitters are often placed fairly close to the trunk.

Do not place emitters so that they will get the trunk wet. This is especially important with perennial plants, such as fruit trees and vines, which are susceptible to root rots. A general rule of thumb is to keep emitters 12 to 18 inches away from the trunks of plants.

The area wetted by an emitter will vary according to soil type. Sandy soil allows water to percolate down rapidly, while water on clay soil moves horizontally much further before it goes down below the root zone. On sandy, coarse soil, a one gph emitter will wet an area on the surface about 12 to 15 inches in diameter. On clay soils, the emitter may wet an area 24 inches in diameter. Below the surface, a large onion-shaped area is wetted as water percolates down. Emitters should be placed to cover root zones well.

Emitters themselves should never be buried because it is much easier to check and maintain them when you can see them. Also, buried emitters can have problems with clogging due to root intrusion or back-siphonage of dirt.

Maintenance

It is important to inspect emitters periodically to ensure they aren't clogged.

Filter screens should be flushed out at least once a month. Checking the filter screens for debris after the first few irrigations will help determine how often you will need to clean them. In some systems, the filters should be cleaned after every irrigation.

Tubing lines should be flushed periodically—at least once a year. Again, water quality will determine frequency.

You can always take your design questions to an irrigation supply center.

PART 4

COEXISTING WITH WILDLIFE

CHAPTER 20

WATCHING WILDLIFE

Watching and photographing the wildlife that use your property can be a source of great pleasure to you and your family, even when it is challenging. This pastime, which is rapidly growing in popularity, usually causes little disturbance or harm if it is done responsibly.

When landscaping for wildlife, much of your labor will be for wildlife you never see. Some species are nocturnal, others live underground, and others stay hidden in debris or in dense tangles.

For information on watching fish and other specific types of wildlife, see the chapters in Part 2. For information on watching butterflies, see Chapter 18. For information on watching bats, see Chapter 13.

Etiquette of Wildlife Watching

In most cases, any harm you cause wildlife is due to stress and the energy the animal uses to move away from a perceived danger—you. This is particularly true in winter when animals are forced together in areas that still provide food and shelter. The stress triggered by careless intruders can lead to a variety of food and health problems for mammals and other wildlife.

Even in urban areas, where wildlife are accustomed to people, not disturbing animals may allow you to better observe their behavior. Here's how to keep from disturbing wildlife:

1. Observe wildlife from a distance they consider safe. Get your "closeup" by using binoculars, a spotting scope, a telephoto camera lens, or a wildlife viewing blind. What is a safe distance? You are probably too close if most of the animals are looking at you with heads up and ears pointed toward you, look nervous, or are jumpy when you move or make a noise. If you see these signs, sit quietly or move slowly away until the behavior changes.

2. Move slowly and casually, not directly at wildlife. Allow them to keep you in view; don't sneak up and surprise them. Most wildlife rely on their eyesight and sense of smell to keep them from danger.

3. Never chase wildlife. Don't follow them or behave in any way that might be seen as harassment, which is unlawful. Keep pets at home or on a leash.

4. Using animal behavior as a guide, limit the time you spend with them, just as you would when visiting any friend's home.

To observe wildlife, try these tips:

- Move like molasses: slow, smooth, and steady.
- Crouch behind boulders or vegetation to hide your figure and break up your outline.

Binoculars and How to Use Them

Binoculars come in several sizes and degrees of magnification, such as 7 x 35, 8 x 40, and 10 x 50. The first number refers to how much the animal will be magnified compared to the way it appears to the naked eye. A "7 x" figure, for example, means the animal will appear seven times larger than it would if you viewed it without binoculars.

More magnification is not always better. Greater magnification can amplify hand movements, making wildlife harder to see; a bird in a tree will be harder to find with a 10 x magnification than with a 7 x because your small movements, even your breathing, will cause the image to move.

The second number refers to the diameter of the large end of the lens (the end facing the animal). The greater that number, the greater the amount of light entering the lens—which means better viewing in dim light. This is important in viewing animals because more animals are active in low-light conditions. A 7 x 50 pair of binoculars will produce an image approximately 1.5 times brighter than will a 7 x 35 pair, though the 7 x 50 model will also be heavier. At 100 feet, a 7 x 35 pair of binoculars will allow you to see an animal as if it were just 14 feet away.

There's no perfect pair of binoculars. When purchasing a pair, consider in what conditions and circumstances they will be used. For example, pocket-sized binoculars are small and easy to carry.

continued on next page

However, they may not work well in low light. The binoculars most often chosen by bird-watchers are 7 x 35; these strike a reasonable balance between compact size and amount of light entering the lens, and they provide a useful degree of magnification. Lower-powered binoculars that focus closer than most are also best for butterfly watching.

Binoculars are relatively easy to use:

1. Find the subject with your unaided eyes.
2. Bring the eyepieces just under your eyes.
3. Sight the subject over the tops of the eyepieces.
4. Slowly bring the binoculars to your eyes.

If at first you don't succeed at finding your object in the lens, locate a larger object or landmark close to the animal with your naked eye. Make a mental note of where the animal is in relation to the large object. Bring the binoculars to your eyes again, find the larger object, and then bring the animal into your field of vision.

Spotting scopes, which can be set on tripods, are used for viewing more stationary wildlife at long distances, such as animals tending a nest. Spotting scopes are monocular (having one lens) and feature much higher magnification than binoculars.

For information on hand lenses, dissecting scopes and other lenses, see Chapter 9.

- Look for out-of-place shapes and motions, such as quickly moving grasses.
- Make "mule ears" by cupping your hands around the backs of your ears to amplify natural sounds.
- Watch with binoculars for movement in areas where you suspect mammal traffic, such as along hedges, fences, and bushes besides streams.

Learn as much as you can about the particular wildlife species you are watching. This will help you to locate it and to better understand its behavior. To further enhance your pleasure in wildlife-watching, develope your own species list for your property (Table 1).

Tracking

Studying or following prints left by animals is an ancient activity first practiced by hunting and gathering societies that depended on it for survival. It is now also a fascinating aspect of wildlife watching.

Tracking tips:

- When studying animal tracks, stand with the tracks between you and the sun for best perspective.
- Measure the length and width of several prints. On many mammals, the front feet will be larger since they support more weight.
- Count the number of toes, note any claw marks, and look for a heel.
- Follow the tracks and note any patterns, which vary depending on whether the animal was walking or running.

Blinds for Viewing Wildlife

The secret to observing wildlife behavior is to sit still and keep quiet. If you're hidden from view, so much the better. A useful tool for watching mammals is a wildlife viewing blind. Using a viewing blind not only conceals you but also allows some limited movement—you can fiddle with a spotting scope or camera, take notes, and fidget to your heart's content, while remaining undetected.

Your house is a ready-made blind for observing wildlife. The easiest way to get a close view of birds is to locate your feeder, bird-bath, or pond so it can be viewed from a convenient window. However, for those who want to observe the habits and movements of wildlife, it's worth the effort to construct a simple blind. This need not be expensive or time consuming. Often these projects can involve the whole family.

Let experience be your guide: Before you decide where to build a blind, spend some quiet time alone in the area. You can determine the exact location for the blind by figuring out the lighting requirements at the time of day you plan to use the blind, where you will have easy access to the blind without disturbing wildlife, and how the blind will look in the landscape.

Good places to locate a blind include:

- Near a water source, such as a creek, seep, pond, garden pool, or birdbath.
- At the edge of a woodland, hedgerow, or thicket.
- Near an existing snag, brush shelter, or rock pile.

Table 1. Sample backyard wildlife species list

(Adapted with permission from Jim Pruske, Washington State Department of Fish and Wildlife)

Key
A = 1996, B = 1997, C = 1998, D = 1999
Upper case: Observed birds using feeder or birdhouse.
Lower case: Observed birds and other wildlife in trees, on ground, in flight, or hawks hunting at feeder.

Birds (Sample)	MAR	APR	MAY	JUN	JUL	AUG	SEP	OCT	NOV	DEC	JAN	FEB
Mallard	CD	BCD	D	D	D	D			c		d	d
Cooper's Hawk	bcd	b		cd	c	c	B	bcd	abd		d	
Great Horned Owl						ab	c		d			
Rufous Hummingbird	BCD	BCD	BCD	BC	ABCD	BC						
Downy Woodpecker			d	d			C	BCD	BD	D	CD	CD
Violet-Green Swallow	BCD	BCD	CD	CD	ACD							
Steller's Jay	BCD	BCD	CD	BCD	BCD	BC	ABC	ABC	ABC	B	CD	BCD
Crow	Bd	D	BD	CD	d		C	BC	AC	C	D	Bd
Black-capped Chickadee	BCD	BCD	BCD	CD	ABCD	BC	BC	BC	ABC	ABC	BCD	BCD
Varied Thrush	Cd	C								BCD	CD	CD
Starling	CD	BCD	CD	BCD	ACD	BC	ABC	BC	B	B	CD	C
Evening Grosbeak	d	BD	BCD	D	D	C	BC	B	B			
American Goldfinch		BC	BCD	BC	ABCD	BC	ABC	ABC	A			

Mammals	MAR	APR	MAY	JUN	JUL	AUG	SEP	OCT	NOV	DEC	JAN	FEB
Douglas Squirrel	a	a	a	a	a	a	a	a				
Little Brown Bat	bc	bc	bc	bc	bc	bc	bc					
House Mouse	abcd	abcd	abcd	abcd	abcd	abcd	abcd	abcd	abcd	abcd	abcd	abcd
Raccoon		cd	cd	cd	c	c	c	cd				

Reptiles/Amphibians	MAR	APR	MAY	JUN	JUL	AUG	SEP	OCT	NOV	DEC	JAN	FEB
Northwestern Garter Snake			d	d	d		d					
Tree Frog	d	d	d	d	d	d	d	d	d	d	d	d

Butterflies/Moths	MAR	APR	MAY	JUN	JUL	AUG	SEP	OCT	NOV	DEC	JAN	FEB
Pale Swallowtail				cd	d	cd	d					
Cabbage White		abcd	abcd	abcd	abcd	abcd	abcd	abcd				
Spring Azure			ac	ac								
Painted Lady					abcd	abcd	abcd	abcd				
Woodland Skipper					abc	abcd	abcd					
Polyphemus Moth					abc	abc	abc					

Other	MAR	APR	MAY	JUN	JUL	AUG	SEP	OCT	NOV	DEC	JAN	FEB
Northern Bluet Dragonfly				a	ab							
Orchard Mason Bee	abcd	abcd	abcd	abcd	abcd	abcd	abcd					

Building Blinds

The simplest open blind is a bench, chair, or tree stump placed behind or among existing shrubbery that will camouflage you from visiting wildlife. Another type of open blind is a wooden screen with openings to see through (Figure 1). In wooded areas, a blind can be constructed from freshly cut saplings, tree branches, and weathered lumber (avoid using new materials that stand out or clash with the surroundings). Gather together what you find to form a teepee-like or free-form structure, and add branches where needed for camouflage.

Old canvas tents make good blinds because they can keep out rain, wind, and snow. Most also have room in which to move around. Provide eyeholes at standing and sitting levels, a slit for your camera and spotting scope, and a shelf for a notebook, sandwiches, etc. With chicken wire, tall grasses or rushes, and some ingenuity you can make a lightweight portable blind like those many duck hunters use. Plastic isn't recommended as a covering because it's noisy, doesn't breathe, and doesn't hold up when exposed to sunlight day after day. These types of blinds can be anchored with conventional tent stakes and ropes, or secured to trees and fence posts.

A useful dodge is to arrive at the blind with a friend, who then leaves after a short time; wildlife may think the blind is then empty.

You can build a permanent blind by nailing together a simple boxlike structure with 2 x 4-inch lumber and then covering it with brush. Leave holes for your camera and other stuff. Be sure the structure is large enough. A space 3 feet square is about right for one person, and an average person is comfortable on a seat 18 inches high. Finally, be patient—because it may take wildlife a couple of weeks to become accustomed to the blind.

A gazebo, outbuilding, or tree house can double as a blind. You can place bird feeders and birdbaths nearby for an optimum view from inside. Finally, you might consider painting the blind to help it blend in with the surroundings and installing screens to keep mosquitoes and flies out.

The future of the Pacific Northwest's wildlife depends on our children's commitment to wildlife conservation. So take a child wildlife-watching. Search out bugs and butterflies with your nephew. Explore the deep woods with your daughter. Introduce

Figure 1. A simple blind for viewing.

children to the mysteries and beauties of the natural environment around your home. It's fun; it's also a learning experience. You will be helping a child develop positive feelings toward wildlife that may last a lifetime.

Because many animals are nocturnal or elusive in other ways, often the pleasure of maintaining wildlife habitat must come from the knowledge that you are helping creatures without necessarily getting to see the result of your work. Birds and insects tend to be the exceptions, as both conduct part or all of their daily lives where we can see them.

Trails in Public Areas

Good trail planning can enhance the viewing experience and protect the natural qualities of a wild area. Not every area should include trails, but they are the safest and most comfortable way to direct viewers to a particular viewing place, while at the same time protecting the habitat.

The first thing to do is decide whether access should be encouraged to all and, if so, to what extent? This will be determined by the degree to which the wildlife conservation value of the site may be put at risk by increasing the level of human disturbance. The major concern is physical damage to plant communities and disturbance to breeding or feeding birds or mammals. For birds, human-induced disturbance during the breeding season may lead to nest desertion and increased predation or, outside the breeding season, to a reduction in the level of foraging activity and ultimately the use of a site.

As with all structures, trails should be designed and built in

harmony with the landscape. They should also be easily maintained and have minimum susceptibility to erosion. Minimizing the visual impact of a trail is relatively easy in woodlands, but more difficult in grasslands and wetlands. Direct ascents of hills and straight-line paths should be avoided. Whenever possible, paths on permanent or seasonally water-logged soils should be avoided.

Trails can be made with a variety of materials, each with its own suitability to particular sites. Asphalt and concrete trails are only appropriate for heavily used sites and sites making a special effort to accommodate the physically challenged. Such trails are expensive, but they require little maintenance. Gravel, crushed glass, or crushed concrete are also suitable surfaces for trails with heavy use, but not necessarily for the physically challenged. Moderately used trails and trails on small properties are ideally surfaced with wood chips, bark, or crushed rock. Those with low seasonal use and a fairly even grade can have natural surfaces such as soil and fallen leaves.

Boardwalks can be used over wet or sensitive environments. Boardwalks need to be carefully constructed and well maintained. There is often no physical reason for walkers to stay on boardwalks, except in very wet marshes and bogs, and so the width and layout must be such that walkers are not tempted off them. Screens of evergreen shrubs or hedges of thick or thorny vegetation are good deterrents to wandering off the route. In woodlands, a few small, well-positioned logs at the edge of the path can be effective in reducing wandering from the path.

Think Small: Backyard Wildlife Photography

(Adapted with permission from: National Wildlife Federation, Vienna, VA, (703) 790-4434, www.nwf.org/habitats)

"Think small!" That's hardly what our parents and teachers and other adults in our lives told us when we were growing up. But if you learn to think small in your own backyard wildlife habitat, you'll quickly discover an amazing new world. And with the right kind of photography equipment and know-how, you'll be able to capture this world on film—to the delight of yourself and others.

The basic tools necessary for close-up work are a single lens reflex (SLR) camera body and the appropriate lenses and other gear to go with it. (A "point and shoot" camera is incapable of anything approaching true close-up work.) Today's electronic SLRs are precise and dependable, easy to operate, and relatively inexpensive.

The cheapest way to get started in close-up shooting is to buy some extension tubes to put between the camera body and the lens that may have come with it. (The more extension, the closer view you'll be able to get.) Another choice is to add close-up lenses to the front of a regular lens. But for the best results, buy a lens especially made for close-up work—the macro. Macros come in focal lengths of between 50 and 200 mm, with the longer lengths allowing more "working room" between you and your subject—something to keep in mind when trying to capture an image of a nervous dragonfly or cranky wasp. True macros can be expensive, but

continued on next page

they give you such versatility and such tack-sharp images that they're worth every penny. (Don't confuse a true macro with a zoom lens with a "macro" setting.)

Magnifying your image also magnifies every move you make. A good tripod will take care of that problem, so long as you have enough natural light and a subject, such as a dew-covered spiderweb, that holds perfectly still. But with low light or a moving subject, you'll need an electronic flash. The quick burst of light from the flash will freeze all action, allowing you to move freely through your backyard, stalking and shooting your "prey" wherever you find it.

You'll be amazed by the intricacies that close-up photography will reveal. Soon you'll be seeing in a whole new way—slowly and carefully observing, learning where to look, quietly letting new forms of life introduce themselves. Even the smallest backyard habitat will become an undiscovered wilderness, full of tiny treasures—each with something to show you.

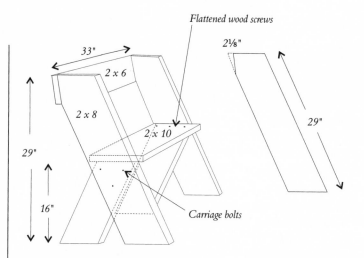

Figure 2. *The Aldo Leopold bench is sturdy and easy to build.*

Although human intrusion into wetlands can be extremely damaging, they need not be totally inaccessible to people. To avoid startling wildlife (especially nesting birds), it's best not to provide access too far into the wetland and not to ring it with a path, unless the wetland is large (5 acres or larger).

Trails into dark woodlands may pose very different problems since these can be intimidating to some people. Trails that have to go under a canopy should be clearly defined by surfacing, such as with wood chips, to encourage use and ensure that people are not unduly concerned about losing their way.

In public areas, the opportunity for well-designed interpretive or educational signs and displays should not be overlooked. Informational signs along high-use trails help educate and increase appreciative use.

Finally, it is important to include resting areas for relaxation and enjoyment. Locate benches or comfortable logs along the trail in places where there is a good view, an opportunity to see a particular wildlife species, or a need to rest (see Figure 2).

CHAPTER 21

HOW TO DEAL WITH WILDLIFE-RELATED PROBLEMS

Our human perspective on wildlife often results in labeling animals "pests" or "nuisance animals." But this label is unfair. Wildlife do not intend to be a nuisance to anyone. They're merely attempting to fulfill their needs—often in the same area where they and their kind have lived for hundreds, if not thousands, of years.

When frustrated, some people seek short-term, frequently lethal solutions to wildlife conflicts. These methods usually fail to address the real problem and just create a void or vacant territory, which is quickly filled by another animal. Fortunately, there are humane, long-term, and effective ways to deal with wildlife-related problems. Ultimately, as long as they're kept out of your home and their interaction with children and pets is limited, you can consider wildlife to be friendly visitors.

Avoid Potential Problems

The best thing you can do for yourself and for wildlife is to prevent potential problems. Do this by removing things that attract animals to places where you don't want them. Typical attractions include uneaten pet food left on your porch, uncovered garbage, a pile of birdseed under a feeder, or any opening to an attic, wall, or crawlspace. If you intentionally or unintentionally provide any of these, you may attract unwanted wild animals.

To prevent problems:
- Inspect your house for any outside holes and other access routes. (Note: A mouse or bat can slip through a hole the size of a dime.) Cover any large openings with heavy-duty wire mesh, boards, bricks, or sheet metal.
- Stuff steel wool around holes where pipes and wires enter buildings.
- Screen off unscreened louvers and air intakes.
- Cap chimneys with a commercial chimney cap, heavy wire mesh, or other cover that an animal can't pry or chew off or gain access through.
- Inspect trellises or vines that may give animals a route to your roof or attic.
- Cover large, ground-level, open-sided porches from the floor level to the ground with wire or wood fencing. Bury the fencing 12 inches underground to prevent wildlife such as skunks from digging under.

Animals Under a Building, Porch, or Deck

Animals most often found under a building, porch, or deck include opossums, skunks, marmots, raccoons, and rats. If you need to verify that something is living in one of these places, first do some detective work. Can you smell, hear, or see anything in a particular area? Listen for baby animals crying or adults talking, scratching, gnawing, or preparing a nest. Look for other signs, including evidence of digging: hair, body-oil stains, gnaw-marks around a hole, or tracks leading into the area.

To get rid of unwanted guests in any of these areas, first eliminate all access except at one point, then seal up the access point after you're sure the animal and any young are gone.

To find the access points, cover suspected access points with loose dirt or some other material. If an animal is in there, it will dig itself out or in during the night. Another technique is to spread a light layer of white flour on the ground outside any suspected holes; animals will leave tracks in the flour if they walk through the area. Cover an area at least one foot by one foot. The next morning look for tracks leading into or out of the holes. Allow three days for either of these tests, as animals often use more than one entrance or may not appear during a cold spell. If nothing is disturbed after three days, seal up the holes with boards, bricks, sheet metal, or wire mesh (⅜-inch hardware cloth is recommended) as a preventive measure.

If you find that an access hole is being used, follow these steps:

Step 1. Seal all but one entry with boards, bricks, sheet metal, or hardware cloth.

Step 2. Drive the animal out of the area using one or more of these techniques:
- Toss rags soaked with white

How to Create and Use a One-Way Flap Door to Exclude Wildlife

A useful technique to exclude bats, raccoons, opossums, squirrels, and rats from an area is to build a one-way flap door. This one-way door allows animals to exit but not to return.

To construct the door, cut out a piece of 3/8-inch wire mesh (hardware cloth) that is 3–4 inches wider, per side, than is the access hole (Figure 1). Place the door over the outside of the hole and create a hinge by attaching a long piece of duct tape or a couple of U-shaped fencing nails to the top to serve as a hinge. If the access hole is larger than 8 inches square, you may need to increase the rigidity and strength of the hardware flap door with a frame of half-inch trim or molding. Make sure the door flap closes flush to prevent the animals from nosing their way back in.

Leave the one-way door in place several nights to ensure that no animals remain inside, especially in rainy or cool weather, when some bats may have entered torpor (similar to a self-induced coma). Once the animals are evicted, you can then permanently seal off the point of entry after determining that no wildlife are still inside.

Note: Do not use one-way flap doors during May or June, when babies may be in the den. If the mother cannot return, the babies will starve and create further problems. Also, watch for signs of any new holes being created near the flap door. A flap door for bats needs to be small. If it is a large hole you are covering, first cover the hole with a piece of plywood with a 2-inch hole drilled in it. Next, place the 4-inch-square flap door over the hole.

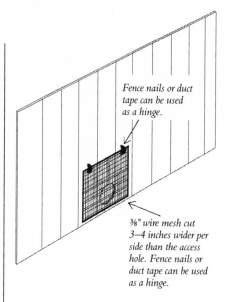

Fence nails or duct tape can be used as a hinge.

⅜" wire mesh cut 3–4 inches wider per side than the access hole. Fence nails or duct tape can be used as a hinge.

Figure 1. *A one-way flap door can be constructed with wire mesh and nails or duct tape to exclude wildlife from a building.*

vinegar into the crawl space or other area near where you expect the animal may be. Check daily to ensure the odor is present and replenish as needed. Place small bowls of white vinegar inside the entry-way and refill as needed.

- Brightly light the space day and night. Lighting can be accomplished by taping a mechanic's drop light or trouble light to a long 2 x 4-inch board and sliding it into the area.
- Put a radio near the opening or slide it into the area and play it day and night.
- Fit the remaining hole with a simple-to-make one-way door, so the animal can escape but not reenter (see Figure 1).

Note: If young are inside, it is kinder and less stressful for the animals to be patient and wait for the mother to move her young before you use these techniques. Also, never approach a mother with her young. Her protective instincts can make her dangerous.

Step 3. Check to see if the animal(s) is (are) gone. Spread a light layer of flour on the ground outside the entrance to record footprints as the animal exits. After sunset, check the flour every half-hour for tracks. When you see tracks leaving the opening, it's time to seal it temporarily.

If you hear noise after closing the entrance, determine where it is coming from. If the sound is coming from inside the enclosure, an animal may still be trapped inside. Re-open the area and repeat the process until all the den residents have departed, then seal the entrance permanently.

Skunks, marmots, and rats may burrow under a foundation or where an enclosure meets the ground. To prevent this re-entry, you'll need to dig a trench 18 inches deep and 6 inches wide along the outside wall. Place wire mesh vertically so that it extends 18 inches down, and bend the bottom 6 inches outward at a 90-degree angle. Backfill the trench with dirt once the wire is positioned. You may first want to see if these animals try to dig back into the area before making this effort.

Animals in Your Attic or Walls

Wildlife that may get into attics and walls include squirrels, bats, rats, and mice. Occasionally a large mammal, such as an opossum or raccoon, will find its way into your attic via an overhanging tree or large shrub. To get rid of any of these unwanted guests, take the same steps described under "Animals Under a Building, Porch, or Deck." Limit their access down to one hole, then seal up the hole after you're sure the animal (and its young) is gone. (Note that older structures may be very difficult to bat-proof.)

Determining where the animal is can take some investigation, however. Here's how to find the entry hole:

Step 1. Look for entry holes under eaves and around vents, ducts, circuit boxes, and unattached washer and dryer hookups. Also, look for openings in walls and floors where pipes or wires enter your house. Since small mammals can squeeze through openings the size of a dime, you may have to look hard and long to find where they're entering. To find where a bat or flying squirrel is gaining access, try watching your house from a distance and from different vantage points around dusk. The surface areas around bat entry holes often have a smooth, polished appearance. The smooth gloss of these rub-marks is due to oils from fur and other bodily secretions deposited there by animals.

Step 2. Close all but one known used hole following the steps outlined under "Animals under Your Building, Porch, or Deck." To identify the holes, you'll probably have to depend on clues other than tracks in flour. These include hair, body oil stains, or gnaw marks around the hole rather than tracks in flour.

Step 3. Drive the animal away by using a nasty odor, bright lights, noise, or a one-way door as described under "Animals under a Building, Porch, or Deck."

Step 4. Seal the one opening.

Keeping Mice and Rats Out of Houses

(Adapted from "Principles of Vertebrate Pest Management" by Dave Pehling, Washington State University Cooperative Extension, Snohomish County. http://snohomish.wsu.edu/vertchap.htm)

The imported house mouse is the most common mouse species to enter houses and other structures in urban areas. In outlying areas you're likely to have the native deer mice. The two imported species of rats found in and around houses are Norway and roof rats. (See Chapter 5 for additional information.)

The spread of these imported species coincided with ship travel, which provided them with transport over water; the construction of houses and barns which provided them shelter; and with the development of agriculture, which provided them with food.

Because they can carry diseases and have the potential to damage your home, you should have a zero tolerance for mice and rats in your living space. If you think you can solve the problem with a house cat or two, remember that cats also kill birds, chipmunks, snakes, lizards, flying squirrels, rabbits, and low-flying butterflies. (For information on cats and wildlife, see Chapter 22.)

Mice and rats are very adaptable and it is unlikely that they will ever be totally eliminated from neighborhoods and other areas where they have access to food, water, and shelter. Their control in wild areas is often carried out by coyotes, hawks, owls, and snakes. However, if mice and rats are a problem

Pet Door Problems

Pet doors installed in your home conveniently allow pets to come and go as they please, but they can also provide easy access for some wildlife. Many reports of raccoons, opossums, and skunks in homes are related to the use of pet doors. If you're willing to risk a wild animal inviting itself in, a two-way pet door can be used. Some pet doors can be set to allow animals to get out but not to get back in. If you prefer to avoid any chance of a wild neighbor getting inside, completely seal or remove pet doors.

For some new pet doors, the cat or small dog wears a magnet-equipped collar that opens the door when the pet gets within a certain distance of it. However, raccoons have been known to follow dogs inside to get pet food.

in or around your immediate home, follow these four important steps for effective control:

Step 1. Elimination of Shelter

Elimination of shelter is often overlooked in a mouse and rat control program. Rats in particular like to inhabit woodpiles and stacks of stored material. Materials that are stored in stacks or piles next to your home should be 24 inches from adjacent walls and stacked on a pallet or other structure at least 8 inches off the ground—double pallets or 16 inches is even better.

Dense vegetation and rubbish piles next to structures should be reduced as much possible. Also, avoid planting ground covers, such as ivy, that can offer shelter to rats next to your home.

Finally, spaces under buildings should be blocked off and old burrows filled in.

Step 2. Mouse- and Rat-proofing Structures

Complete rodent-proofing is usually not possible, especially in barns and older buildings, but if approached properly will go a long way toward reducing an infestation.

To keep rats out of buildings, any opening they can get their teeth into (that is, over ¼ inch in diameter) must be closed up. Mice can often squeeze through openings as small as ⅜ inch. Be sure to seal up openings around pipes and floor drains and keep the drains tightly covered. Edges subject to gnawing, such as door bottoms, should be covered with heavy sheet metal or wire. Place metal rat guards on pipes, wires, and other places rats climb. Other openings can be covered with hardware cloth no larger than ¼ inch.

Step 3. Elimination of Food and Water

Elimination of food and water is the third step for effective mouse and rat control.

Strict control of food materials is essential. Do not put food of any kind in open compost piles; bury food waste in an underground composter or use a lidded worm box instead. It is best not to feed cats and dogs outside, but if you have no other alternative, pick up bowls and leftovers promptly. Also, clean up pet droppings—rats can do very well on a straight diet of feces. If you feed birds, put only as much seed in feeders as will be eaten in a day, and clean up spilled seed and hulls before nightfall.

Step 4. Killing Mice and Rats

Along with general cleanup and controlling food sources, direct reduction of mice and rats in and under homes can be obtained with the careful use of poison and/or traps. Remember, killing existing rodents will not solve the problem. You'll have to remove attractions and repair the areas where the animals are entering the structure.

There are several effective poisons on the market. They are all effective and safe to use as long as the label directions are followed. **Always read the label before using any poison.**

To be most effective, bait stations must be placed where the rodents can easily get to them. At the same time, they must be protected from dampness, pets, and children. This can be done by placing pans of bait in runways and covering each with a box which has a hole in each end.

Traps can also be used effectively to help control rat and mouse infestations. They are especially useful when poisons are not wanted or to catch bait-shy rodents left after a baiting program. When traps are used for rats, it is most effective to set one or two every 15 to 20 feet, the same spacing you'd use for bait, wherever there are signs of activity. For mice, place traps every 5 to 10 feet.

Use caution when setting a mouse or rat trap outside where there are ground-feeding birds. These birds are easily caught when they peck at the bait. Peanut butter mixed with rolled oats is a good bait for all traps.

You may have to try several baits to find what works best in your situation. Whichever bait you choose, be sure it is fixed securely to the trap so it cannot be licked off. Being very cautious by nature, rats will sometimes avoid a trap or strange bait for up to 14 days. Mice, on the other hand, are generally unwary and easily caught. Remember to check traps daily to remove dead rodents.

Animals Down Your Chimney

Animals that may be found in chimneys include bats, raccoons, squirrels, and occasionally birds. If an animal is in your chimney, here are some ways to remove it:

Place a large bowl of white vinegar in the fireplace at the base of the chimney and set a barrier across the opening to contain odors and prevent the animal from exiting into your house.

If you don't have access to the base of the chimney because of a stove insert, lower a sock partially filled with mothballs or rags soaked in white vinegar from the top. Be sure to leave the top open so the animal can climb or fly out. Make sure that animals too young to climb or fly are not left in the chimney; given the opportunity, an adult mammal will move the young to a new den site.

To prevent entry, cap the chimney with a commercial chimney cap, heavy wire mesh, or other cover that an animal can't pry or chew off or gain access through.

Birds Nesting or Roosting on or in Your House

The birds most likely to nest on or in your house include house sparrows, starlings, pigeons, and swallows, although woodpeckers occasionally will attempt to nest in the walls of a house. All but the swallows and woodpeckers were introduced to the Pacific Northwest from Europe and have adapted to finding shelter sites on and in buildings.

Allowing pigeons, starlings, and other birds to nest in your house can result in an accumulation of droppings and other material, which is messy and creates an environment for the growth of molds, which are potential disease risks.

Take precautions when working around areas where birds are roosting. Anyone working directly with large amounts of bird droppings and other waste material should wear a respirator approved for protection from microbiological agents.

The following sections detail ways to deal with specific types of birds:

House Sparrows and Starlings

On existing buildings, seal all existing entry holes, and replace loose shingles and siding. If the birds have already nested and are caring for babies, wait a couple of weeks until the young fly from the nest. Then remove all nesting materials and close the opening.

In new construction, avoid creating small cavities or spaces with access from the exterior. (For more information on managing these species see Chapters 12 and 14.)

Pigeons (Rock Doves)

Pigeons prefer to nest and roost on flat surfaces. To make it difficult for the birds to land, install sheet metal or wood at a 45-degree or greater angle on all ledges and flat areas. You can also create an unstable surface by placing a rolled-up piece of plastic mesh netting or chicken wire over these areas. Tightly strung wire placed 2 inches above a roosting area will also discourage birds from landing. Some hardware stores and feed stores sell "porcupine spine" strips with sharp wire points on them to prevent birds from landing in the area.

If you can't reach a ledge from inside a building, attach netting or wire mesh to the roof and lower it to drape across the front of the building, then tightly secure it at the base. Attach netting under beams, supports, or girders to create a "false wall," blocking birds' access to roosting areas. (Many netting and wire-mesh products are almost invisible and don't detract from a building's appearance.) Always use care when

working high above the ground and ensure that pigeon barriers can't fall and injure a passerby.

Routine removal of all nests and newly laid eggs will discourage pigeons from returning. Pigeons will give up after several unsuccessful attempts to build nests. Nests with babies should not be touched until the young have flown away. This is typically three weeks after the eggs have hatched.

Swallows

Barn and cliff swallows build nests from mud in protected places, such as walls, where they're safe from predators. If swallows start to build a mud nest in a spot that will create a problem for you or for them, such as near a busy entry, knock the nest down with a pole or a strong hose spray. Do this at the first sign of nest building but never when a nest is in use.

To prevent birds from rebuilding or starting a new nest where you don't want it, staple a piece of heavy plastic over the nest-building area. Mud doesn't stick to plastic. Even newly painted areas are sometimes unsuccessful sites for swallow nests because mud won't adhere to new paint, particularly glossy kinds.

If you decide to let swallows nest on your building, you can deal with the droppings (a good fertilizer) by placing newspaper or a similar material under the nest. This material can be carefully added to a compost pile or dug into the ground.

Woodpeckers

When you hear or see a woodpecker drumming or excavating a hole on your house, it could be doing one of three things. You need to know which before you can take appropriate action:

Flocks of Crows and Starlings in a Tree

Where crows and starlings are roosting in trees in large numbers, loud noises have been found to be temporarily effective. Propane guns, which create loud booms repeatedly, can be used in locations and at times when people will not be affected (let your neighbors know well ahead of time what you're up to). Recorded cries of frightened starlings are also reported to work. Tapes and streamers of mylar or other bright reflecting material are effective.

The trouble with these techniques is that the effectiveness wears off after the birds get used to the sounds or movements. If the perching area is still available, the birds usually come back. One suggestion is to modify the perching area to make it less inviting. In the case of a tree, open up the structure of the tree by removing up to one-third of the total canopy.

1. The woodpecker could be drumming to proclaim its territory and attract a mate. This typically happens during the breeding season (February to May). To discourage drumming, try one or more of the following:
   - Be patient and realize that the behavior won't last forever.
   - Cover or wrap the gutter, downspout, section of siding, or other drumming site with fabric, a tarp, plastic netting, or some other material.
   - Cover any ledges or cracks the bird uses as a foothold while drumming.
   - Persistently shout at, bang pans at, chase away, or squirt the bird with a garden hose as soon as it starts drumming.
   - Hang strips of foil, lightweight fabric, or commercially available bird-scaring tape (generally mylar tape) from the eaves.
2. If woodpecker activity is not restricted to one location on your house, and if it occurs throughout the year, the bird is probably finding insects to eat. You'll need to control the insects if you have an infestation, and then make any necessary repairs.
3. If you find a round opening in the siding or other boards, it means the woodpecker is probably drilling a cavity for nesting, roosting, or storing food. In the spring or early summer, assume there is an active nest with eggs or chicks inside. Wait until you're sure the young birds can fly and have left the nest on their own, and then immediately repair the opening.

A Bat or Bird Inside a House

A bat or bird may accidentally enter a house or other building through an open window or door. If the intruder is a bat, close off all doorways to the room and open a window. The bat will usually depart on its own. If the bat does not exit, it may be diseased, possibly with rabies, and you should contact your local doctor, health department, or fish and wildlife agency. (See Chapter 5 for information on rabies.)

If a bird enters your house, darken all windows with curtains or sheets and then leave one door or window open. The bird will likely move toward the light and leave. You can also keep the bird flying until it tires and has to land. Then, gently toss a towel over it to catch it. Place the bird on a ledge where it will be safe from predators, such as a house cats, and observe it until it takes off. If after 30 minutes the bird still does not fly, call a wildlife rehabilitator for information on what to do next. Your local Department of Fish and Wildlife office will have a list of rehabilitators in your area. (Note: Don't toss a bird into the air to release it; it may be hurt or disoriented.)

Animals in Your Garbage, Compost, or Pet Food

Animals that may be attracted to garbage, compost, and pet food include raccoons, opossums, coyotes, skunks, foxes, rats, and crows. The simplest way to keep animals from eating these things is to eliminate their access to them. To do this:

Use garbage cans with tight-fitting lids. To keep a lid on tight, secure it with rope, chain, bungee cords, or weights, or buy cans with clamps or other mechanisms.

To prevent tipping, secure the side handles to metal or wooden stakes driven into the ground, or keep your cans in tight-fitting bins, a shed, or a garage.

Don't put food in open compost piles where rats and other wildlife are a problem. Instead, bury food waste underground in an underground composter, or use a lidded compost or worm box.

If you feed a pet outside, pick up food and water bowls, as well as leftover and spilled food, as soon as the pet has finished eating. Don't leave bowls or food scraps outside at night.

Wildlife Problems in Your Pond

Animals that may eat the fish or disturb the plants in your pond include raccoons, great blue herons, river otters, opossums, dogs, and house cats (see Chapter 22).

Having the opportunity to watch wildlife search for food may be worth losing some fish from your pond. However, if you want to protect the area:

- Place a 2-foot-wide strip of chicken wire horizontally around the inside of the pond edge and just under the water. Lightly secure it with rocks. Animals won't be able to reach over the wire, and tend not to stand on it because it is unstable.

- Install an electric fence about 8 inches above the ground around the perimeter of the pond. The fence can be hooked up to a switch for discretionary use; when you want to work near the fence, turn the system off. Be sure to monitor the plant growth under the fence to prevent the fence from shorting out when touched by vegetation.
- Construct your pond with steep sides to prevent great blue herons and other animals from wading in. They will, however, hunt from the shore. (Note: Steep sides present safety problems for small children and the elderly.)
- Install a thick planting around the sides of the pond where animals gain access; you can keep the portion designed with steep sides open.
- Use a heron decoy on small ponds—herons generally prefer to hunt alone.
- Install a motion-detector light to help reduce night activity around the pond. A decoy or motion detector will probably help only if it is installed before animals discover your pond.
- To discourage animals from accessing the pond from the air, run some brightly colored weed-eater line or mylar tape above your pond.

For additional information on protecting fish, see Chapter 8.

Wildlife Eating Your Plants

Some of the wildlife species that may eat landscape plants include deer, raccoons, opossums, rabbits, mountain beavers, squirrels, geese, and ducks. The following are ways to work with specific wildlife:

Deer

Many people are willing to trade the opportunity to view deer for moderate consumption of plants around their homes. However, in areas where hunting isn't permitted and natural predators no longer keep populations in balance, deer may heavily browse plants they used to leave alone. You can tell whether a plant has been browsed by deer by looking for a characteristically rough, jagged cut. Rodents and rabbits leave a clean cut when they browse. Also, deer browse-marks will often extend higher up on plants than marks left by smaller animals, such as rabbits.

Ways to prevent deer from eating plants include fencing large areas, protecting individual plants with barriers and repellents, and using various home remedies. There are also plants that deer favor more than others. (See "Short Plant Lists" in Appendix B for a list of landscape plants that ordinarily won't be ruined by browsing deer.)

Fencing deer out of a large area usually means fencing them onto someone else's property. This only concentrates deer and deer problems. Fencing is recommended only when replanting can't be successful without it, or when the plants are of high value, such as in a nursery, vegetable garden, or orchard.

Deer-proof fences are generally 6–8 feet high and made of woven wire. Mule deer are more likely to jump a 6-foot fence than are the smaller white-tail deer. A 6-foot-high wire fence can be used for mule deer if it is topped with two strands of single wire spaced approximately a foot apart. A solid board fence need be only 5-feet

Deer Repellents and Home Remedies

There are several types of commercial deer repellents that work either through taste or smell aversion. Like home remedies, all work on some deer but none seem to work on all. Most repellents must be constantly replenished, especially after rain or heavy watering. Always use commercial repellents according to the directions provided by the manufacturer.

Home remedies that may keep deer away include:

- The family dog humanely tethered in or near the planted area.
- Human hair, deodorant soap, or blood meal placed in old stockings or cheesecloth bags and hung along plant rows or in the foliage of plants.
- Five old eggs blended in a gallon of water and placed in small, vented containers hung from limbs.
- A portable radio turned on at night. You can connect the radio to a motion sensor, so it will come on only when something moves into the area.
- Low-wattage light bulbs hung in vines and plugged into a Christmas tree blinker so that when they blink, they cast lines of shadows.
- A motion detector that triggers a sprinkler to come on and spray the area.

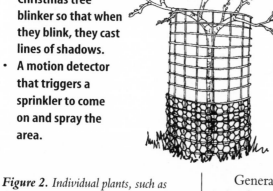

Figure 2. Individual plants, such as fruit trees and conifers, can be protected with a mini-deer fence. Additional wire mesh can be added to the bottom to keep rabbits and mountain beaver out.

high, as deer will not jump over objects when they can't see what's on the other side. If a fence is located on sloping ground, it may be necessary to build it taller to guard against deer jumping from above.

Several types of electric fences are also available and can be used successfully. Consult your local Extension Agent, Department of Fish and Wildlife officer, or farm supply center for current design information.

Deer are persistent in trying to gain access to preferred areas. Thus, fences need to be carefully designed, constructed, and maintained to keep deer out. For example, a deer will often try to go under a fence before attempting to jump over it and can work its way under a loose fence. Large stones or logs placed at the bottom of the fence, or stakes pounded in the ground, can be used to hold the wire down in areas where the fence is loose. Also, when building a fence, keep in mind that you must enclose the entire area. Deer will simply walk around one-, two-, or three-sided fences.

You can protect individual trees and shrubs with barriers and eliminate having to fence an entire area. The type of barrier depends on the size of the plant you want to protect. To protect large trees, you can use mini-deer fences made from any of a variety of materials. Generally, 4–6-foot woven-wire fence material is cut and attached to form a circular fence (see Figure 2). A fence like this can be installed around each tree

or shrub, allowing plenty of space between the plant and fence to prevent deer from nipping growing tips. Mulch the plants in these hoops to decrease the need to remove them for weeding.

Commercially available barrier tubes can provide individual protection for seedling trees and shrubs. Tubes made from a variety of materials, including rigid polypropylene mesh, are available in a variety of height and diameter combinations.

Sink the tube a couple of inches in the hole around the tree when it is planted. Attach the tube to a support stake to hold it upright. Tubes should fit loosely to allow for growth and should extend at least 2 feet above the normal snow line.

When you design a new landscape in an area where deer live, walk around your neighborhood and see what plants appear to be doing well without protection. Most native plants are equipped to survive moderate or even occasional heavy browsing. Also, plant any new plants away from known deer trails.

Rabbits

Rabbit browse can be identified by a clean, angled cut on the end of the branch or twig. Also, browsing and debarking by rabbits usually does not extend more than 2 feet above ground.

A wire mesh fence two feet high will keep rabbits out of a planted area. The wire mesh should be 1 inch or smaller to exclude baby rabbits. Make sure the bottom is secure or buried a few inches underground to prevent animals from scooting under. Existing fences can be rabbit-proofed by adding wire mesh to the bottom and making sure that

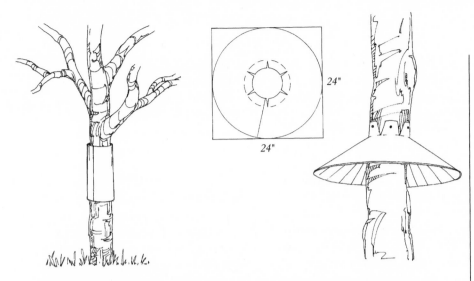

Figure 3. Stove pipe or sheet metal can be placed around a tree or post to keep mammals from gaining access to a feeder, nest box, or your house. A baffle can also be cut out of a piece of sheet metal.

the fence remains secure at the bottom. You can protect individual small plants and seedling trees with the commercially available plastic tubing described in the deer section.

Mountain Beavers

If a mountain beaver is eating your landscape plants, you may want to increase the quantity of plants you install. In addition, several methods may be used to discourage them:

- Use individual plant tubes on seedlings or mini-deer fences on larger plants as described for deer.
- Locate new plants away from known dens.
- If mountain beaver are climbing trees, install a predator guard (Figure 3).

Beavers

Beavers eat twigs, branches, and bark of young deciduous trees and shrubs. Large trees and shrubs can be protected with the mini-deer fence described in the deer section, or by wrapping the trunks with chicken wire, hardware cloth, or metal flashing (see Chapter 4). Trunks should be wrapped at least

3 feet above ground, or at least 2 feet above the high-water mark in areas that are occasionally flooded. Leave room for tree growth, and plan on adjusting the wrapping every couple of years.

It's pointless to destroy a beaver dam because beavers will rebuild immediately. However, it's possible to modify a dam with a trickle tube. The trickle tube masks water flow and the sound of flowing water—the beaver's two most noted stimuli. When these two things are undetectable, beavers are less compelled to build dams. Contact your local Department of Fish and Wildlife office for more information on beavers and trickle tubes.

Moles and Gophers

You'll rarely see a mole or gopher. The most reliable form of identification is their mound. A gopher mound is generally made up of fine soil and is fan-shaped, with an obvious plug in one end; a molehill is made up of clumpy soil and is circular with the plug at its center. Much of the burrowing by moles is done close to the surface and it often raises a visible ridge.

Predator Guards

A predator guard can be constructed to keep animals from climbing a tree and gaining access to your attic, bird feeder, nest box, or fruit trees. An easy predator guard to install uses two sections of 24-inch aluminum or galvanized vent-pipe, which is available at most hardware stores. These sections can be interlocked around any tree or post that can fit inside them (see Figure 3). Tack the top and bottom to the tree or post and always leave enough room (2–3-inches) for tree growth and air circulation.

If the tree is too large, you can buy sheet metal in a roll, 24 inches wide. Measure two pieces long enough to go around the tree, leaving a few inches between the guard and the trunk and an inch overlap for the seam. Cut with tin snips. Nail the two lengths to the tree along the seams, overlapping the top piece over the bottom. Jiggle or pry the metal guard away from the trunk so that air and water can move freely around the trunk. You can also use a 24 inch by 24 inch piece of sheet metal to make a baffle to deter climbing wildlife.

Some athletic house cats can jump 6 feet, so it's best that these guards be placed at least that high.

Although mole and gopher activity in the garden or lawn may be unsightly, these animals are among nature's rototillers; their tunnels allow air and water to get deeper into the ground. Moles and gophers are here to stay. Extermination is impracticable.

Moles are territorial and don't like having other moles around except during breeding. Having one mole around may keep others away. Moles rarely eat plants, and their diet is made up of insects, including grubs, found in your lawn.

Most plant damage blamed on moles is actually caused by meadow voles, which often use the mole's tunnel systems. The only real damage caused by moles is when roots dry out as a result of their shallow tunnels lifting up the turf or garden soil. To prevent this, just press the soil down with your foot and water the area if it's dry.

If you feel it's necessary to keep moles out of an area, you can install a barrier. Bury ½" hardware cloth 18 inches below the soil and leave a couple of inches above ground. All pieces should overlap and be tightly secured.

As with moles, the most effective way to manage gophers is to "design the animal out." Sometimes gophers can be excluded from a garden by burying hardware cloth as described for moles. Bulb beds or individual shrubs or small trees can be protected by half-inch mesh wire if it's laid on the bottom and sides of the planting hole. Be sure to place the wire deep enough so that it doesn't restrict root growth.

Because gophers occasionally feed on the bark of certain trees, particularly fruit trees, it's wise to protect the trunks of these trees during planting with cylinders of ½" galvanized hardware cloth sunk 12 inches underground and rising 6 inches above the surface.

Canada Geese and Ducks

If geese and ducks are eating your plants, leaving messes, or polluting your pond, the first thing to do is to stop feeding them. Not only does feeding attract them to an area, but it can also cause problems for other wildlife. Nutritionally deficient food such as bread can weaken the animals, making them more susceptible to diseases, which are then spread among visiting wild populations.

To keep geese and ducks out of an area:

- Spray grazing areas with a commercially available nontoxic chemical that repels geese. Contact your local Department of Fish and Wildlife office or Animal Damage Control office for current recommendations.
- Plant trees and tall shrubs in the line of flight between the water and the grazing area. Geese want to see an escape route into water and will be less likely to use a grazing area that is not within visual range of water.
- Keep a balloon or mylar streamers flying above geese and ducks. Mylar streamers can be hung overhead from a wire. This will keep them nervous, and they will be less likely to use the area.

To keep geese and ducks from rooting out plants while these are getting their roots established:

- Cut newly planted waterside plants such as rushes and sedges almost to the ground, and place large stones next to them.
- Insert a large metal staple (used to hold down jute netting) over each new plant.

- Place wood lath around three sides of the plant and secure it with metal staples or rocks.
- Plant new plants around large shrubs where the geese and ducks will feel insecure grazing.

Squirrels

Squirrels will sometimes be a problem in newly planted flower beds. Repellents such as diluted Tabasco sauce or cayenne pepper sprinkled on soil will help keep them out. To protect bulbs, place chicken wire flat across the top of the flower beds before the plants break the surface. Make sure that the chicken wire is attached to something or held down with rocks so that squirrels cannot remove it.

Wildlife Eating Garden and Orchard Fruits

Wildlife that are often found eating garden or orchard fruit include birds, raccoons, and opossums.

The best way to protect your garden and orchard fruit from birds is to create a barrier between them and the fruits you wish to protect. Flexible mesh netting can be used to cover vines, berry bushes, and small fruit trees (semi-dwarf or dwarf varieties). Netting can be suspended from a makeshift trellis to avoid interfering with plant growth, or it can be placed over the plant or tree just before fruits ripen. Be sure to secure netting at the base so birds can't get underneath. Also, pull netting tight to avoid entangling and injuring birds.

Loud noises, recorded alarms of birds, and high-pitched sounds may temporarily frighten birds away. Visual repellents such as scarecrows and reflecting surfaces may be effective at first, but their impact decreases as the birds become accustomed to them.

In agricultural areas, crops left in fields to lure birds from other areas and tactics designed to scare birds away are also only temporarily effective. For local solutions, contact agencies that work with farms and ranches in your area.

Raccoons and opossums are good climbers who relish fruit, and they may break branches. To prevent them from getting started raiding your fruit tree, keep fallen fruit picked up or install a predator guard (see Figure 3) to prevent access up into the tree.

An area can also be protected by erecting a two-strand electric fence with the first wire 6 inches from the ground and the second 6 inches higher.

Animals Eating Your Poultry or Pets

Coyotes, foxes, mountain lions, bobcats, skunks, opossums, hawks, and owls are the wildlife most likely to eat farm animals or pets. The best defense is to fence in your animals and carefully manage your pets.

Ducks and chickens need to be completely enclosed, especially at night. Make certain that the coop or pen is secure, with well-fitting doors and floor. Use heavy-gauge mesh wire or wood to cover up any holes in the structures. The outdoor portion of your coop will need to be completely covered with wire mesh, such as chicken wire. Extend the mesh at least 12 inches underground to prevent skunks from digging under.

Although a hungry coyote, mountain lion, or bobcat may prey on a cat or dog at any time of year, a particularly dangerous time is between June and August.

This is when these animals may be trying to feed their growing young. To prevent potential conflicts, keep companion animals indoors, especially from dusk to dawn. Dogs are less apt to fall prey due to their larger size and because they are often supervised by owners, at least in the city. However, even a large dog can fall prey to several coyotes working together.

Live-Trapping Troublesome Animals

If the techniques described here have not completely stopped troublesome animal visits to your home or property, live-trapping is another, albeit short-term, solution. Live-trapping isn't usually recommended, for several reasons.

- The released animal that has been trapped and released is unlikely to survive in the new location.
- Relocated animals don't know where to find the safest shelter, or where the best places are to forage for food and water. They must compete with resident animals of the same species who know the territory well.
- Both the relocated animal and animals resident in the area to which it is relocated are likely to be stressed by the need to adjust social structures around the new animal.
- Live-trapping and relocation can spread diseases.
- Live-trapping may not stop the problem, because other animals may move into the vacancy created when the previous animal was removed.

If you think you need to live trap an animal:

- Consult your local Fish and Wildlife office to determine

whether trapping is allowed under law in your area.
- Be aware that animals can't be trapped while on another person's property without permission from the landowner.
- Don't trap between October and March, because relocated animals may be unable to find suitable dens and will die from exposure or lack of food.
- Take care not to relocate a female raising young, because the young are highly unlikely to survive the loss of their mother.
- Check the traps morning and evening because wildlife are susceptible to heat and hypothermia and can die in the trap. To reduce chances of this you can cover the trap with a piece of plastic and a board.

If you decide to enlist the services of a professional wildlife-removal company, make sure you have a firm agreement with the company about their responsibility to monitor traps frequently and remove the animals in a suitable way. Give clear instructions about whether the animals will be relocated or euthanized. Relocation is not usually recommended unless the animals are to be released near where they were trapped.

The best bait to use depends upon the animal. Peanut butter on a cracker works well for attracting raccoons, opossums, rats, and squirrels. Apple or peach slices, a raw unbroken egg, or marshmallows will also attract raccoons and opossums. Squirrels prefer birdseed or sunflower seeds. Mountain beaver are attracted by fennel, which smells like licorice.

Live traps may be available from your local Fish and Wildlife office, rented from rental outfits, or purchased from a hardware store.

CHAPTER 22

THE IMPACT OF DOMESTIC CATS AND DOGS ON WILDLIFE

Often unbeknownst to their loving owners, pet dogs and cats that are allowed to roam free represent a serious threat to wild animals. Free-ranging domestic cats represent a major threat to wildlife in urban, suburban, and especially in rural areas. Domestic dogs also harass wildlife.

With forethought and preventive measures, however, people and their pets can coexist with wildlife.

The following questions and answers address some misunderstandings about free-ranging domestic cats and dogs. They are intended to provide helpful information to people who also care about wildlife. If you are not a pet owner, keep in mind the emotional attachment your pet-owning neighbors have for their animals when sharing this information with them.

Domestic Cats

Do cats really injure and kill a significant number of wild animals?

Yes. Wildlife rehabilitators in Washington State have reported that, of all the animals treated in wildlife rehabilitation centers, 17 percent were injured by cats. Cat attacks are one of the most common causes of injury and death to wild animals brought to HOWL (Help Our Wildlife), the largest of Washington's rehabilitation centers. Studies in other states have shown that cats are a major threat to bird populations, especially ground-nesting species; a Wisconsin study tallied 19 million songbirds and 140,000 game birds killed by cats each year in that state.

Can I teach my cats not to hunt wildlife?

No. Hunting is an instinctive behavior that cannot be modified by training. Even cats that have been declawed are still able to bat down and kill wildlife. You can prevent attacks by keeping your cats indoors or safely confined in an outdoor enclosure.

Predatory animals play an important role in keeping wildlife populations balanced and healthy. Since cats are predators, don't they help preserve the balance of nature?

No. Domestic cats are not a part of the natural ecosystem; they are predators introduced and maintained by humans in high numbers. (Thirty-five thousand domestic cats are born every day in the United States—more than three times the human birthrate.) Populations of other predators such as owls, foxes, and snakes are kept in balance by factors such as disease, human interaction, and competition with other species. Meanwhile, the number of cats continues to grow. A study by the Department of Fish and Wildlife found that there are at least 30 cats per block in some Seattle neighborhoods. This represents greatly disproportionate predatory pressure on wildlife if cats roam freely.

While the "kept" cat is a problem if it is allowed outside, the animals that are allowed to reproduce over and over are an even bigger problem. This is one of the reasons it is important to spay or neuter your cat. Many local shelters now offer low-cost spaying and neutering for dogs and cats.

Since cats are hunters, don't they fill a niche in the food chain?

No. The cats that do the most damage to wildlife are usually well-fed pets that do not need to hunt to survive. The urge to hunt is independent of the urge to eat. Cats that are well-fed, healthy, and strong will focus hunting activities on a broader array of wild animals.

Cats are nonselective predators and capture healthy as well as sick and injured wildlife. They capture not only problem species, such as non-native mice, but also many species of native wildlife including snakes, lizards, meadow voles, flying squirrels, rabbits, chipmunks, and butterflies. Only about half of the animals killed by pet cats are brought home.

Can cats live happily indoors?

Yes. Indoor cats lead long and healthy lives. The Humane Society of the United States reports that the expected life span of an indoor cat is at least triple that of cats that spend their lives outside. Indoor cats avoid the dangers of traffic, diseases, parasites, poisoning, and other animals. Pet owners benefit from lower health-care expenses and the knowledge that their cats are safe.

Even outdoor cats can make the change to indoor living. It is easier to start kittens on a completely indoor living regime, but you can change an outdoor cat into an indoor one. Try to make the indoor environment as inviting as possible.

Also, have your cat neutered or spayed. This reduces aggression, wandering, and territorial behavior (including the urge to spray) that can result in injuries from fighting with other cats, dogs, and animals.

Place identification and license tags on your cat's collar. If it accidentally gets outdoors, this may enable your cat to be returned home safely.

Feel good about your indoor arrangement. You're doing both your cat—and wildlife—a favor by giving them the protection they deserve!

If you cannot or prefer not to offer your cat a run or enclosure, consider leash-training the cat so you can supervise her time outside. Attach the leash to a harness. Your cat may resist leash-training at first, but she will eventually accept the leash.

Some cats may develop behavioral problems when they are no longer allowed outside. Most of these problems can be attributed to a change in routine that is too abrupt or to lack of attention and stimulation inside. If your cat becomes destructive or un-house-trained, consult a veterinarian or animal behaviorist to find ways to solve the problem. Remember that these symptoms can also be attributed to boredom and loneliness.

Does putting a bell or bells on a cat really do any good?

No. This is a myth. There is no evidence to suggest that wild animals equate the sound of a bell with a predator. Ground-nesting baby birds, baby mammals, or slow-moving species such as amphibians have little chance of escape anyway.

I keep my cats indoors but my neighbors let theirs roam. Can I do anything to keep their cats from killing birds in my yard?

Yes. First, try talking with your neighbors, sharing your concerns, and giving them a copy of this chapter. Many people are unaware of the pressures their pets put on wildlife populations or of the benefits of indoor life for cats.

It is usually the responsibility of the owner to control the cat's movements. In most areas, cats can be live-trapped and either returned to the owner or turned over to authorities if they wander onto other peoples' property. Many municipalities have leash laws and require vaccination and neutering of pet cats. Because laws vary, one should check local ordinances for the appropriate way to deal with stray cats.

You might want to take measures on your own to discourage offending cats. Scent and sound repellents that are available at garden and pet supply stores may help keep animals out of marked

How to Have a Happy Indoor Cat

Although it takes patience, an outdoor cat can be turned into a perfectly content indoor pet. The key is to make the conversion gradually and to provide lots of attention and stimulation while the cat is indoors.

Cats are creatures of habit, so you must be careful to slowly replace your cat's old routine of going outside with the new exciting routine of staying in. If your cat is outdoors most of the time, bring your cat inside for increasingly longer stays. Gradually shorten the length of time the cat is outside until you no longer let him or her out at all.

Cats need human companionship to be happy, and when they spend all their time out of doors, they get very little attention. An outdoor cat may welcome the indoors if he or she gets more love, attention, and play.

To make life more enjoyable indoors, try the following:

- Build or buy resting shelves or high perches so your cat can watch the outside from a window.
- Provide a scratching post (one with different levels is good), corrugated cardboard, or sisal rope for your cat to scratch, and praise your cat for using it. Trim your cat's claws regularly.
- Place a bird feeder where it can be seen from the window inside. (There are also videos of birds for cats to keep them occupied inside.)
- Give your cat its own protected, screened outside enclosure that it can access through a window or

continued on next page

pet door. Make the enclosure at least 5 feet wide and 12 feet long.

- Bring in a pot of fresh green grass or catnip for the cats to chew.
- Clean your cat's litter box daily and have fresh drinking water available at all times.
- Amuse your cat with safe toys such as boxes, paper bags, soft rope, or a ping-pong ball in a tub of water. Some cats enjoy searching for toys. If your cat likes to explore the house looking for "prey," hide his toys at various places so he can find them throughout the day.
- Spend quality time loving your cat: it's dependent on you for care and companionship.
- Consider adopting a playmate for your cat. Two cats are just as easy to care for as one and they keep each other company.

garden areas. You may have limited success at discouraging cats from entering your yard by spraying them with water from a hose or squirt gun. A gentle blast of water will deter, but not injure, cats. However, you will have to be persistent.

Place your bird feeder at the top of a tall pole—the taller the better. This will be especially effective if the pole is equipped with a squirrel baffle, which will prevent not only squirrels but also cats from climbing it. You can also scatter something difficult for a cat to walk on—for example, plastic forks with the tines up—on the ground around the feeding area. This is not to hurt the cat but to surprise and confuse it.

A 2-foot-high wire fence can be placed around a feeder, a birdbath, or other areas where there is a concentration of birds at ground level. A low fence won't keep cats out, but will give birds enough time to escape. A higher fence will eliminate the problem. (See Chapter 11 for an example of a fence around a birdbath.)

You can also install an inexpensive electric fence around your feeding or nesting area, or run a single line on top of your existing fence. Large hardware stores or farm supply centers have systems that are easy to install. Keeping the area around bird feeders and birdbaths free of shrubbery prevents cats from ambushing birds at these locations.

If all else fails, can I get help from my local animal control agency in controlling cats in my neighborhood? Are there any laws regulating cats?

The answer depends on where you live. Legal restrictions on cats vary by jurisdiction. As human and cat populations expand, legislators are writing new laws and police are enforcing old ones that govern cats.

If you have talked with your neighbors about their cats, and if the aversion tactics outlined above have failed, ask your local animal control agency for help. Authorities may suggest using a humane trap to catch free-roaming cats on your property. If the cats have proper identification on their collars, they can be returned unharmed to their owners by you, an animal control agent, or the local animal shelter.

What can I do if I find wildlife that has been injured by a cat?

Any wild animal that has been caught by a cat probably needs medical care to survive. Because birds have high metabolisms, they are particularly susceptible to bacteria from cat bites and scratches, and infections set in quickly. Never attempt to care for injured animals yourself. Local Department of Fish and Wildlife offices can refer you to the closest wildlife rehabilitation center. (For more information see Chapter 23.)

Finally, don't dispose of unwanted cats by releasing them in rural areas. This practice enlarges rural cat populations and is an inhumane way of dealing with unwanted cats. Cats suffer in an unfamiliar setting, even if they are good predators. Contact your local animal welfare organization for help.

Don't feed stray cats. Feeding strays maintains high densities of cats that kill and compete with native wildlife populations.

Domestic Dogs

Do dogs really injure and kill a significant number of wild animals?

Yes. Dogs that are allowed to roam free will often harass, injure, or kill deer, raccoons, skunks, squirrels, chipmunks, rabbits, grouse, quail, and other wildlife. Even the most obedient dog may harass deer or other wildlife if given the opportunity. Two or more dogs working together can relentlessly pursue an animal, causing a loss of valuable energy that is especially detrimental in cold weather. In deep snow crusted over with ice, deer and elk punch through with their small, sharp hooves, while dogs race along the top on their big paws, snowshoe-style. A dog-chased deer or elk can be rapidly stressed in these conditions, allowing the dog to catch it and slowly tear it apart. Washington Department of Fish and Wildlife offices, particularly on the east side of the state, receive hundreds of complaints about such situations every severe winter.

What can happen to my dog if it is allowed to run free?

Dogs running at large are in danger of being hit by a car or of being bitten or attacked by another dog or by wild animals, including raccoons, foxes, and coyotes. Although rabies is rare in the Pacific Northwest, dogs can contract rabies from an infected animal. They can also develop bacterial infections from bite or scratch wounds.

Many local jurisdictions consider free-ranging dogs to be feral and subject to being shot, especially if they are harassing livestock or otherwise being a nuisance or threat. Dogs chasing deer or elk in Washington are legally considered a nuisance under wildlife law and owners can be cited with a misdemeanor, subject to minimum fines of about $150. In severe winter conditions, Washington Department of Fish and Wildlife law enforcement officers are often authorized to remove or even destroy offending dogs.

What can deter dog/wildlife conflicts?

Keep dogs under restraint, either kenneled or leashed, whether you live in town or in the country. Suburban or rural living does not entitle your dog to roam free; in fact, the greater your proximity to wildlife, the greater your responsibility to keep dogs confined.

Neuter or spay your dog to help the animal's overall health and decrease the desire to roam. Neutering also reduces aggression and territorial behavior that results in injuries from fighting with other dogs and other animals.

Keeping vaccinations current will protect both wildlife and your pet by preventing the spread of viruses and infectious disease. Both pet cats and dogs should be closely monitored during the nesting seasons of wildlife, generally in the spring and summer months.

CHAPTER 23

WHAT TO DO WITH ORPHANED, SICK, OR INJURED WILDLIFE

In areas where human and wildlife activities overlap, it is not unusual for people to find wild animals that are orphaned, sick, or injured. These animals are often in need of care by licensed wildlife rehabilitators. Phone numbers for wildlife rehabilitation centers in your area are available at Fish and Wildlife offices. If a rehabilitator isn't available, it's best to leave the animal in the wild and let nature take over. Regardless of your best intentions, if you offer first aid without the proper training, you may do more harm than good.

Before you decide to remove an animal from the wild, remember that most wild populations have both high birth and death rates. Many animals die, in a variety of ways, to make room for many more to be born. It's nature's way of keeping things in balance.

Orphaned Wildlife

When you find a young wild animal that you think is orphaned, reconsider.

Many infant animals received at wildlife rehabilitation centers are not actually orphaned. Wild animals, including deer, often leave their young alone for long periods of time as a protection measure. Young animals often do not have much body scent, so parent animals stay away from them between feeding times to avoid drawing predators. Do not automatically assume young animals found alone need help.

If a dead parent or sibling is found nearby, or the animal is listless, cold, or thin, it may be an orphan. Unfortunately there isn't much you can do but let nature take its course. Rehabilitators don't work with animals unless they are sick or injured.

Fledglings, or young birds just learning to fly, spend a lot of time on the ground. These birds are still protected and fed by their parents and should be left alone unless they are injured. Safeguard the area by keeping cats and dogs away until the bird is able to fly and escape danger.

If you find a baby bird that has fallen from the nest and is uninjured, you can gently return it to the nest. Because birds can't smell human scent, mother birds don't abandon their babies if people touch them. Wash your hands immediately after handling the bird.

If the nest is out of reach, put the bird in a small, open box and secure it on the highest branch possible. Look for a site protected by other branches so that the box will be sheltered from wind, rain, and sun, and will be less obvious to predators. If a nest has blown down, you can put it back in the tree.

Sick or Injured Wildlife

When you find a sick or injured animal, you need to know the following:

Wild animals are protected by state and federal laws. It is illegal to keep wildlife, even for a short period of time. Wild animals should be taken to a rehabilitation center as quickly as possible. If a licensed rehabilitator is not available or is unable to handle the animal, it should be left alone.

Even when sick or injured, wild animals can be dangerous to handle. Do not attempt to handle a sick or injured animal without first calling a rehabilitation center to find out how to proceed safely.

Time is critical. The faster an animal gets help at a rehabilitation center, the greater its chance of survival.

Wild animals are very sensitive to stress caused by human contact. Handle wildlife as little as possible to avoid stress as much as possible.

Small birds may be placed in a covered cardboard box or large paper bag to protect them from predators and to keep them from escaping. Maintaining the bird's body heat is essential because small animals can quickly go into shock when they are injured and their body temperature drops. Place the box in a warm, dark, or partially lit quiet place to minimize stress. A securely plugged hot-water bottle, wrapped in a towel, will keep the bird warm while it is being transported. Large birds, such as ducks, owls, hawks, crows, and gulls, can be placed in a box as well, but call your local wildlife rehabilitator for handling instructions. Always wash your hands after handling any wildlife.

Do not feed the animal or give it water. Inappropriate food

can further harm, or even kill, wildlife.

An animal that has been captured by a cat, even if it appears uninjured, needs immediate care at a wildlife rehabilitation center. Infection from cat bites and scratches spreads very quickly in small animals.

Birds and Windows: Tips to Ensure Safe Flight

Many birds are killed each year by flying into windows. This is generally because the windows are reflecting open sky or an object they regard as a landing area. Problems with window collisions may increase after a bird has indulged in its annual binge of fermented berries, or if there is a predator, such as a Cooper's hawk, chasing birds in the area. During breeding season, male birds (robins, woodpeckers, hummingbirds) may "fight" their own reflections in windows and car mirrors. This activity generally stops after the breeding season.

Catalogs and stores selling bird-feeding supplies often offer silhouettes of falcons or owls to be attached to windows, supposedly to frighten birds away from them. Placing one silhouette on a window is usually ineffective. If it does work it's because the image breaks up the reflection on the glass, not because of the image itself. It's better to use several silhouettes. Whatever you place on the window should be on the outside surface; anything on the inside of the glass will lose its effect because it won't cut the reflection.

Here are some other ways to deal with the problem of window collisions:

- Rub soap over the outside of the window to create a dull appearance.
- Leave windows dirty.
- Place a pattern of small stick-on decals, 8 inches apart, over the entire window surface.
- Install screens, gauze, clear plastic, twine, streamers, or closely spaced adhesive strips on the outside of the window.
- Install black plastic garden-protection netting mounted on frames or thumb-tacked from below the eaves to below the windows. This will act as a trampoline-like surface to soften the force of impact.
- Close the curtains on one side of a large corner window to prevent birds from seeing through a corner of the house and attempting to fly through. (Collisions may even happen when people have uncurtained windows directly opposite one another on different sides of the house.)
- If feeders are attracting birds to the area around your windows, either move the feeders farther from windows (20 feet or more) or place them next to windows (3 feet or closer). If feeders are close to windows, birds leaving feeders don't get up to flight speed before impact.

Important: A bird that hits a window and falls to the ground may simply be unconscious. You may be able to save it: pick it up right away, especially if cats are in your area. Place the bird upright in the palm of your hand, cup your other hand over the bird and hold it for about five minutes. When the bird starts moving around, lift your hand and release it. Wash your hands immediately.

If the bird is a large bird or doesn't revive within five minutes, place it in a brown paper bag or container with air holes and put it in a quiet place. Then, if you hear the bird moving, open the container outside and give it a chance to fly away.

If you find a dead bird with a band on its leg, you can provide a vital service by removing the band and returning it to the U.S. Fish and Wildlife Service. If the bird has colored bands, note the colors and their order, reading from the top to the bottom of the leg. Also note which leg the bands are on. (It is very important to record this information accurately.) Next, flatten the band tape and attach the flattened band to a letter that states the date of your find, location (nearest town), and the condition of the bird. Don't forget to include your name, address, and phone number. Send the information to the U.S. Fish and Wildlife Service, Bird Banding Lab, Route 197, Laurel MD 20811.

Finally, researchers or your local Audubon chapter may benefit from the specimen. To preserve the bird, wrap it in a plastic bag with your name, date, and the location of the find, and place it in the freezer. (Wear rubber gloves when handling dead birds.) Then contact the Audubon chapter, university, or state wildlife agency in your area to see who accepts bird specimens.

PART 5

APPENDICES

PACIFIC NORTHWEST HABITATS

WOODLANDS AND WOODLAND LANDSCAPES

Much of the Pacific Northwest is naturally covered with trees, and trees are often planted in areas where they might not ordinarily grow. All these areas can loosely be described as woodlands. Woodlands offer many benefits, including shade, beauty, and wind protection. Trees filter air, muffle noise, and protect and enrich the soil with their roots, leaves, needles, and twigs. In addition, a woodland around a home creates an ideal living area for wildlife, as well as opportunities for you to observe various creatures.

The ideal woodland wildlife landscape is made up of different vegetative layers: tall trees with shorter trees growing under them; tall and short shrubs below the trees; and ferns and flowers growing under the shrubs (see Figure 1). With these different layers, wildlife have many opportunities to find enough food, shelter, and safe places to breed and raise their young.

Creating Woodlands

Woodland habitat can be created almost anywhere. The size and type of woodland you create will depend on the conditions of the site, where it's located, and your intentions. Examples include a buffer next to a pond, wetland, or creek, and a naturalistic woodland landscape around a home or other building. Forestry projects also involve wildlife considerations, and these are described later in this Appendix.

A woodland landscape around your home will rarely satisfy all of the life requirements for feeding, nesting, and resting of wildlife species. However, it can serve as an oasis in the midst of a developed area, especially if it connects with another wooded area or a large undeveloped area, particularly one with year-round water.

The benefits of a new woodland can be increased by creating a simple landscape plan. Chapter 2 describes how to create such a plan; Chapter 3 explains how to choose trees, shrubs, and other woodland plants.

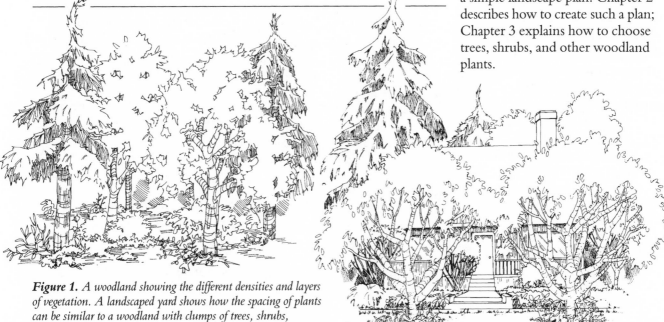

Figure 1. *A woodland showing the different densities and layers of vegetation. A landscaped yard shows how the spacing of plants can be similar to a woodland with clumps of trees, shrubs, and a sun-dappled ground layer.*

Restoring Native Woodland Habitat

Woodland restoration is the process of returning a disturbed woodland habitat to a close approximation of its pre-disturbed condition. Successful woodland restoration requires knowledge of the woodland conditions that existed prior to disturbance and a clear understanding of the site's existing conditions.

A woodland that has been severely degraded by logging, grazing, or invasion by non-native plant species is a good candidate for restoration. The goal for such a project may be to restore the structure, function, diversity, and dynamics of the woodland so that the woodland will eventually sustain itself and grow without continued human management.

Woodland restoration projects are generally large in scale and involve consultants, large budgets, and permits. Because of this they are beyond the scope of this book. However, you can use the vision of a restored woodland as a goal to move toward with any size woodland project. Large steps in the direction of woodland restoration can be made utilizing the suggestions described in this Appendix.

Naturalistic Woodland Landscapes

A naturalistic landscape is one that seeks to emulate—in a cultivated form—the appearance of a wild area, borrowing design ideas from natural landscapes. For instance, a naturalistic woodland landscape may utilize the most characteristic tree canopy, middle-story, and ground layer species of a woodland type, such as an oak woodland or Ponderosa pine savanna. In these cases it is possible to have the key plant species, spacing, and similar distribution patterns capture the essence of the woodland model.

To keep maintenance to a minimum, benefit local wildlife, and help create a naturalistic-looking landscape, plant native woodland plants that naturally grow together in your area. This is particularly important in Area 3 as described in Chapter 2, or if your aim is to partially restore the site to its pre-disturbed condition.

However, naturalistic woodland landscapes can include non-native plants which simulate the form and community function of natives. For instance, new plantings directly around your home and in other high-use areas (Areas 1 and 2 as described in Chapter 2) may include non-native species because they are known to attract certain wildlife, meet certain site requirements, and are personal favorites. In addition, native woodland trees (particularly conifers) often get too large for woodland landscapes around homes. In such case some non-native firs, pines, or spruces (see "Coniferous Trees" in Appendix C) may be more appropriate. A combination of native and non-native species can still provide the various layers used by woodland wildlife.

Whether they are all native plants or a mixture of natives and non-natives, grow your plants in groups of species that require similar site and soil conditions. Woodland plantings next to wild areas such as ravines, greenbelts, or wetlands should never include non-native plants that are known to be aggressive, such as English ivy. For more information on native and non-native plants, see Chapters 1 and 3.

You can learn a lot about what plants to use in a new woodland landscape by observing woodlands growing nearby. State, national, or provincial parks are good places to find the most natural woodlands. However, even these areas have probably been modified by human activity in some way: by management for timber, firewood, grazing animals, or the introduction of non-native plants such as Himalayan blackberry. Even so, look to see what the dominant trees and shrubs are, what their ultimate size is, and where these plants are located in the woodland.

To become familiar with native wetland plants that grow in your area and to help you select plants for your enhancement project, contact your local Native Plant Society, County Conservation District, or other private or public group in your area that works with wetlands. These groups often lead plant identification walks, host planting projects, or have staff and literature that can assist you in plant selection.

Appendix B contains a list of common woodland plants found in the Pacific Northwest; most of these are described in Appendix C. For information on obtaining plants, see Chapter 3.

Planting a Woodland Landscape

Here are some different approaches to planting new wooded habitat. Consider the advantages and disadvantages of each and choose one(s) that will work best for you and the site you're working on:

1. Plant trees and shrubs in the desired density and proportion you want them to be at landscape maturity. (Landscape maturity means 3–5 years for most shrubs and 10–15 years for most trees.) Mulch the spaces between the newly planted trees and shrubs or plant the areas with sun-loving ground-layer plants. Shade-tolerant plants can be added as the tree and shrub canopies develop.

   This is a common landscaping approach used when creating a new landscape with plants purchased in 1, 2, and 5 gallon containers. It is also a good method to use when replacing a lawn with a naturalistic woodland. This technique is most successful when aggressive weeds are eliminated at least one growing season prior to planting. For additional information, see "How to Transform a Lawn into a Wildlife-Friendly Landscape" in Chapter 3.

2. Plant trees and shrubs using the above approach, but mulch only around the newly planted trees and shrubs and let plants, including any weeds, grow in the ground layer. Keep the ground layer mowed with a weed-eater or other tool and eliminate any extremely aggressive species. After the tree and shrub canopy fills in you can begin to add permanent ground-layer plants and/or mulch.

   This approach is used in large and out-of-the-way areas. It works best if you can plant large trees and shrubs (a minimum of 24 inches tall) so they aren't quickly overcome by or lost in the weeds. While new plants fill in, the ground layer provides excellent wildlife habitat. Be prepared to carefully mulch, weed, and water the individual trees and shrubs for several years. You may also want to flag plants with bright nursery tape so they aren't lost or cut down.

3. Plant short-lived, fast-growing trees (alder) and tall shrubs (elderberry) as a temporary planting. As the canopy develops, under-plant with shade-tolerant trees (western hemlock, red-cedar) that later will become dominant. Thin the temporary planting species as necessary to reduce competition with the eventual dominants.

   This approach is used for reclaiming disturbed land, quickly screening an unattractive view, or working in poor or wet soil. This method is much more successful when large quantities of bare-root or potted plants are available and when aggressive weeds are eliminated the year prior to planting. Native white or red alder trees are fast growers that improve the soil for later successional plants such as conifers.

4. Do nothing and allow trees and shrubs to naturally invade the area. Selectively remove those which are not desired and encourage the preferred species. Treat the ground layer in the same way.

   Although this is an inexpensive approach, tree species may not get established if there is heavy competition from

Breaking New Ground in Wooded Areas

(Adapted with permission from Elliott Menashe, Greenbelt Consulting, PO Box 601, Clinton WA 98236)

Owners of rural woodland property who value and want to maintain the land in a natural condition are faced with a bewildering array of possibilities as they consider how best to blend comfort, safety, financial returns, and environmental concerns. Aside from the central requirements of access, construction, utilities, and cost, they should also consider the long-term effects of their actions.

To best protect your interests and the health of the land, carefully consider the results of what you do. The best way to protect and preserve your property's special qualities is through common sense and sound planning.

When clearing a wooded area remember:

- If possible, plan where your house and driveway will go *after* you've identified the trees and other vegetation and have selected the ones you most want to save.
- Leave undisturbed buffers of vegetation as large as possible for sight and sound barriers.
- Identify areas of wildlife habitat and avoid disturbing them.
- Comply with special ordinances to preserve streamside and other habitat.
- Don't plan to clear more than you are able or willing to maintain.
- Consider the effects of grading and utilities, such as septic tanks and underground telephone lines, upon existing and future vegetation.

ground-layer plants (especially grasses), or if mother seed-plants of the species you prefer don't occur in the area.

Salvaging Native Plants

(Adapted with permission from King County Department of Natural Resources, 1994)

Salvaging plants from an area that's slated for development is an excellent way to conserve a natural resource, obtain local native plants, and save money. To locate potential salvage sites, contact construction companies or contractors and talk to the surveyors of your local Planning Department. School sites and other public construction sites are good sources for salvage because these projects are often large in scale and in the planning stage for a long time. Contact your local Sheriffs Department to see if you'll need to get a free native plant transportation permit.

Be sure to salvage only after all permits have been received for the construction project and you have permission from the land-owner. Also, salvage only in the areas of the site that will actually be developed.

Salvaging is most successful on wet, cloudy days from November to March. However, with care at the time of transplanting and superb aftercare, you can salvage plants at any time, although probably with more mortality.

Here are some transplanting tools you'll need and steps to follow:

- A sharp, flat-bladed shovel (sharpen with a metal file before using in the field).
- Hand clippers (for pruning).
- Burlap bags.
- Two 2 x 4 inch boards that are 8 feet long.
- A wheelbarrow.

Step 1. Find the Right Plants

Select plants that can stand transplanting and will regrow quickly after being replanted (see list below). Good candidate plants—including Pacific rhododendron, vine maple, and sword fern—have shallow roots, and can adapt to disturbance. Try to find a small shrub or tree seedling (4 feet or less in height) growing by itself. Trees and shrubs growing in clumps connected by underground runners are not as likely to survive transplanting.

Step 2. Prepare the Plants

Check that the plant looks healthy. Clear the area around the base of the plant of leaves and twigs. Shrubs can be pruned back somewhat if they have a few long (more than 4 feet) branches. Sword fern fronds can be cut off, pruned to half their length or less, or tied with a string.

Step 3. Dig the Plants

Hold the shovel vertically and dig in a circle at least 8 inches from the stem, or 1 foot if it is a large shrub or seedling (3–4 feet high). Gently work the shovel under the plant's roots. If you encounter a root that the shovel will not slice through, cut it with clippers.

Lift the root ball out with the shovel blade using the strength in your legs, not with your back, and get someone to help if the plant is heavy. Put the root ball into your burlap sack to move it, or cover it with moist leaves or soil if it is to be stored for any period.

Native Plants That Transplant Successfully

Trees

White fir, *Abies concolor*
Vine maple, *Acer circinatum*
Douglas maple, *Acer glabrum* var. *douglasii*
Rocky Mountain juniper, *Juniperus scopulorum*
Douglas-fir, *Pseudotsuga menziesii*
Cascara, *Rhamnus purshiana*
Western red-cedar, *Thuja plicata*
Western hemlock, *Tsuga heterophylla*
Bay (Oregon-myrtle), *Umbellularia californica*

Shrubs, Ferns, and Ground Covers

Lady-fern, *Athyrium filix-femina*
Deer fern, *Blechnum spicant*
Twinberry, *Lonicera involucrata*
Osoberry (Indian plum), *Oemleria cerasiformis*
Oxalis (Wood sorrel), *Oxalis oregona*
Pacific ninebark, *Physocarpus capitatus*
Sword fern, *Polystichum munitum*
Pacific rhododendron, *Rhododendron macrophyllum*
Red-flowering currant, *Ribes sanguineum*
Wild rose, *Rosa* spp.
Thimbleberry, *Rubus parviflorus*
Elderberry, *Sambucus* spp.
Snowberry, *Symphoricarpos albus*

Step 4. Transport the Plants

Plants in burlap sacks can be carried to a truck by hand, in a wheelbarrow, or with a homemade stretcher. A stretcher can be constructed with burlap sacks and two 2 x 4 inch boards that are 8 feet long.

Step 5. Hold the Plants

Plants can be installed immediately or held for later use. However, most plants, particularly small ones, will develop roots more quickly if they are first planted in containers.

If necessary, large plants can be stored in mulch in a special raised bed. This can be a wooden frame about 1 foot deep and approximately 4 feet wide by 8–10 feet long. Line the bed with heavy-mil plastic and half-fill it with soil or mulch that you'll keep saturated with water. Punch holes in the corners halfway up the liner so the entire bed will not fill with water. Place the plants on the bottom layer of saturated mulch and cover their roots with mulch. Capillary action provides the plants with sufficient water and they can survive for a year or more with very little maintenance.

Enhancing Existing Woodlands

Existing woodlands are ready-made woodland wildlife landscapes. Mature vegetation with a mass of leaves to support insect larvae, dense branches and stumps to attract nesting birds, and a deep leaf litter for small animals and insects all add up to a rich living world.

To improve an existing woodland area for wildlife, you can start by removing, aggressive non-native weeds such as English ivy and Himalayan blackberry. These displace other plants by competing for space, light, nutrients, and water. Because these plants spread so quickly, remove them as soon as possible where they begin to invade, especially in areas next to wild places. For information on how to manage these and other plants, see Chapter 4.

Other things you can do for wildlife in existing wooded areas are:

- Link a wooded habitat to another habitat by planting trees and shrubs to create a corridor between them. A corridor is an area of continuous habitat that permits animals to travel securely from one area to another. A corridor allows wildlife to find islands of habitat that have been separated by fragmentation such as by housing development. (See Figure 1 in Chapter 1.)
- Plant a vegetative layer where one doesn't exist. For example, add shade-tolerant shrub-layer plants under some existing trees that lack this layer. You'll need to water and mulch the new plants because they'll be competing for moisture.
- Create a snag out of an existing woodland tree that's creating a hazard with a forked top, weak wood, or disease. Other snag candidates are trees that are shading an area, have invasive roots, or are in a group of trees that needs thinning out.
- Girdle a large limb or a forked trunk in an existing tree to create a perch for birds and potential habitat for other wildlife.
- Expand the plant diversity in an existing woodland edge. Woodland edges create edge habitat that benefits both woodland and open-country wildlife. They offer excellent escape cover for rabbits and grassland songbirds, good feeding sites for grouse, deer, and woodland birds, and good hunting conditions for hawks, owls, foxes, and coyotes. The best edge is one where ground-layer plants, short and tall shrubs, and short and tall trees are present. An edge that creates a gradual transition between two habitats is better than one creating a sudden change. Enhance aesthetic qualities and wildlife habitat by including trees and shrubs that flower and fruit at different times of the year. These can be concentrated along access routes for maximum visual impact. Many of the woodland plants listed in Appendix B can be included in an edge planting.
- Add rock and brush shelters near the woodland edge and decaying logs in the interior of the woodland.
- Maintain nest boxes for flying squirrels, bats, songbirds, owls, and other wildlife.
- Install a small pond or a birdbath to attract a wider array of wildlife.

Forestry Management and Wildlife

The forests throughout the Pacific Northwest can be improved using forestry techniques that benefit both trees and wildlife. There is no single best way to maintain a working forest. Goals vary widely among landowners and may include having a continuing source of revenue, clearing a portion for a home, managing for future income, and ensuring good wildlife habitat.

If the forest is developing well, the best long-term option for forest management may be to leave it alone, re-surveying it every five years to monitor its quality. Under natural conditions it can remain as wooded habitat for hundreds of years. However, you should be ready to provide necessary maintenance if forest development starts in a direction you don't desire—for example, because of overgrazing pressure, the formation of a very dense, closed canopy, or threat of domination by invasive non-native plant species.

Here are some guidelines to follow to lessen a forestry operation's impact on wildlife habitat:

- Minimize disrupting wildlife by limiting activities to small areas in order to complete them in the shortest time.
- Avoid harvesting or thinning trees during nesting or calving season (March 15th to July 15th).
- Maintain wide buffers around springs, streamsides, wetlands, and high-quality wildlife habitat.
- Preserve or create snags and logs.
- Keep firebreaks and roads as narrow as possible, while still meeting objectives.
- Keep roads out of areas near streams and creeks.
- Pile or spread slash to provide cover for wildlife.

Timber Harvesting and Marketing Assistance

There is much more to a successful timber sale than meets the eye. Consider retaining the services of a professional consulting forester. Consulting foresters offer a wide range of independent forest-management services. Their fees may be based on the acreage of timberland involved, a percentage of revenues from sales, dollars per thousand-board-foot of product, amount of time required to perform the job, or, in the case of tree planting, dollars per thousand seedlings planted.

As in any business arrangement, be aware of competing interests. The consultant should not be the one who physically removes your trees or does other woodland work for you. Also, once you've selected a consultant, sign a written agreement or contract. This document should include a list of services, how they are to be performed, who will perform them, and by what date.

Working with a Logging Contractor

(Adapted with permission from Terra Verde, Consulting Foresters, Vancouver, WA)

When considering logging a portion of your property, take your time and be careful how you proceed. Logging is traumatic and initially pretty ugly. Once you cut trees down, it is too late to change your mind. Don't be bullied into doing something you are not sure of, and beware of claims that this is your last chance to sell your trees because of new regulations or falling prices. Remember, contrary to what someone may tell you, your trees are still growing larger, in most cases increasing in value.

The following guidelines should help eliminate most of the undesirable players vying for your resource:

- Don't work with a word-of-mouth agreement. A properly written contract eliminates all surprises. Include a list of *all* fees to be charged and determine who will pay any timber taxes.
- Ask for professional references, a list of clients, and a written description of work experience in areas similar to yours. Call all of the references.
- Ask to be shown an example of a recently completed job. Verify that the person showing you the place actually did the work.

- Check with your local University Extension Forester, or with a resources office listed in Appendix E. Ask the foresters if they know the operator and if there have been any complaints or citations issued against him.
- If you have any questions about the value of your timber or costs associated with getting it to the market, contact a consulting forester. As with any professional, ask the forester for references.
- Often a logger or timber buyer will want a percentage agreement. Ask a consulting forester to provide you with standard logging costs based on operational conditions.
- Prior to entering a contract, photograph all areas of concern to establish their condition prior to logging. These areas may include sensitive leave-tree zones, roads, and buffers.

Make sure buffers and trees that are to remain are identified with bright spray paint, and the form of identification is stated in the contract.

Make certain you have all of the responsibilities (for both parties) identified in the written contract. Issues to address include condition of roads, slash disposal, "leave" trees, trespass authority, cleanup, duration of the job, payment schedules, costs, taxes, and verification of log volume removed. Have the contract reviewed by an attorney. The money you invest in a legal review will go a long way in preventing later problems. No reasonable operator should object to providing you with a sample contract. Always get a second opinion.

For additional sources of information on woodlands, see Appendix E.

FIELDS, MEADOWS, AND GRASSLAND LANDSCAPES

In general, a grassland is any area dominated by grasses. Grasslands in the Pacific Northwest range from manicured lawns, containing one or two grass species, to natural meadows containing a variety of grasses and wildflowers. Between these two extremes are grasslands you can create and manage for yourself and for wildlife.

Informal grasslands can support a variety of different plants and wildlife, and they are good settings for people of all ages to discover and appreciate nature. They display unique aesthetic qualities, including color, texture, sound, movement, and visual changes throughout the year.

As economic, environmental, and aesthetic factors favor more natural landscapes, the popularity of grasses and wildflowers is increasing. Alternative mowing techniques are also becoming popular.

A variety of grassland plants can thrive in your local climate and soils. They don't require fertilizer or long-term irrigation, and many attract butterflies, hummingbirds, songbirds, birds of prey, and larger animals. Whether you live on acreage or on a suburban lot, you can tailor grassland plants to fit into your landscape design.

This section tells how to create, improve, and maintain grassland landscapes. For more on lawns, see Chapters 3 and 4.

Pacific Northwest. These changes have also meant the suppression of wildfires. Fire stimulates the growth of grassland plants and keeps encroaching trees and shrubs from taking over.

About one-third of all urban land is grassland, two-thirds of which is closely mowed lawn.

Almost all of the plants in recently created grasslands are non-native species. Many are drought-tolerant, mowing-tolerant, and trample-resistant plants that have adapted to your lawn, field, clearing, or pasture. Those commonly found growing in lawns include English (oxeye) daisy, plantain, dandelion and its many relatives. Non-native plants growing in fields and clearings include scots broom, Queen Anne's lace, chicory, and evening primrose. Common non-native grasses include wild oats, cheatgrass, orchard grass, crabgrass, reed-canary grass, and Kentucky bluegrass.

Pacific Northwest Grasslands

Native grasslands in the Pacific Northwest formed where the soil was either too wet or too dry to support trees and where frequent fires kept trees and shrubs in check. They include wet meadows, dry prairies, and the shrub-steppe of sagebrush country.

Wet meadows, also called wet prairies, once blanketed much of Oregon's Willamette Valley with tufted hairgrass, sedges, and wildflowers. Drier native prairies, found in patches in the Puget Sound lowlands west of the Cascade mountains, were covered with native fescue and wildflowers such as camas, balsamroot, shooting star, and lupine.

The most widespread native grassland is the shrub-steppe or sagelands. In areas east of the Cascade mountains in Oregon and Washington, and in British Columbia's Basin and Range Province, precipitation too meager to support trees maintains a conspicuous community of grassland plants. It's a grassland sparsely carpeted with bunch grasses and wildflowers, and dotted with sagebrush, bitterbrush, and rabbitbrush.

Two hundred years ago native grasslands covered a much larger area. Over time, they've been converted to farms, pastures, rangelands, and developments, all of which have resulted in the introduction of many non-native plants now common throughout the

How Grasslands Support Wildlife

Grasslands harbor a surprising number of animal species, from tiny soil organisms to large mammals, including grazers, burrowers, and their predators.

A chemical-free lawn or other mowed grassland is home to many small organisms. Spiders live near the tops of grass blades and avoid danger by rapid escape down into the thatch layer or soil. Ants and other invertebrates spend most of their lives in the thatch layer to avoid mower blades, predators, and dehydration.

Mowed grasslands offer rabbits and birds access to food and flat, open spaces where the view of approaching danger is unob-

Three Grassland Birds

(Adapted from *Wild Flyer*, Oregon Fish and Wildlife, Spring 1998)

Horned larks like short grass (less than 6 inches high) and bare ground, including gravel roads and shoulders of lightly traveled paved roads, recently plowed or burned fields, and young Christmas tree farms with lots of open grassland.

Vesper sparrows occur almost exclusively in two habitat types—Christmas tree farms and lightly grazed pastures with scattered shrubs where grass is less than 2 feet high. In Christmas tree farms, they are common in unmanicured areas. They seem to prefer areas that have furrows, stumps, weeds, and many large clods of dirt between rows of trees.

Western meadowlarks favor grass-dominated fields—especially lightly grazed pastures or fallow fields with grass less than 2 feet high and a few shrubs. Shrubs seem to be important as singing perches for Oregon's state bird. In cultivated taller grass fields, meadowlarks occur only where fences or telephone poles offer singing perches.

structed. Here, mourning doves and finches find seeds, robins hunt for worms, and flickers search for ants.

Unmowed or occasionally mowed grasslands can be home to mice, voles, and gophers, which attract their predators: coyotes, hawks, owls, and snakes. Deer feed and bed down in unmowed fields of tall grass and bracken fern. Meadowlarks, savanna sparrows, song sparrows, and other ground-nesting birds may nest and raise their young in a field left uncut until mid-July. Even a small area of flowering grassland plants can be a refuge for butterflies, crickets, grasshoppers, and bumblebees.

Deep, well-drained shrub-steppe soils are ideal for burrowing animals such as kangaroo rats, pocket mice, and ground squirrels. Burrows provide thermal and hiding cover, as well as places for hibernation.

Creating a Grassland Landscape

Grassland landscapes come in all shapes and sizes: a lawn, a roadside, a hayfield, an orchard, a patch of wildflowers in a front yard, and a shady glade among large trees. While the ambitious habitat landscaper may create a grassland of dazzling diversity, the less energetic can maintain an existing grassy area with modest effort.

It's important to match the grassland planting to the site. Be sure the soil, light, and water conditions are well understood before trying to establish new plants in the area. Knowing what plants are already growing on the site can give you an idea of the growing conditions that exist there.

Where soil and other site conditions are suitable, a naturalistic-style grassland can be created using a prairie, shrub-steppe, wet meadow, or other native grassland as a model. A naturalistic grassland landscape is a simplification of the natural model, and is almost always smaller. The aim is to capture the essence of a certain grassland type by using the dominant plant species.

Creating a grassland, especially one containing Pacific Northwest native plants, isn't easy, cheap, or quick. It requires more attention to planning, site preparation, and maintenance than any other area in the landscape. It is not a project for the novice gardener.

There are several things you should know about creating a grassland landscape:

- Don't underestimate the difficulty of getting grassland seeds and small transplants established before the inevitable invasion of fast-growing weeds.
- The area may need to be cut several times with different machinery than you use on your lawn. A weed-eater (string trimmer) with thin line is often the most practical tool.
- Grasses and wildflowers will look brown and rather untidy in the winter when they remain in place to reseed and become sources of food and cover for wildlife. If you're planting a large area within city limits, consult your planning department to determine whether local codes place any restrictions on the appearance of your property.
- Establish a small area first, then gradually increase its size. This will let you experiment with soil preparation methods and discover what does and doesn't

grow well before you commit to managing a large area.

Site Preparation

Improper site preparation and establishment procedures are the main reasons why newly created grassland landscapes containing a variety of grasses and wildflowers fail from the start.

If you want to sustain native grasses and wildflowers, it is best to avoid site preparation in an area with fertile soil. Non-native grasses flourish in fertile soil but are less able to handle stressful growing conditions and often can't compete with native grasses and wildflowers, which are accustomed to growing in soil with low fertility.

The following describes various ways to prepare and plant a 500- to 3,000-square-foot, naturalistic grassland landscape:

After you've roughly removed unwanted plants, use one or more of the following techniques to further eliminate existing seeds and vegetation in preparation for new grassland seeds and plants. Any of the following techniques can also be used to prepare other areas for trees, shrubs, or other landscape plants.

Solarization is a simple technique for killing seeds and vegetation that requires minimal equipment. In places where the air temperature regularly exceeds 90 degrees, tightly cover the ground with clear plastic. The heat of the sun will trap moisture beneath the plastic and turn it to steam. The heat and steam will kill both weeds and weed seeds. Weight the plastic down around the edges so it seals in the heat and doesn't blow away. To make sure even the toughest weeds are killed, keep the soil moist during the hottest period of smothering.

To use this technique over a large area, set up a rotation that treats each area several times over the summer. Keep the plastic in one place for four or five sunny days and then move it to the next area. Water the areas that have been treated, so remaining viable seeds will germinate before you do the plastic treatment again.

Smothering is an effective way to kill existing plants and prevent seeds from germinating. Cover an area with several layers of newspaper, cardboard, or leaves, and then 3 inches of weed-free topsoil or mulch. Different size planting pockets can be opened up to receive live plants and seeds.

Stripping is a way to remove the top 2–3 inches of vegetation, roots, and soil. For a small front lawn, a mattock or a grub hoe works well (see Figure 1). Detach strips of turf by cutting through their roots and using the chopping motion of the blade to roll the turf toward you. For areas larger than 600 square feet, consider renting a motorized sod cutter. With a sod cutter the sod comes off in long strips about 18 inches wide and 3 inches thick, and usually creates a nearly weed-free site ready for seeds or transplants. Be aware that the area stripped will be lower than adjacent ground after sod removal.

You can compost the turf you dig up by piling it upside down and keeping it moist (the consistency of a damp sponge). For faster composting, layer in blood meal or fresh manure. Turning the pile will also speed decay. In about a year you should have a rich loam.

Cultivating is a way to incorporate existing vegetation into the soil and deplete the supply of weed seeds. First, turn existing plant matter and newly sprouting weeds

Managing Existing Grasslands for Wildlife

You can improve an existing field, lawn, roadside, or other grassland for wildlife by modifying your mowing technique, adding desirable plants and removing undesirable plants, and installing nest boxes and other enhancements. A field or other grassland can be improved for wildlife by appropriate mowing. The goal is to reduce the harmful effects that mowing can have on wildlife and to create different habitats within the area. For information on mowing lawns and other grasslands, see "Mowing Options to Improve Wildlife Habitat" in Chapter 4. Surrounding trees and fence posts are ideal places to attach nest boxes for cavity-nesting birds (see Chapter 12). Large grasslands can also be ideal places to install perching poles for the birds that use them (see Chapter 6).

into the soil. In a larger area this is generally done with a rototiller; in a smaller area or over a long period of time, this can be done by hand. Weeds should be allowed to resprout and then be turned into the soil to a depth of 4–5 inches. The number of times you have to cultivate will depend on the abundance of weed seeds, the amount of "seed rain" that falls on the area during preparation, and your expectations for a weed-free planting site.

Normally, you should expect to cultivate three times, with three weeks between cultivations. If you cultivate in the dry season, after cultivating, run a sprinkler for 15 minutes, three days in a row and wait a week to see what germinates. Cultivating is least effective against aggressive grasses and other plants that re-sprout from pieces left in the soil, but it can prepare a smooth area for planting.

Herbiciding with a glyphosate herbicide such as Roundup™ requires the least work and is very effective at killing invasive grasses, but it won't kill weed seeds. (For information on how to safely apply an herbicide, see Chapter 4.) Spray the area when new weed growth is vigorous (March 1st–September 15th) and again (if necessary) two weeks later to kill newly sprouted seeds and plants that haven't succumbed to the previous spraying.

Note: It may take a full growing season, from early spring to late fall, to carefully prepare a site for new grassland plants. Even though the area may look desolate, the longer you can wait to make sure weeds aren't present, the easier it will be to establish new plants. In addition, a newly cleared and/or graded area may take a year to fully reveal its bank of weed

Figure 1. A mattock has two blades and a short handle. It's used for chopping out roots as well as stripping turf. A grub hoe is a lighter tool with one blade and a longer handle.

seeds. If you intend to establish grassland plants in such an area, it may save you a lot of time and money to wait and see what germinates in the spring before planting the area. Bare ground can be covered with a mulch or cover crop.

In an area prone to erosion, you can plant a temporary cover crop to hold the soil through the winter or while other species get established. A common seed choice is annual ryegrass. (Don't use perennial ryegrass, which is used in permanent grass mixes.) Annual ryegrass will germinate in early fall, root into the soil, but die before your spring planting. Regreen™ is a sterile hybrid wheat that can be used for temporary erosion control in the spring and summer.

Plants for Grassland Landscapes

Wildflowers, shrubs, and even trees can be included in grassland landscapes, but grasses are the main component. Grasses help stabilize the soil, add seasonal color and texture, and provide food and cover for wildlife.

Large grassy areas are often dominated by vigorous sod-forming grasses such as Kentucky bluegrass and agricultural varieties such as orchard grass. Their aggressive, dense root-mats create barriers that other plants can't grow through and thus they aren't recommended for small grassland creation projects where a diversity of flowering plants is desired. However, there are many opportunities to manage these existing grasslands for wildlife (see "Mowing Options to Improve Wildlife Habitat" in Chapter 4).

Native bunch grasses occur throughout the Pacific Northwest and are well-suited for small grassland landscapes and larger projects where competition from aggressive non-native grass species can be carefully controlled. Rather than spreading out into a sod-like cover, bunch grasses grow in clumps—leaving room between them for wildflowers to become established and for birds to search for insects and seeds. They are not, however, suited for areas that you want to be smooth, like a lawn for croquet or volleyball.

Bunch grasses will survive in areas with little or no supplemental irrigation after they are established, and they stay green as long as moisture is available in the soil. In a typical inland Northwest summer, this is generally until July or August. About this time, plants go dormant and turn a straw color. They are not dead or dying, they are just conserving moisture to survive the hot, dry period. They will green up when fall rains come before they go dormant again for winter. When watered, bunch grasses can look fresh all summer.

Most native bunch grasses are small to mid-size at maturity. However, some bunch grasses (tufted hairgrass for example) can get quite large so choose your species carefully. See Appendix C for descriptions of individual species and recommendations for your area.

Selecting Grassland Plants

Height of mature plants, flower color, and season of bloom are things to consider when selecting plants for your project. Use the information in Appendices B and C to help compile a list of possibilities. You can also contact your local native plant society or one or more growers of grassland plants (see Appendix E) who may suggest plants that could meet specific site conditions.

A long species list for your grassland landscape is unnecessary and generally expensive; an attractive grassland can be created with relatively few species, chosen to suit the site conditions.

Beware that many non-native grasses and wildflowers are aggressive and can quickly displace native species. Because of this, they should never be scattered in or next to a truly wild area.

Planting the Grassland Landscape

There are two ways to establish a grassland landscape with new plants: by seeding and by transplanting. Large areas are often seeded because of the expense and scarcity of transplant stock, and the labor involved in planting numerous small plants. Small areas are often planted only with transplants (pots, plugs, divisions) when the desired species are available in these forms. An area can also be planted with both seeds and transplants.

Most perennial wildflowers and grasses are slow to grow from seed; an Idaho fescue seedling may take three years in the field to reach the size of a one-year nursery transplant. While initially more expensive, transplants are quicker to establish and give complete control over the placement of species by height, color, and other design requirements you have. In addition, many transplants will flower the first growing season.

Transplants are also preferred where weeds are a critical problem. Weed control is made easier by mulching around the newly planted transplants or by planting after a weed-free mulch has successfully smothered the weed competition.

Planting seeds or small transplants directly into an area dominated by aggressive grasses is a waste of seeds, plants, and time. Transplants have little chance of getting established among existing grasses because of competition for light, water, and minerals. The grass cover also may keep germinating seeds from penetrating into the ground.

To help make establishment possible, use one of the techniques described in the section "Site Preparation." Plant the seed or transplants, and control aggressive grasses for the first couple of years.

When to Plant

The grassland can be planted with seeds and/or transplants in spring (March through May) if the area can be watered through the growing season. Although rain is generally reliable in spring, hot, dry weather can kill new plants if they are under-watered.

Fall planting (September and October) is recommended for large

Wildflower Seed Mixes

Seed mixes for "eco-lawns," wildflower meadows, and other themes are now available from a variety of commercial outlets. Although a seed mix may seem a simple, no-fuss way to plant, you need to be cautious when choosing a mix. Although some seed suppliers now blend mixes for specific habitats, these mixes may be designed for broad geographic regions and may include mostly aggressive, non-native plants, or plants not adapted to your local area.

However, local mixes are becoming more common. They are collected by commercial collectors and blended for a specific geographic region and type of habitat, such as a Puget Sound prairie or a shrub-steppe area in eastern Oregon. The optimal match of seed to a specific area is worth the extra expense. To locate a seed nursery, call a local nursery (see Appendix E) and ask for a recommendation.

Whether you use a commercially available seed mixture or create a custom local mix, here are some guidelines:

- Buy from a reputable source to prevent potentially weedy contaminants.
- Check the seed list and make sure you are planting the things you want. If the mix doesn't have a list, request one from the seed company.
- Check with your county Extension Office, Native Plant Society chapter, Nature Conservancy, or County Weed Board to see if any species are a problem in your area.
- Include a mixture of species that bloom at different times to

continued on next page

Table 1. Non-native plant species found in mixes and their origins

Four-o'clock, *Mirabilis jalapa* (Peru)

Bachelor buttons, *Centaurea cyanus* (Europe)

Chicory, *Cichorium intybus* (Mediterranean)

Queen Anne's lace, *Daucus carota* (Eurasia)

Foxglove, *Digitalis* spp. (Europe)

Candytuft, *Iberis sempervirens* (Europe, Asia)

Tansy, *Tanacetum vulgare* (Europe)

Oxeye daisy, *Leucanthemum vulgare* (Europe)

Sweet alyssum, *Lobularia maritima* (Europe, Asia)

---

provide nectar and pollen for a variety of insects. Aster, pearly everlasting, and goldenrod can extend the bloom period into late summer.

- Don't include species that are difficult to grow from seed, such as kinnikinnik or camas. Also, avoid including rare species in your mix; they're expensive and generally have a high rate of failure. These are best added as transplants.

- Be sure the seeds in the mix are packed individually by species to allow you to sow them separately or together.

- When ordering seed mixes, write "no substitutions will be accepted without prior notification" on the order form. Otherwise, suppliers may substitute one plant for another if the one you want is out of stock.

areas, where supplemental watering is not practical, and when the seed has germination requirements. It's also recommended for dry and clay soils because young seedlings can become established before the clay dries in the summer and restricts downward root growth.

Don't attempt fall seeding or transplanting on sites subject to soil erosion. Runoff from snow melt or heavy rains can cause extensive erosion in areas where the soil has been prepared for planting.

Consult with your seed supplier about the optimum time to sow their seed and any special requirements (stratification, scarification) of individual species.

How to Seed a Grassland

Future grassland sites under 3,000 square feet can be seeded by hand. To seed by hand, prepare the seed bed as best you can and broadcast seeds (distribute seeds by hand or with a handheld broadcaster) on a windless day. Because many wildflower seeds are tiny, sow any large seeds first and then after they've been lightly raked in, plant the smaller seeds (about the size of coarse pepper seasoning). To help define your design before you sow the seeds, outline the edges of different planting areas with lime or white flour and plant each area separately.

To provide more uniform seeding, keep the seeds from clumping together, and make it easier to see what areas have been seeded, mix the seed with slightly moist peat moss, sawdust, or potting soil. The proportions don't matter. Broadcast half the seed in a north-south direction, and then half in an east-west direction to reduce bare spots.

After you broadcast the seed, lightly rake it into the soil for good seed-soil contact, or mulch the area lightly with a *weed-free* mulch. If possible, roll the site with a roller (available at most rental stores) to press in seed and stabilize the mulch.

To help you later distinguish a weed from a new grassland plant, cover a small area of soil with newspaper or a sheet of plastic before you sow the seeds, and remove the barrier after seeding. Anything that sprouts in that area is likely a weed.

How to Plant a Grassland with Transplants

After a planting area has been prepared, transplants can be placed in sweeps (broad, meandering groups) or random arrangements according to your design. Spacing for grasses and wildflowers should range from 1–3 feet. Any shrubs or trees should be planted first so as not to disturb smaller transplants later. Disturb the soil as little as possible when you transplant to prevent bringing dormant weed seed to the surface.

A layer of weed-free mulch will help keep weeds down. Baled field hay and even "clean" straw are not recommended because they contain weed seeds. Fine wood chips or bark are good choices. Either of these can be spread over the soil or over layers of cardboard or newspaper.

Maintaining a New Grassland Landscape

A newly planted grassland needs watering and frequent weeding for the first few years. Spring and summer plantings need regular watering through the growing season. Water every few days for the first four to six weeks to encourage higher seed germination and plant survival. Use a fine mist at first, so you don't create runoff and dislodge seeds. After four to six weeks, water longer but less frequently.

Your new planting will inevitably be invaded by weed seeds left in the soil, or from surrounding grasses and other plants. *Accept that some weeds will always be present, decide which weeds need to be controlled, and focus on managing those.* You will need to be diligent with any target weed species as it may only take one (or a few) recurrences to reestablish the population.

Maintain small areas by persistent hand-pulling, digging, or smothering with layers of newspaper and a covering of mulch. Weeds pull easier a day or two after rain or watering (when soil is soft but not muddy). Another control option is to clip weeds near the ground with pruning shears. As with transplanting, be careful not to spread soil around when hand-pulling. This exposes dormant seeds that germinate and form a new weed crop. For more information on weed control, see Chapter 4.

Burning a Grassland

Burning a grassland is a way to control unwanted plant species and encourage preferred species. Burning is a natural process that removes the accumulated plant litter from the previous year's growth and exposes the soil's surface to the warming rays of the sun. Most grassland plants need warm soil temperatures to grow efficiently. Therefore, burning encourages early soil-warming and typically increases growth, flowering, and seed production of native flowers and grasses.

The benefits of fire will only last one year when the grassy area is surrounded by weedy species, which will quickly move into the area. Rotational burning, burning just one-half or one-third of a grassland each year, is recommended.

Using fire as a management tool requires designing firebreaks into your landscape. Always plan fire safety into plantings, even if you are not going to use burn management. A closely mowed grass strip 5–10 feet wide can serve this purpose. Driveways, sidewalks, lawns, ponds, and streams also serve as firebreaks.

Even under carefully controlled conditions, burning a grassland can cause accidents. A permit to burn is necessary almost everywhere, and burning is not allowed at all in some places. Before burning or planning your firebreaks, get advice and approval from local fire-prevention authorities.

Burning is not recommended in urban areas or in landscapes where mulch has been used to control weeds.

Mowing

If you want to try and replicate a natural burn in a small grassland and burning is not an option, you can substitute a combination of mowing and raking. Although not as effective as burning, mowing followed by raking can have a similar effect by removing the previous year's vegetation. Raking keeps soil fertility low, which discourages the growth of non-native plant species, and prevents clippings from smothering small plants. (Raking is not recom-

Collecting Seeds

Collecting seed by hand is time consuming, but suited to small-scale projects because you can choose exactly what is to be planted, gradually building up a stock of seed for a variety of species.

Even where plants are full of ripe seed, collect no more than one-tenth of the total seed present for any one species. Taking more reduces that species' chances of regeneration in the next growing season. Be aware that others may be collecting from the same area. In addition:

- Get permission from the landowner or manager before collecting, especially in parks and nature reserves.
- Collect dry seed and place it in non-airtight containers such as boxes, trays, or stout paper bags.
- If wet seed is collected, spread it to dry as soon as possible (damp seed may become moldy or may germinate prematurely).
- Avoid collecting seeds of different species in the same container.
- Keep a record of the species collected, place collected, date, and amount collected.
- Collect seeds in your immediate area when possible. These are best adapted to local conditions.

Advertise Your Grassland Project

To prevent problems with neighbors it's important to explain your plans to help them understand and accept your grassland landscape project. A sign that shows the intent of the project will improve viewer acceptance. On a private property this could be a small sign that designates the area as a wildlife sanctuary; in a public area it may be a sign that explains what is being done, why it is being done, and what people can expect to see over time. If the area is signed and there is evidence of maintenance, such as a mowed boundary line (see Figure 3 in Chapter 4), people will see the area being managed in some way, and they will be more likely to accept its wild look.

mended where your weed control strategy relies on a mulch.)

If the site or number of weeds is too large for control by hand, you'll need to regularly cut the weeds. During the first growing season the area should be mowed or cut to 4 or 5 inches whenever weeds have reached a height of 8 inches, or before they set seed.

A flail-type mower is preferable for large areas because it chops cuttings into small pieces. A weed-eater can also achieve this if it is carefully used. If a sickle-bar or rotary-type mower is used, mow more frequently so cuttings will not become large enough to smother the seedlings. If you miss a seed head, cut it off by hand and carefully remove it from the area.

Small trees and shrubs that could be damaged by mowing should have colored flagging tape attached to them so they can be avoided. You should expect to mow the grassland once or twice a month during the first growing season.

Continue mowing the second growing season, but raise the cut height to 10 inches, and eventually to 12 inches by the end of the growing season. After the first two years, any area that remains thick in weeds can continue to be mowed to approximately 12 inches. Adjust mowing heights according to the response observed in your grassland, and your own preferences. For additional information on mowing, see "Mowing Options to Improve Wildlife Habitat" in Chapter 4.

Some plants may be so small the first year that you can barely see them. At times weeds may seem to be taking over and you may think the project is a failure. However, your persistence at weed control will pay off. After a few years, the new grassland plants will cover the ground, and their dense roots will make it difficult for new weeds to become established.

For additional sources of information on grasslands, see Appendix E.

WETLANDS AND WETLAND GARDENS

Until recently, wetlands have been considered worthless, bug-infested wastelands. As a result, phenomenal numbers of wetlands in North America have been drained for agriculture and filled for development. In the Pacific Northwest, nearly 90 percent of presettlement wetlands have been degraded; about 50 percent have been lost outright.

Only recently have wetlands been recognized as a valuable resource. Wetlands help control floods, filter and purify water, and provide rich and diverse habitat for fish and wildlife. Wetlands also offer visual interest by displaying unique populations of plants and animals. For these reasons it's important to carefully consider how landscape projects could affect wetland habitat.

What Is a Wetland?

A wetland is just that: wet land. It is typically a low area saturated or covered with shallow water for at least part of the year. Wetlands are shallow and some dry up late in the summer. However, even when wetlands look dry, their soil is often saturated with water, at or just below the surface.

Wetlands have many names, such as bogs, swamps, marshes, and wet meadows (Table 1). They vary in size from many acres to a year-round damp area in a low corner of your property. Whatever the water

Table 1. General characteristics of three types of wetlands. For a list of additional plants found in these wetland types, see the "Native Plant Occurrence Chart" in Appendix B.

Type of Wetland	Description	Wildlife Information
Shrub and Forested Wetlands	Shrub and forested wetlands are often found next to rivers and streams, even in urban areas. They are partially or totally shaded by plants such as red-cedar, Sitka spruce, black cottonwood, Oregon ash, red-twig dogwood, willow, and hardhack. Grasses and grass-like plants include hard-stem bulrush and slough sedge.	Dense vegetation under trees provides an abundance of food and cover for wildlife including ducks, geese, herons, nesting songbirds, small mammals, and amphibians. Snags (dead trees) in these areas are very important for woodpeckers and the many bird and mammal species that use tree cavities for nesting and roosting.
Fresh-water Marshes	Marshes typically have standing water throughout the year and support aquatic plants such as cattail, soft rush, sedge, and duckweed. Marsh-like habitat is sometimes created in urban areas where stormwater is managed by a city or county.	Marshes receive high use by wildlife. Where connected to a river or stream they may provide critical winter habitat for young salmon. Typical wildlife species found in marshes include muskrats, marsh wrens, red-winged blackbirds, dabbling ducks, geese, frogs, salamanders, and garter snakes.
Wet Meadows	Wet meadows may be flooded all winter and be dry enough to walk through in late summer. They often contain grasses such as, rice cutgrass, small fruited bulrush, tufted hairgrass, and woolly sedge; also Oregon ash, Douglas hawthorn, and nootka rose.	Wet meadows may be used by Northwestern salamanders, treefrogs, and toads. These species may breed in wet meadows where shaded puddles remain full through late spring. Large ant mounds may be built up to avoid winter flooding.

Wetland Buffers

To perform its functions for people and wildlife, a wetland needs to be surrounded by a buffer. This is an undeveloped or carefully landscaped area next to the wetland that helps protect the wetland from human intrusion. A buffer also helps a wetland maintain water quality by filtering out sediments and pollutants through its vegetation and soils.

Buffers are part of the wetland system and are equally as important as the wetland itself. Buffers are essential, no matter how large or small a wetland is. Often, small wetlands need more protection because they can't recover from harm as quickly as large wetlands can.

Wildlife depend on both the wetland and the adjacent buffer to meet essential life needs. Ducks and geese, for example, feed mostly in wetlands, but many species nest on dry ground where nests won't be flooded. In some areas, mowing or heavy grazing by livestock next to wetlands removes buffer vegetation and reduces nesting sites for waterfowl and shelter for other wildlife. A wetland may be preserved, but if the nesting habitat and shelter in the adjacent buffer is lost, wildlife usage will decrease substantially.

As the width of the buffer increases, direct human impacts such as dumped debris and trampled vegetation will decrease. As the width of the buffer increases, its effectiveness in removing pollutants from surface water and the numbers and types of wetland wildlife will increase. To maintain the functions of a wetland for wildlife you need a buffer of 50 to 300 feet, depending on the type of wetland, the surrounding land use, and the types

continued on next page

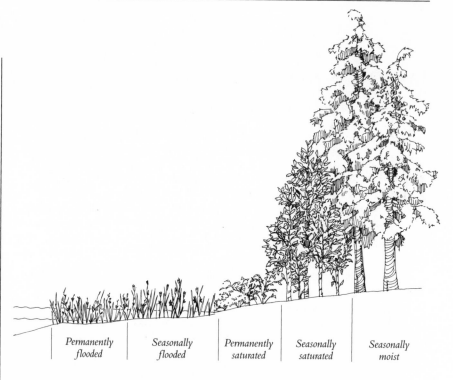

Permanently flooded   Seasonally flooded   Permanently saturated   Seasonally saturated   Seasonally moist

Figure 1. *A cross-section of a typical wetland showing how wetland plants range from those that are permanently flooded, to those that are seasonally moist.*

source—a high water table, rainwater that remains on impenetrable layers in the soil, frequent flooding, or groundwater seeps—prolonged saturation of the soil defines the wetland.

Why Wetlands Are Vital to Wildlife

As a complex combination of water, plants, and rich soils, a wetland (including its adjacent buffer) offers protection where water- and land-based wildlife can feed, breed, and rear their young.

Wetlands don't have to be big to sustain large and diverse wildlife populations. For instance, ducks can be particularly abundant in a wetland that is less than one-half acre where shallow, open water is available. Even a puddle 15 feet wide may host a pair of mallard ducks through winter. In addition, many frogs and salamanders achieve their highest densities in small wetlands because they can't survive in the presence of fish and

bullfrog populations found in the larger ones. A year-round moist area in your backyard may attract butterflies, dragonflies, treefrogs, and certain songbirds. Instructions on how to create a mini-wetland are described later in this Appendix.

Many wetlands are wet in winter and dry in summer. These seasonal wetlands are critical for wildlife because they are available when wildlife need them the most. For example, ducks need open water when breeding in the spring and migrating in the fall. Treefrogs, Northwest salamanders, and redback salamanders need water in the winter and spring to breed and grow in before they move into the surrounding upland area.

Because bullfrogs (an aggressive non-native species that preys on native amphibians) require water year-round, they can't survive in a wetland that "disappears" in the summer. Although a limiting factor for bullfrogs, this dry period is often the only thing that makes it possible for native amphibians to flourish in areas where bullfrogs are common.

Working in and Around Wetlands

Wetlands are complex systems and we are still learning what they do and how they do it. Tinkering with them must be done cautiously, with humility, respect, and an understanding of how the wetland fits into the surrounding watershed.

As a general rule, you shouldn't alter a wetland for any reason. Instead, efforts should be taken to support the unique populations of plants and animals that already are in balance with existing conditions. Introducing animals to a wetland to establish or increase populations is not a good idea, and in most cases it is illegal. Most animals will find the wetland without your help.

Wetland Regulations

For activities including filling, excavating, cutting trees, grading, draining, and construction work in or next to a wetland, expert opinions should be obtained. Talk to local agencies and wetland specialists about the potential problems, and possible alternatives to the activity being considered. Consultants can be found by calling your local government planning office, the Society of Wetland Scientists, the Department of Ecology, the Environmental Protection Agency, or your local Fish and Wildlife office. In larger cities and towns, look under "Environmental" in your phone directory. See Appendix E for addresses and telephone numbers.

Because permits may be required, you should begin the process as soon as possible. Contact your city or county planning office for permit information.

Wetland Restoration

Wetland restoration is the process of returning disturbed wetland habitat to a close approximation of its pre-disturbed condition. Successful wetland restoration requires knowledge of the wetland conditions that existed prior to disturbance and a clear understanding of the site's existing conditions.

A wetland that has been severely degraded by clearing, filling, or invasion by non-native plant species is a good candidate for restoration. The goal for such a restoration project may be to restore the structure, function, diversity, and dynamics of the wetland so it will operate without continued human management or reliance on engineered structures.

Wetland restoration projects generally involve consultants, large budgets, and permits from local, state, or federal agencies. Because of this they are beyond the scope of this book. However, you can use the vision of a restored wetland as a goal to move toward with your wetland project. Large steps in the direction of wetland restoration can be made utilizing the suggestions described in this Appendix.

For references on how to carry out a large wetland project, see Appendix E.

Enhancing a Wetland for Wildlife

Wetland enhancement projects can increase the quality of wildlife habitat around a wetland of any

of activity in it. Anything under 50 feet is generally ineffective in protecting a wetland. Your local Fish and Wildlife office can provide you with recommendations.

A half-acre wetland can provide feeding and breeding habitat for:

- One pair of bitterns.
- One pair of marsh wrens.
- Two pairs of red-winged blackbirds.
- A small flock of green-winged teals.
- Ten Townsend's voles.
- Six garter snakes.

If it's associated with a stream it can also provide wintering habitat for cutthroat trout and young coho salmon. Visitors could include great blue herons, ospreys, deer, raccoons, weasels, dragonflies, and other animals. The soft mud around the edge is used for nest-building by robins and barn swallows, and clouds of newly hatched midges, gnats, and mosquitoes provide food for nighthawks, swallows, and bats.

Pole Plantings

The use of black cottonwood and willow polewood in wetland and riparian restoration has shown promise in areas dominated by the aggressive reed-canary grass. These poles are essentially gigantic live stakes or dormant hardwood cuttings. The large size of the poles allows rooting below the reed-canary grass root-mass, and shooting above the dense reed-canary grass thatch. Thus, plants obtained from rooted polewood cuttings have a strong competitive edge over the reed-canary grass and can assist in shading out this nuisance weed.

The polewood material is typically taken from young sapling trees or from larger branches and suckers on more mature trees in existing wetland and riparian habitats.

The native black cottonwood should be used, not the hybrid cottonwoods commonly being planted for pulpwood production. Nearly all native willows will propagate by this method. Material is collected and planted during the dormant season (November through February). Although the dimensions can vary, preferred polewood material is between 2 and 4 inches in diameter at the base, 1 to 3 inches in diameter at the top, and 6 to 8 feet long. Poles are sawn off from the donor plant and all branches are removed with a saw or pruning shears. A 12 to 18 inch long lateral branch can be retained for use by perching birds.

Once a batch of poles has been prepared and bundled, spray-paint the tops of all the poles with tree-marking paint. This provides an

continued on next page

size. Enhancement projects may be appropriate within a wetland that lacks certain wildlife features or plants that might have naturally occurred there. Most enhancement projects won't require a permit, but you should contact your local planning office to make sure.

An example of a wetland enhancement project is to increase the amount of food and shelter for wildlife by expanding or adding native plants to the wetland buffer.

Other examples include:
- Remove or regularly control aggressive plants that threaten to dominate the wetland and buffer area.
- Add appropriate wetland plants to areas after aggressive, non-native plant species are thoroughly eliminated. Add plants to increase the number of vegetative layers in the wetland, create a screen, or attract certain wildlife.
- Create a wildlife corridor that connects the wetland with another nearby habitat, such as a wooded area. This enables land-based wildlife in the woodland to more easily use the wetland for food, water, cover, and breeding areas. For additional information on corridors, see Chapter 1.
- Add snags, perching poles, roost boxes for bats, and nest boxes for cavity-nesting birds.
- Add rock shelters, brush shelters, and decaying wood around the edges of the wetland. To enhance these areas for different species of wildlife, install these enhancement features within the wetland and within the wetland buffer.
- Control human access by adding fences, footpaths, benches, or observation areas. Fence off

livestock pastures or any place where extra protection is needed. A low (3 foot) fence will discourage entry by small children, dogs, and casual intruders, while allowing access by older children and adults who are interested in the wetland.

Wetland Planting Projects

Each wetland plant species has its own water requirements and tolerances, and the familiar adage "right plant, right place" applies at least as much to a wetland enhancement project as it does to plantings around your house. Because of this you'll need to understand the fluctuating water level of the wetland to determine where plants should be located. Failure to anticipate the receding water in summer is the most frequent cause of plant failure. For this reason it is important to observe the wetland you are working on for at least a year before undertaking a major planting. The time prior to planting can be spent eliminating aggressive weeds, designing a planting plan, getting familiar with wetland plants, and locating their sources.

Any enhancement project that involves installing a community of plants will need a planting plan. The planning process will help you discover aspects of the wetland and its surroundings that will influence where other enhancement features should go. For information on how to create various types of plans, see Chapters 2 and 3.

As part of the planning process, visit a wetland or wetlands that occur in your area to learn more about wetland plant communities. State, national, or provincial parks are good places to find the most natural wetlands. Wetlands you visit in a developed area probably have been modified by human activity in several ways. The most obvious will be the introduction and abundance of aggressive non-native plants. Even so, note what plants are growing there, whether they are grouped or growing alone, where in the wetland they are growing, and their size.

To become familiar with native wetland plants that grow in your area and to help you select plants for your enhancement project, contact your local Native Plant Society, County Conservation District, or other private or public group in your area that works with wetlands. These groups often lead plant identification walks, host planting projects, or have staff and literature that can assist you in plant selection.

As a general rule, anything other than a small, ornamental wetland garden should contain only native plants that naturally grow together in your area. For information on using native plants, see Chapter 3. Appendix B contains a list of native plants that are common in wetlands in the Pacific Northwest.

Be aware that some non-native plants that grow in wetland settings can quickly displace native species. Two species that should never be planted are reed-canary grass and purple loosestrife. Yellow iris (yellow flag) is another non-native that can quickly spread into a wild area; an alternative iris species, such as Siberian iris, should be used in ornamental wetlands.

Planting

The approach to planting in a wetland depends on the conditions of the site. If a portion of the wetland has been disturbed, you can sometimes rely on existing wetland seeds and plant parts to regrow into the disturbed site. Often in wetlands it is these volunteers that grow most vigorously and reseed because they are able to quickly adapt to the site conditions. Nearby wetland plants may also generate seeds that will be transported via air, water, or wildlife to the disturbed area. In both cases, your main job will be to control unwanted plants.

Wetland plants can be purchased from nurseries that specialize in wetland plants; some retail nurseries are starting to carry common native wetland plants. For a list of nurseries, see Appendix E. For information on obtaining plants from other sources, see Chapter 3.

If you want to vegetate a portion of a wetland but can't reach an area because of water depth, you can introduce plants by putting dormant roots, or tubers, and rocks in a loosely woven sack (burlap or other natural fiber). Tie the sack and toss it gently into the water where the plants are to grow. The bag will sink to the bottom where the roots and new shoots will quickly grow through the material to colonize the area.

If an existing area is infested with an aggressive, fast-growing plant such as reed-canary grass, it can make establishing new plants difficult or impossible. Even wetland trees and shrubs will have extreme difficulty getting established if competing plants aren't first thoroughly removed. Because of this, plant establishment is most successful when aggressive weeds are eliminated at least one growing season (preferably an entire year) prior to planting.

Fast and tall-growing trees and shrubs such as Sitka spruce, red-twig dogwood, and willow

easy way to identify the tops of the poles from the bottoms and also prevents the poles from being lost or becoming a safety problem after being planted.

The poles are then planted in holes prepared using post-hole diggers, soil augers, or shovels. Poles should never be driven into the ground using hammers or sledges. The idea is to plant one-half to-two thirds of the bottom end of the pole in the ground. The base of each pole can be wounded with a hatchet to enhance rooting, but that's generally not necessary. Make sure the bottom of the pole is planted, not the top of the pole.

can sometimes be established if 4-foot planting circles are first prepared within the reed-canary grass or other competing plant. Weed mats installed around individual plants at planting time will also help new plants get established in areas where competition from weeds is high. Plants will benefit if the planting circle is kept weed-free or at least kept mowed down during the first two or three growing seasons. A mower or weed-eater can do a lot of harm to new plants and it is important to flag small plants and protect the trunks of larger plants in the area (see Figure 1 in Chapter 4).

Create Your Own Mini-Wetland Garden

Most wetlands have developed naturally over centuries and many of them possess distinctive collections of plants and animals. Wetland creation is a new science that requires considerable expertise, and even then projects are often unsuccessful because of the many factors involved in their creation. If done haphazardly, such attempts can damage surrounding natural systems by altering surface water or groundwater flows and creating weed problems.

If a small wet area exists on your property as a result of a "drainage problem," it may be an ideal place to grow some beautiful, moisture-loving plants that also provide food and shelter for wildlife. The planting design could be inspired by a natural wetland, creating a naturalistic landscape that captures the essence of a local wetland setting. Before you make any changes make sure the existing plants in that area aren't already serving an important function for wildlife.

On a small scale, some gardeners combine moisture-loving plants together to form a mini-wetland garden, sometimes called a "bog garden." (A bog is one type of wetland; bog gardens rarely resemble natural bogs.) A totally artificial mini-wetland can be easily constructed with the same type of rubber or vinyl liner used to create a pond. Even used-plastic sheeting and tarp material can be recycled in this way. Small holes and tears won't pose a problem because, unlike a pond that is designed to hold water, a mini wetland garden merely requires a reduction in drainage.

Families with young children may want to consider constructing a mini-wetland until children are old enough to understand the danger associated with a pond. A small, leaky concrete pond can be filled in with dirt to create an area that simulates a wetland.

To create a mini-wetland garden using an artificial pond liner, first outline the desired shape and remove 18 to 24 inches of soil. Keep this soil next to the excavation. Next, lay down the liner making sure it covers all the sides, much like it would for a shallow pond. Punch a ring of ½-inch holes about 6 inches up from the bottom and spaced about 24 inches apart around the pond. (These holes will prevent too much water from collecting in the bottom.) Then refill the area with the original topsoil or a modification of it to meet the needs of the plants you choose. Next, fill the new wetland with water to settle the soil and add additional soil if necessary. Finally, trim the excess liner, leaving a small lip around the edge that can be tucked under the surrounding soil or mulch.

Plant the mini-wetland garden with species that grow well in wet soil (see Appendix B). These gardens tend to look most natural in partial shade with a background of trees and shrubs.

To expand planting possibilities around an existing pond, you can design the mini-wetland garden next to it. Direct the overflow from the pond into the wetland area to provide a moist environment for plants and wildlife. (Butterflies and other invertebrates will congregate in the moist mud, swallows and robins will use the mud to construct their nests.) If the overflow is not sufficient to keep your mini-wetland moist, you will need to water it periodically to keep it thriving. For additional information on mini-wetlands, see Chapter 10.

Adding Enhancement Features

Adding or creating snags, floating logs, stumps, and brush and rock piles is beneficial to wetlands and their wild inhabitants. Snags may naturally develop over time or can be created around the edge of the wetland and in the buffer.

Logs and stumps can also be installed in saturated soil, or in a buffer area, to provide wildlife cover. Logs that extend from the shore into open water provide basking sites for water birds, reptiles, and amphibians. They also provide travelways in and out of the wetland for animals, and cover for fish. Logs or stumps located in areas where they could cause problems if they float away may need to be anchored.

One of the most pleasurable aspects of a wetland is the opportunity to observe different birds. You can help attract them by installing various types of nest boxes built and located to attract specific wetland birds such as wood ducks, swallows, and wrens. For information on how to choose and build appropriate nest boxes, see Chapter 12.

Since some wetlands contain shallow standing water at some time during the year, mosquitoes can probably breed there. However, these areas may also be nurseries for desirable insects such as dragonflies and damselflies, which eat both larval and adult mosquitoes. Unless the mosquitoes are a health problem, tolerating them is the recommended approach. In lieu of using a pesticide, a more natural control method is to install nest boxes for swallows and bats, both of which consume large quantities of adult mosquitoes.

Maintaining a Wetland

When considering what long-term maintenance is appropriate to a wetland, you need to remember that a wetland is an evolving system. Some open-water wetlands will change as the vegetation proceeds from marsh, to woods, to dry land over many centuries. Wetlands also change with the seasons. You can use these visual changes as a way to appreciate the subtleties of the seasons, and any maintenance you do should take these changes into account.

For the long-term health of the wetland, you will need to protect both the wetland and the buffer area from activities that would impair their ability to function normally. Often, restricting access by grazing animals and humans will be the single most effective maintenance action. Once a disturbance is removed, native plant communities can be

encouraged by keeping aggressive plants in check. Controlling domestic cats and dogs is also important. Cats prey on small birds, reptiles, and amphibians that are concentrated around the edges of wetlands. Dogs can scare off water birds and disturb amphibian egg masses.

No matter how you install new wetland plants, they will require maintenance, especially frequent weeding because aggressive weeds can quickly take over an area. Trees and shrubs may also need basal wrap to prevent damage from moles, voles, and other wildlife (see Figure 1 in Chapter 4). (For more information on wildlife damaging plants, see Chapter 21.) The need for careful maintenance applies even more to wetland plants because they have special needs and, once they are installed, they typically are left to fend for themselves.

For general information on maintenance, see Chapter 4.

Removing Aggressive Plants and Weeds

Both native and non-native plants can overwhelm a wetland, choking out other vegetation, reducing diversity, and destroying its natural balance. Despite its popular status as the unofficial symbol of wetlands, the common native cattail can quickly grow into large, dense stands in which few other plants will grow. If a majority of the wetland is between 1 and 2 feet deep and doesn't already contain cattail, it may be best to prevent it from getting established. A good alternative is hardstem bulrush. This plant occupies similar habitat, is less aggressive, and provides food and cover for many species of wildlife. Hardhack, also called Douglas spirea, is another aggres-

sive native species that can overrun enhancement sites by prolifically seeding onto areas of bare soil and forming dense thickets.

For suggestions on how to remove aggressive plants, see Chapter 4.

Stormwater Control

The urbanization of land has a direct impact on wildlife. Construction of buildings, roads, and parking lots, for example, removes extensive acreage of productive wildlife habitat. These effects are obvious. However, a more subtle, secondary effect is the management of runoff water from the hard, impervious surfaces created by development. These "storm" waters increase in volume with increased development. Past practices were simply designed to remove these waters as quickly as possible by shunting them into the nearest natural drainage way. However, this increased downstream flooding, stream channelization, degraded wildlife habitat, increased erosion and sedimentation, and diminished groundwater levels.

Recently, planners, environmental engineers, and others have recognized the negative impacts of traditional stormwater management. Emphasis has been shifting to capturing precipitation near where it falls. This is generally done by creating detention and retention basins to slow runoff. Detention basins detain runoff for a short time, releasing it slowly, and are often without water between storm events. Retention basins retain a permanent pool.

In the design stage of most basins, the concern for improved water quality, increased wildlife use, and aesthetics, have received little attention. However, studies by the National Institute for Urban Wildlife showed that urban residents overwhelmingly favored design and management of basins for fish and wildlife as well as stormwater management. The potential benefits to urban residents for better management practices involving stormwater seem great.

Design

Research has shown that retention basins, with their permanent water pool, provide the most benefit to both wildlife and improved water quality. A permanent pool assures greater security to a wider variety of wildlife, especially during the breeding season. The pool improves water quality by holding sediments, nutrients, and other chemicals better than retention basins. Retention ponds are, therefore, preferred. However, detention basins can be of great value.

To optimize the value of either retention ponds or detention basins for wildlife, water quality, and aesthetics, the following guidelines are recommended:

1. Side slopes should be as gentle as possible. Often created at 3:1, anything greater is an improvement, with 10:1 preferred. Gentle slopes encourage desirable vegetation as wildlife habitat and water quality enhancement. They are also safer than steep sides.

2. Water depths should be approximately 2 feet for up to 50 percent of surface area. The remainder can be deeper.

3. Larger ponds should be designed with islands. Islands provide secure nesting and undisturbed nesting areas for wildlife.

4. Stormwater ponds should be designed with the capability to regulate water levels up and down, including complete drainage. Drainage every five to ten years may be necessary for maintenance and cleaning and periodic drainage will often improve wildlife values. By occasionally "aerating" wetland bottoms through exposure to air, productivity is increased.

Maintenance

Proper planning and construction with wildlife in mind are an important first half of stormwater management. Following construction, site maintenance is the second important factor influencing wildlife use. In basins originally constructed without wildlife use as an objective, it is probably foremost in importance. To further optimize wildlife benefits in a stormwater management system, the following maintenance guidelines are recommended.

1. Plant woody vegetation around the perimeter of the detention and retention basins. A dense screen of shrubs, for example, will buffer the wetland from the noisy and busy urban environment. Wildlife will, therefore, enjoy a more peaceful habitat. In storm collection basins that are less urban, e.g., within natural habitats, this is probably unnecessary. Simply refraining from regular mowing of shorelines (as is often practiced) will encourage wildlife use.

2. Vegetation management within the wetland basin is also very important. Too much or too little can be a problem. For example, a 50:50 ratio of open water to emergent vegetation is preferred by most ducks. This also provides maximum "edge" between differing habitats, thereby further improving wildlife habitat diversity. As they regularly dry and allow vehicle access, most detention basins are mowed within their entirety. Many shallow retention ponds often become choked with vegetation. Neither condition is desired. Best management practice for detention basins is long-term rotational mowing (e.g., once every three to five years) of 50 percent of the basin bottom. The central 50 percent of the basin should remain unmowed, as it will provide an "island" of secure escape cover.

For retention ponds, design phase is most important in creating open water. If the pond is already constructed, but filling rapidly with emergent vegetation throughout, dredging might be the only solution unless the pond can be drained and mowed. If dredged, the spoils should be used to create a central nesting island.

3. Fencing of the wetlands should be considered, especially in high use areas such as public parks. A low (3-foot) fence will prevent small children and discourage pets and casual intruders from entry. However, older children and adults interested in the wetland can still gain access. The fence will, therefore improve wildlife security. It may also act as a "litter" barrier. Planting clinging, vine-type vegetation to grow onto it will soften its image, further improve sight and sound buffers, and provide increased wildlife habitat.

4. The opportunity for interpretive, educational signs and displays should not be overlooked. Informational signs near wetlands within parks, or along high-use paths, sidewalks, roads, etc., will educate and increase appreciative use.

STREAMS, DITCHES, AND OTHER WATERWAYS

Wildlife seek out streams, creeks and other waterways for their rich food supply, cool water, and dependable shelter. In urban areas and other places where natural space is often fragmented, creeks and ravines provide valuable wildlife corridors, reserves, and temporary refuges.

Not only do these areas support wildlife, but they also benefit our communities directly; they purify water, control floods, and offer shade, pleasant scenery, and wildlife viewing opportunities.

For these reasons it's important to carefully consider how landscaping for wildlife could improve and affect wildlife habitat in and adjacent to waterways.

For information on seeps and springs, see "Water Developments for Birds (and other wildlife) in Arid Areas" in Chapter 6.

Wildlife in and Along Waterways

Waterways are particularly suitable as breeding sites for wildlife because they provide food, cover, and water all in close proximity. The milder climate they create offers plentiful insects and early forage in the spring when birds and other wildlife have young. These areas can also be vital shelter during harsh winters and hot, dry summers.

Over half of the migratory birds in the Pacific Northwest use waterways during migration and breeding. These include the beautiful neotropical species, such as warblers, tanagers, and orioles that need places to rest and feed both in the spring and the fall.

Streams and creeks can also support a variety of large and small mammals. Deer and elk seek shade and rear young there. The mere presence of drinking water makes these areas preferred habitat, especially in hot, arid areas. In the winter, animals seek running water to avoid spending the extra energy needed to eat (and internally melt) snow. Fish and many species of amphibians require waterways for survival.

Enhancement Projects in and Along Waterways

Several enhancements can be made to streams and other waterways. These are often done to improve fish habitat, but they can also boost habitat quality for other species and add to the landscape aesthetics of a site. Examples include placement of logs, rootwads, and boulders in the stream channel and along the banks, adding and/or cleaning spawning gravel, and removing culverts or designing them to allow fish to navigate more easily upstream (see Figure 1, Chapter 8).

People from agencies or private consulting biologists can advise you as to the best activities for improving your stream habitat. Consultants can be found by contacting your local government planning office or your local Fish and Wildlife office. In larger cities and towns, look under "Environmental" in your phone directory. See Appendix E for addresses and telephone numbers.

Because permits may be required, you should begin the process well in advance of your intended start date. First call your city or county planning office for permit information.

Smaller enhancement projects are as rewarding as engineered projects, whether pursued as individual or small-group efforts. There may already be an organization in your area, such as the Adopt-A-Stream Foundation, Stream Team, Tribes, a watershed advisory committee, or a local club that participates in stream cleanup and habitat improvement. These are generally wonderful groups with lots of information on local stream ecology. To learn more, call one of the environmental groups in your area.

Planting streamside plants and removing trash and garden debris from a stream and its banks are examples of how you can enhance/improve wildlife habitat. (Large trash items which are not toxic and are embedded in a stream may have to remain if removal could cause significant erosion, or a release of silt and sediments.)

Evaluating a Stream

A degraded stream may look pretty from a distance, but up close problems become apparent. Often, a lack of plant cover results in few places for fish and other wildlife to live. Because there isn't a protective mass of plants and roots, banks are broken down and open to erosion. In addition, the stream itself lacks diversity; it's flat and shallow and has few areas of pools.

The Value of Large Woody Debris (LWD)

Trees and other forms of large woody debris that fall or are placed into a stream perform many functions.

In additional to forming falls, pools, and riffles that are essential to fish, they:

- **Provide hiding and resting cover for aquatic wildlife.**
- **Change the velocity of the stream, causing gravel, cobble, and sand to settle out in different areas, thus creating spawning and rearing habitat.**
- **Reduce and redirect the current, thus lessening channel cutting and erosion.**
- **Aerate the water.**
- **Facilitate fish passage in streams with rapids by providing stair-steps and resting pools in the channel.**
- **Create pleasant sounds and provide opportunities for wildlife watching.**

To determine if a stream needs help, ask:

- Have the banks been damaged by livestock, heavy equipment, or off-road vehicles? Do they show signs of bare earth crumbling into the stream?
- Was the stream channelized or "cleaned up"? Is the bank rip-rapped?
- Does the stream channel lack fallen logs or other woody debris? Does it lack shade?
- Has the streamside vegetation been logged, cleared, or eaten down by livestock, ducks, or geese?

- Is the water always muddy?
- Is there only one plant species growing on its banks?

The more times you answered "yes" to these questions, the more the stream needs help. If you're unsure of what to do, start by calling the state, federal, or provincial agencies in charge of the waterways in your area. See Appendix E for a list of agencies.

Ditches and Wildlife

An irrigation or drainage ditch can provide wildlife with some of the same benefits as a creek or stream (see Figures 1 and 2). Ditches are

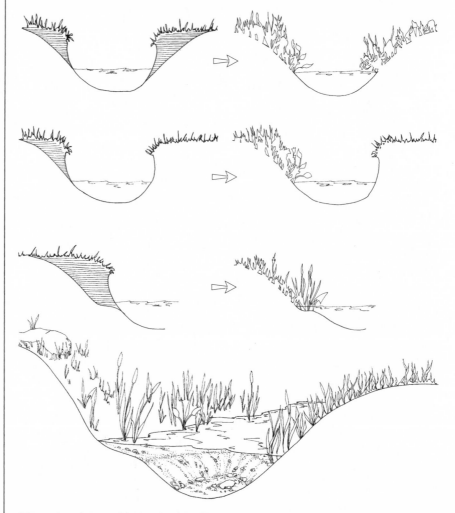

Figure 1. Side view of ditches that have been modified and maintained for wildlife. Note that one side can be terraced to help establish plants.

Figure 2. *An example of how fish habitat can be created over time with plants. A healthy streambank is securely held by a dense mass of plants and roots, and the bank overhangs the stream, providing cover for fish and aquatic animals. There are a variety of trees, shrubs, and ground covers, some of which hang over the stream and provide shade to keep the water cool. The stream itself has variety in its depth and the pattern in which it flows.*

and other wildlife can live in relative security.

Any size ditch that holds water for part of the year should be retained with some natural banks and beds and should not be confined to underground piping. Fencing off a ditch will keep livestock from breaking down the banks, damaging bankside plants, and polluting the water. Varying the depth of a ditch in short stretches will create deeper areas needed by fish, frogs, and other wildlife (Figure 1).

Periodically a ditch may need to be dredged. Dredging is very disruptive to aquatic wildlife and some wildlife species will inevitably be disturbed. It's best to dredge only small sections in any one year (for example, alternate sides of a ditch each year or dredge just one-third of a ditch a year). This leaves some vegetation and undisturbed mud as a refuge from which wildlife can recolonize the area.

If access is needed for regular ditch maintenance, vegetation can be planted on the side that will shade the water. For example, the south side of east-west ditches can be planted and the north side kept open for access. In north-south ditches, the west side can be planted and the east side can be left open for access.

No permit is required for maintenance of totally artificial ditches, but a permit is required when a stream has been channelized into a ditch. The rule of thumb is: if the ditch wiggles any-place upstream, maintenance work requires a permit.

Finally, avoid spreading fertilizers next to ditches. The rich nutrient supply from fertilizer runoff encourages the growth of algae, which depletes the water of oxygen as it decays. Try to leave

often the only available wildlife habitat in urban and agricultural areas. For homeowner and agricultural purposes, the priority is to maintain a ditch's ability to drain; for wildlife the aim is to maintain habitat diversity. With some planning both these priorities can usually be satisfied.

Ditches, particularly ones containing water, create corridors of habitat and can provide water, cover, and food for insects, birds, amphibians, reptiles, and mammals. Birds such as great blue heron and red-winged blackbirds eat insects and other aquatic organisms that live in ditches. Coho salmon and other fish may spend the winter in these ditches hiding in submerged vegetation. If wide enough, ditches can also act as a natural barrier to livestock and the public, behind which birds

Table 1. Erosion control abilities of native streamside plants

Coniferous Trees

Fir, *Abies* spp.	Good
Incense cedar, *Calocedrus decurrens*	Good
Sitka spruce, *Picea sitchensis*	Good
Ponderosa pine, *Pinus ponderosa*	Good
Douglas-fir, *Pseudotsuga menziesii*	Good
Western red-cedar, *Thuja plicata*	Good
Western hemlock, *Tsuga heterophylla*	Good
Pacific yew, *Taxus brevifolia*	Excellent

Broadleaf Evergreen Trees

Bay (Oregon-myrtle), *Umbellularia californica*	Good

Deciduous Trees

Big-leaf maple, *Acer macrophyllum*	Excellent
Vine maple, *Acer circinatum*	Fair
Alder, *Alnus* spp.	Fair
Birch, *Betula* spp.	Good
Pacific dogwood, *Cornus nuttallii*	Fair
Oregon ash, *Fraxinus latifolia*	Good
Bitter cherry, *Prunus emarginata*	Fair
Chokecherry, *Prunus virginiana*	Fair
Quaking aspen, *Populus tremuloides*	Good
Cottonwood, *Populus balsamifera* var. *tricocarpa*	Good
Willow, *Salix* spp.	Excellent

Evergreen Shrubs

Salal, *Gaultheria shallon*	Good
Oregon-grape, *Mahonia* spp.	Good
Pacific rhododendron, *R. macrophyllum*	Good
Evergreen huckleberry, *Vaccinium ovatum*	Good

Deciduous Shrubs

Wild rose, *Rosa* spp.	Good
Oceanspray, *Holodiscus discolor*	Good
Snowberry, *Symphoricarpos albus*	Excellent
Hardhack, *Spiraea douglasii*	Excellent
Thimbleberry, *Rubus parviflorus*	Good
Salmonberry, *Rubus spectablis*	Fair
Serviceberry, *Amelanchier alnifolia*	Excellent
Red-twig dogwood, *Cornus stolonifera*	Excellent
Hazelnut, *Corylus cornuta* var. *california*	Good
Twinberry, *Lonicera involucrata*	Good
Osoberry, *Oemleria cerasiformis*	Good
Western ninebark, *Physocarpus capitatus*	Good
Red-flowering currant, *Ribes sanguineum*	Good
Elderberry, *Sambucus* spp.	Fair

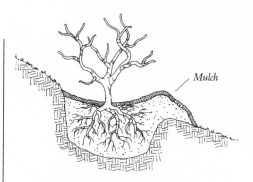

Figure 3. *One example of how to plant on a slope.*

as wide a buffer zone (area of untreated vegetation) as possible along the ditch to prevent the possibility of contaminating the water.

Planting Areas Along Waterways

Vegetation along waterways directly and profoundly influences the life within the channel of water. Plants that grow along the water's edge or hang over the water are particularly essential because they shade out summer sunlight, keeping water temperatures cool for heat-sensitive fish, amphibians, and aquatic insects. The cool water also allows the stream to hold more oxygen, which is essential for healthy fish populations during hot summer days. During the winter, overhanging trees add to the insulating barrier, protecting the water and its aquatic life from extreme cold.

Much of the food for in-water wildlife species also comes from the plants growing next to the water. Leaves, needles, flower petals, fruit, twigs, tree insects, and droppings fall into the water, and eventually sink to the bottom. This decaying material creates a nutrient-rich blend called "detritus." Detritus provides nutrients for bacteria and phytoplankton that in turn serve as a food source for invertebrates and eventually fish, frogs, amphibians, and birds.

Before people knew better, they tidied up their property by removing plants and plant debris that "cluttered" streambanks. Now that people recognize how vital habitat next to streams and creeks is, they're managing these areas more carefully and replanting disturbed sites. Since streamside plants often produce rapid growth, water quality and wildlife habitat can often be improved in a couple of years.

Like all efforts to improve wildlife habitat, a stream- or creekside planting project must be well-planned, carefully planted, and faithfully maintained. When designing a planting project, site evaluation is the most important step to ensure success. In addition to soil and light conditions, it is important to consider what kinds of vegetation already exist on the site. This can provide clues to what other native plants to plant; more importantly, if the area is infested with aggressive non-native plants such as reed-canary grass, blackberries, or English ivy, establishing new plants will be difficult or impossible.

It is possible to establish a diversity of plants in an area dominated by one or more of these aggressive species, but only after the competing plants are eliminated. Any regrowth from the major competitors should be removed immediately. Weed mats installed around individual plants at planting time may help new plants get established in areas where competition from weeds is high. For information on how to manage certain weed species, see Chapter 4. For related information see "Removing Unwanted

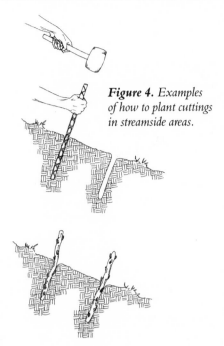

Figure 4. *Examples of how to plant cuttings in streamside areas.*

Vegetation in a Wildlife-Friendly Way" in Chapter 3.

Pay careful attention to any areas where there is the potential for erosion as a result of plant removal or ground preparation. Areas can be seeded with sterile wheat, annual rye, or a custom seed mix (see Appendix A "Fields, Meadows, and Grassland Landscapes"). They can also be planted with fast-growing trees and shrubs such as Douglas-fir, western hemlock, willow, red-twig dogwood, and others that are known for their erosion control abilities (see Table 1). Areas should be carefully mulched with a material that won't blow or be washed away.

Planting Design

Any waterway planting project that involves installing a community of plants will benefit from a planting plan. A planting plan can be a simple sketch that takes 15 minutes, or a detailed drawing created by a professional. For information on planting plans, see Chapters 2 and 3.

Because streams periodically flood or change course, it's impor-

Establishing Willows from Cuttings

Willows are fast-growing trees and shrubs that naturally occur along waterways. They are easily planted as hardwood cuttings and can be installed during the dormant season as soon as they lose their leaves and the ground is moist enough to insert them. To establish a single plant or a stand of willows from cuttings, use the following technique:

1. Select dormant cuttings from willows growing as near the site to be planted as possible. Willow species which are adapted to streamside areas root very readily from cuttings, whereas those species adapted to upland areas are less likely to root by these methods. Select only healthy stems (no cankers or insect infestations, etc.).

2. Prepare the willow cuttings by removing stems (1/2 to 7/8 inch in diameter) with pruning shears or loppers. Trim off all side growth and cut the stems into 18 to 30-inch lengths. To later identify the top of the cutting from the bottom, always make cuttings with the bottom cut slanted and the top cut straight across.

3. Press cuttings directly into the soil, or insert into guide holes made with a metal spike. (Rebar hammered into hard soil can be difficult to pull out, so be sure to use a long piece and leave enough sticking out to keep a good grip. A smooth metal rod constructed from a 5/8-inch steel reinforcement bar is easier to remove.) Plant so that the lower 12 to 24 inches of the cutting will be in close contact with

continued on next page

the groundwater table during the driest part of the growing season. If the bank has been previously rip-rapped, plant cuttings between the rocks in soil. Approximately four to six buds should remain exposed above the ground surface. Following insertion, the soil should be tamped around the cutting (see Figure 4).

Willow cuttings are most successful when planted vertically. However, planting further up a bank at an angle of up to 45 degrees is acceptable, as long as the lower rooted stem portion remains moist throughout the summer (see Figure 6).

For planting densities, a good rule of thumb is to place cuttings 2–4 feet apart. It's often wise to plant additional cuttings, since some may not survive.

New growth from willow cuttings may be eaten by beaver, deer, and other browsers. For the most part this causes no long-term ill effects. Although it's dependent on the willow species selected and the site conditions, first-year growth can be 2–4 feet.

continued on next page

tant to plan for trees and shrubs to be planted according to the existing and possible future characteristics of the creek or stream. Planting in relatively stable areas will increase the likelihood of long-term survival of the new plants.

Establishing a variety of plants is desirable for wildlife use, for effectiveness in anchoring the soil, and for esthetics. For a naturalistic landscape, plant an assortment of native plants that naturally grow together. You can get ideas by studying a streamside plant community growing near your property. However, understand that this example probably can't be described with certainty as totally natural. It's probably been modified by human activity in some way: by logging, grazing animals, or the introduction of aggressive non-native plants.

For information on how to find out what native plants grow in your particular area, see "Wetlands and Wetland Gar-

dens" in Appendix A. Appendix B lists native Pacific Northwest plant species that grow in streamside areas. For additional information on native plants, see Chapter 3.

To stabilize erosion at the water's edge, plant species that establish quickly, have a tenacious root system, and are flood-tolerant. Small-stemmed, flexible shrubs such as red-twig dogwood, willow, and Sitka alder are all adapted to such situations. Salmonberry, thimbleberry, and wild rose are other good choices for areas where they can spread freely.

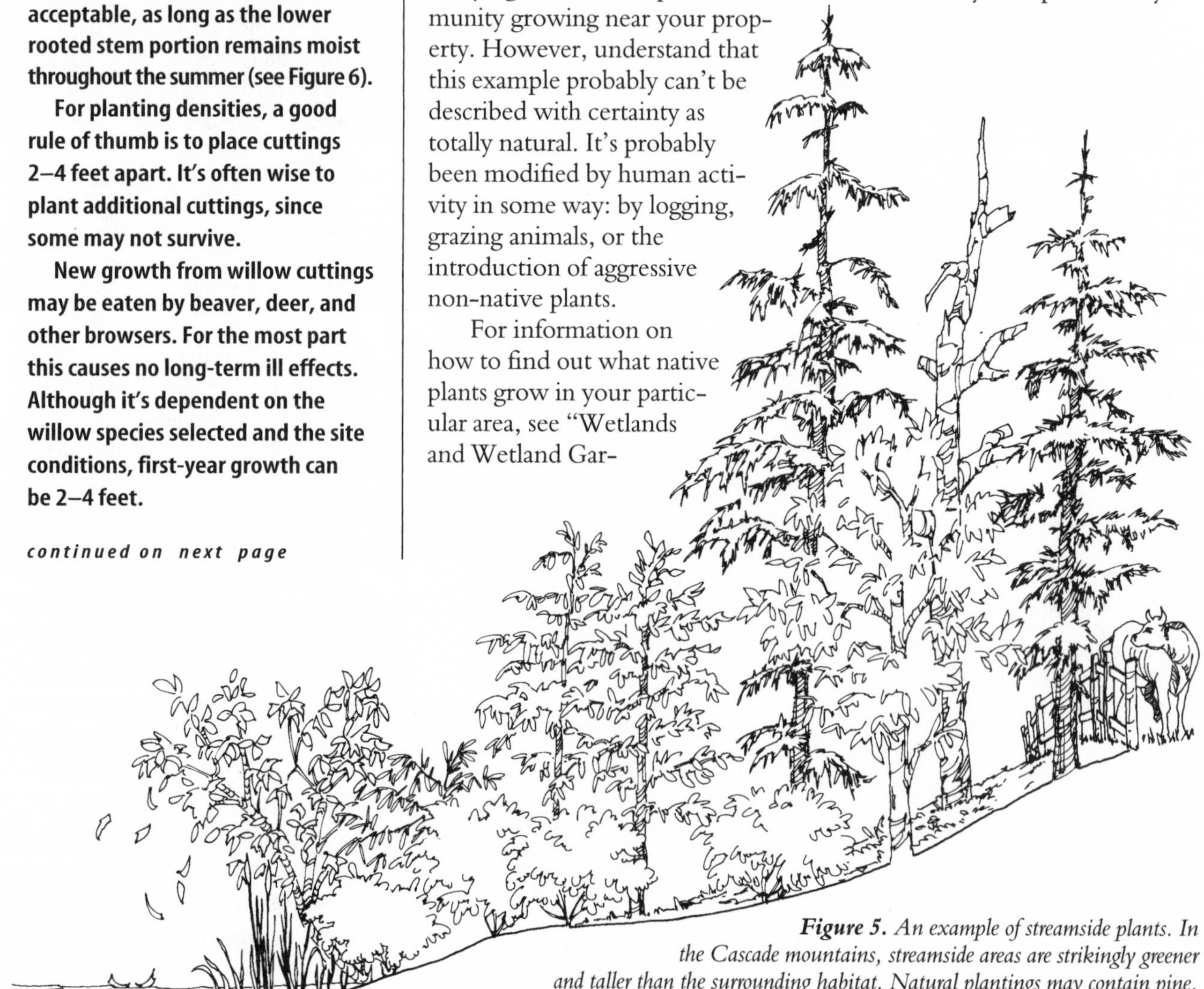

Figure 5. An example of streamside plants. In the Cascade mountains, streamside areas are strikingly greener and taller than the surrounding habitat. Natural plantings may contain pine, cottonwood, and aspen trees with willows, shrubs, grasses, and sedges underneath. West of the Cascade mountains, streamside areas are usually younger and shorter than the surrounding vegetation. Natural plantings may contain Oregon ash, red alder, red-cedar, spruce trees, salmonberry, and sword fern.

The plants should be arranged along the stream so that each is in its preferred condition. For instance, plants which need lots of water should be planted near the water line. Plants which require full sunlight should not be planted next to plants that will quickly shade them out. Shade-tolerant shrubs and ferns can be interplanted among the trees to increase diversity.

The direction each streambank is facing will influence the conditions for the plants; south-facing banks are typically more dry and sunny than north-facing banks. If you have limited time and funding, plant first on the south and west banks as these are the most important for moderating the water temperature.

To keep part of the stream in view, you can plant trees and shrubs in clumps and include ground covers in-between. For plant spacing, see Chapter 3.

Developing a Buffer

Buffers are areas of natural or planted vegetation that parallel a streamside area (Figure 5). Generally, you will need at least 50 feet of buffer between any developed areas and the water. However, wildlife species that are sensitive to human disturbance, such as the American bittern and greenbacked heron, may need up to ten times that width. For fish, buffer widths also vary depending on the size and geography of the stream—major rivers and streams with large floodplains may need more than 300 feet; smaller year-round flowing streams may need a 100-foot buffer, and seasonal creeks about 50 feet. Your local fish and wildlife agency can provide you with recommendations.

Any buffer is better than nothing. However, you may not be rewarded with fish and other wildlife with less than the minimum buffer area. In areas where buffer widths are narrow, try to create height and diversity by planting trees with an understory of evergreen and deciduous shrubs.

For related information, see "Pole Plantings" in "Wetlands and Wetland Gardens."

Maintaining and Protecting Existing Streamside Habitat

To care for new plants, the all-important thing to do is to weed frequently, keeping ahead of weeds before they get out of control. Watering is also extremely important, especially in hot areas, in areas with weed competition, and if plants are installed in late spring or in the summer. New plants may need protection from mice, voles, mountain beaver, beaver, and deer. For information on protecting plants from wildlife damage, see Chapter 21.

Protecting streamsides from undesirable change is always easier than trying to fix a damaged area. Human-related disturbances that need careful observation include logging, road-building, farming, grazing, dumping, and landscaping activities such as removing buffer plants to plant lawn next to the stream. Removing any plants from a streamside area makes the area prone to erosion until regrowth occurs. Furthermore, the stream may collect silt, and if the shade-producing canopy is removed, water temperatures may increase, stressing or even killing fish.

For additional sources of information on waterways, see Appendix E.

Note: Willows can also be constructed into living sculptures. Long whips can be inserted in the ground to form domes, tunnels, or any shape desired. These will root and bud out, making a living, growing structure that can be further woven as it grows. This is an ideal material for kids to work with, requiring little skill and lots of imagination. These willow structures, apart from play, can act as a springboard for cross-curricular activities incorporating art, craft, and ecology. Later these structures contribute to wildlife habitat, as willows are used by a myriad of wildlife.

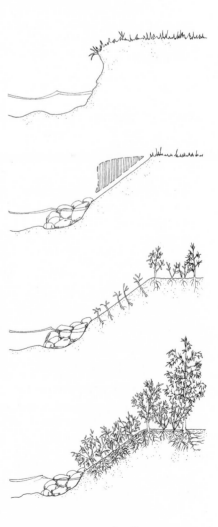

Figure 6. *One example of how cuttings can be planted on a newly cut slope.*

Table 2. Streamside care guidelines

The Situation	What You Can Do	What to Avoid
Streamside vegetation provides food and shelter for wildlife. Snags are important for birds and mammals. Root systems stabilize streambanks, guarding against erosion.	Leave vegetation along sides of streamside areas. Plant a diversity of vegetation.	Do not "park out" the streamside or remove overhanging trees or shrubs from the banks. Trees and shrubs shade the streamside area and provide leaf litter which forms the base of the aquatic food web (leaves–insects–fish–humans). Plant native plants.
Streamside trees die and fall into the stream, become embedded and form pools that are important to fish.	Leave the streambanks and channel in their natural conditions as much as possible.	Do not remove embedded logs from the stream or streambank.
Heavy equipment in the stream can ruin spawning gravel, destroy fish habitat, damage streambanks, and pollute.	Obtain a permit from the proper authority before beginning any work below the high-water mark.	Do not place rip-rap or fill on stream banks without a permit. Do not dig, dredge, or reconstruct the stream channel without a permit. Do not cross the stream with motorbikes or other vehicles.
Fertilizers promote algae and weed growth in streams and lakes. Pesticides and herbicides can be toxic to people and fish.	Use garden and farm chemicals sparingly and according to the label. Learn to garden and farm using organic methods.	Do not spray streamside vegetation or dispose of chemical-laden garden refuse near water.
Construction sites are a major source of sediment, which can ruin spawning gravel and kill fish.	Take precautions to avoid excessive runoff when clearing land. Follow erosion-control guidelines and regulations. Reestablish vegetation as soon as possible. Clear as little as possible.	Do not disturb soils during the rainy season. Do not allow sediment-laden runoff to enter a stream. The responsible party can be subject to expensive penalties and the cost of replacing damaged resources.
Domestic animal wastes cause serious water quality problems. Bank trampling causes sedimentation and destroys native vegetation.	Restrict livestock use in the streamside area. Provide watering away from streams. Keep pets away from streams.	Do not allow banks to be trampled or denuded.
Litter and junk in the stream can degrade water quality and endanger fish, wildlife and recreationists. All litter is an eyesore.	Remove litter (lawn clippings, trash) and junk from the stream and vicinity. Start a stream clean-up club.	Do not tolerate litterbugs.
Increasing urbanization requires innovative regulations and water-quality control programs. Our cities can grow without altering our streams.	Support legislation that benefits water quality (wastewater management, erosion control, organic farming). Work with local planners to preserve our streams.	Do not be apathetic or uninformed on water-quality issues.

WILDLIFE PLANT LISTS, TABLES, AND MAPS

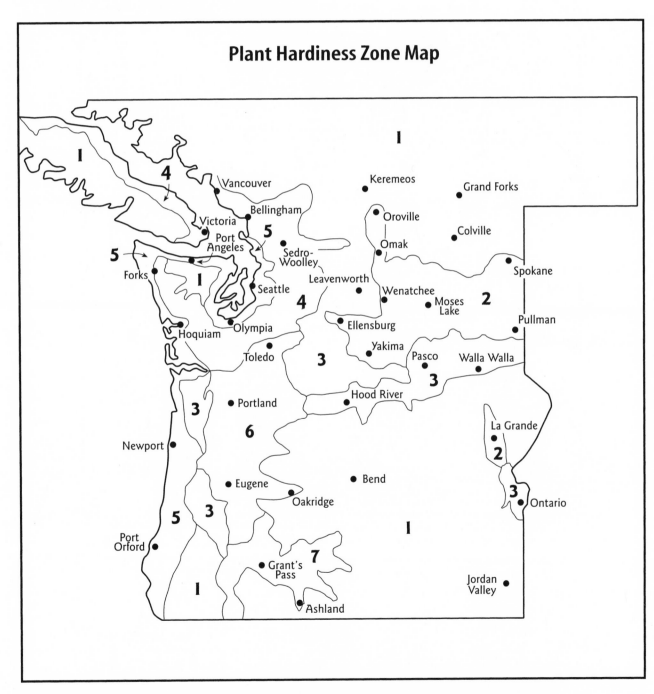

Plant Hardiness Zone Map

Adapted from *Sunset Western Garden Book,* 6th ed. Sunset Publishing, Menlo Park, CA, 1995.

Native Plant Occurrence Map

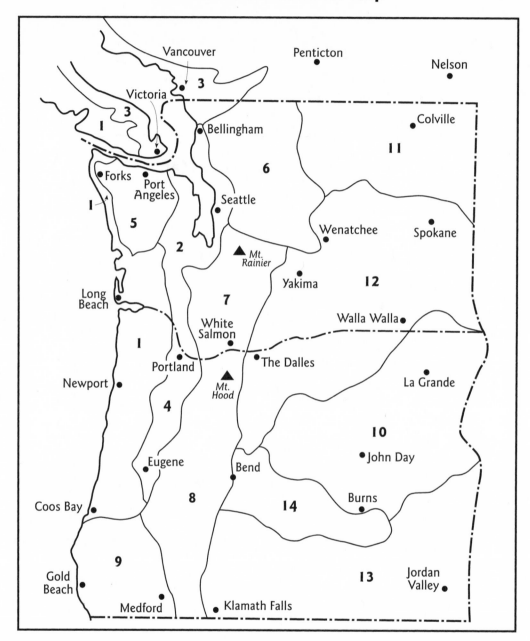

Adapted from *Natural Vegetation of Oregon and Washington,* Oregon State University Press, 1988.

1 Coast Range
2 Puget Trough
3 Georgia Depression
4 Western Oregon Interior Valley
5 Olympic Mountains
6 Northern Cascades
7 Southern Washington Cascades

8 Oregon Cascades
9 Siskiyou Mountains
10 Ochoco, Blue, and Wallowa Mountains
11 Okanogan Highlands
12 Columbia Basin
13 Basin and Range
14 High Lava Plains

NATIVE PLANT OCCURRENCE CHART

The following chart is designed to help you get a *general* idea of what native plants occur (or occurred) naturally in a region, and in what general setting. To locate regions, see the "Native Plant Occurrence Map," page 214. For a list of plants that occur in a specific area within one of the regions, such as your neighborhood, contact your local chapter of the Native Plant Society. For woodland, waterway, grassland, and wetland definitions, see Appendix A. For landscape and wildlife information, see Appendix C.

✓ Common name, *Botanical name*	Region/Notes	Woodlands	Waterways (Riparian)	Grasslands/Shrub-steppe	Shrub/Forested Wetlands	Fresh-water Marshes	Wet Meadows
Coniferous Trees							
Pacific silver fir, *Abies amabilis*	1, 5, 6–8 (west side)	•					
White fir, *Abies concolor*	8, 9, 13, 14 (most common in southwest and south central Oregon)	•					
Grand fir, *Abies grandis*	1–10, 11, 13, 14 (sea level to 6,000 ft.)	•	•				
Noble fir, *Abies procera*	1, 7–9 (mostly west side)	•	•				
Incense-cedar, *Calocedrus decurrens*	8–9, 13, 14 (southwest Oregon and east side of mountains)	•	•				
Lawson cypress (Port Orford cedar), *Chamaecyparis lawsoniana*	9, 1 (southern Oregon only)	•	•				
Western juniper, *Juniperus occidentalis*	8 (east side), 10, 14	•		•			
Rocky Mountain juniper, *J. scopulorum*	2, 10	•		•			
Brewer's spruce, *Picea breweriana*	9, usually in groves at 4,000 to 8,000 ft.	•					
Engelmann spruce, *Picea engelmannii*	5 (in rain shadow), 6–8 (seldom on west side), 10, 11, 13 (high elevations)	•	•				
Sitka spruce, *Picea sitchensis*	1–3, 5, 6–8 (west side), 9	•	•		•		
Shore pine, *Pinus contorta* var. *contorta*	1–5, 9, coastal areas and lower elevations	•					
Lodgepole pine, *Pinus contorta* var. *latifolia*	All	•	•	•	•		
Western white pine, *Pinus monticola*	All but 4 and 12	•					
Ponderosa pine, *Pinus ponderosa*	4, 6–8 (mostly east side) 9–14	•	•	•			
Douglas-fir, *Pseudotsuga menziesii*	1–12	•	•	•			
Western red-cedar, *Thuja plicata*	1–3, 5–9, 11	•	•				
Western hemlock, *Tsuga heterophylla*	1–3, 5–9, 11, less common on east side.	•	•		•		
Mountain hemlock, *Tsuga mertensiana*	1 (BC only), 5–10 (high elevations)	•					

Deciduous Trees

✓	Common name, *Botanical name*	Regions/Notes	Woodlands	Waterways (Riparian)	Grasslands/Shrub-steppe	Shrub/Forested Wetlands	Fresh-water Marshes	Wet Meadows
	Vine maple, *Acer circinatum*	1-11 (not on Whidbey Island)	•	•		•		
	Douglas maple, *Acer glabrum* var. *douglasi*	All (not as common on west side as the vine maple)	•					
	Big-leaf maple, *Acer macrophyllum*	1-9	•	•				
	Mountain alder, *Alnus incana*	8, 10-14	•	•		•		
	White alder, *Alnus rhombifolia*	4, 8-14	•	•		•		
	Red alder, *Alnus rubra*	1-9	•	•		•		
	Sitka alder, *Alnus viridis* var. *sinuata*	5-8, 10, 11 (above 2,500 feet)	•	•				
	Water birch, *Betula occidentalis*	6-8 (east side), 10-14		•		•		
	Paper birch, *Betula papyrifera*	2, 3, 6, 10-12	•	•		•		
	Pacific dogwood, *Cornus nuttallii*	1-9	•	•				
	Columbia hawthorn, *Crataegus columbiana*	6, 10-13		•				
	Black hawthorn, *Crataegus douglasii*	All	•	•	•	•		•
	Oregon ash, *Fraxinus latifolia*	1-9	•	•		•		•
	Pacific crabapple, *Malus fusca*	1-9 (also called *Pyrus fusca*)	•	•		•		
	Quaking aspen, *Populus tremuloides*	All (many disjunct stands)	•	•		•		
	Black cottonwood, *Populus balsamifera* var. *trichocarpa*	All (also called *Populus trichocarpa*)	•	•		•		
	Bitter cherry, *Prunus emarginata* var. *mollis*	1-9 (the shrubby east side form of bitter cherry is *Prunus emarginata* var. *emarginata*)	•	•		•		
	Oregon white oak, *Quercus garryana*	1, 2, 4, 5, 7-9, 12 (also called Garry oak)	•		•			
	California black oak, *Quercus kelloggii*	1, 4, 8, 9 (southwest Oregon only)	•		•			
	Cascara (buckthorn), *Rhamnus purshiana*	1-12	•	•		•		
	Sumac, *Rhus glabra*	6 (east side of mountains), 10-14		•	•			
	Hooker willow, *Salix hookeriana*	1-6, 9 (includes *Salix piperi*)		•		•		
	Pacific willow, *Salix lucida* var. *lasiandra*	1-5, 7-9, 11		•		•		•
	Scouler willow, *Salix scouleriana*	All	•	•		•		

✓	Common name, *Botanical name*	Regions/Notes	Woodlands	Waterways (Riparian)	Grasslands/Shrub-steppe	Shrub/Forested Wetlands	Fresh-water Marshes	Wet Meadows
Broadleaf Evergreen Trees								
	Madrona, *Arbutus menziesii*	1–9	•		•			
	Chinquapin, *Chrysolepis chrysophylla*	1, 4, 8, 9	•	•				
	Tan-oak, *Lithocarpus densiflorus*	1, 4, 8, 9	•					
	Canyon live oak, *Quercus chrysolepis*	1, 4, 8, 9	•	•				
	Bay (Oregon-myrtle), *Umbellularia californica*	1, 4, 8, 9	•	•				
Deciduous Shrubs								
	Serviceberry, *Amelanchier alnifolia*	All	•	•	•	•		•
	Bog birch, *Betula glandulosa*	1–3, 8, 10, 13		•		•	•	•
	Deer brush, *Ceanothus integerrimus*	7–8 (east side), 10, 11, 13			•			
	Red-twig dogwood, *Cornus sericea*	All (also called *Cornus stolonifera*)	•	•		•	•	
	Hazelnut, *Corylus cornuta* var. *californica*	1–4, 6–9 (lowlands to mountains)	•	•				
	Winterfat, *Eurotia lanata*	10–13 (alkaline flats)			•			
	Oceanspray, *Holodiscus discolor*	1–13			•			
	Oceanspray, *Holodiscus dumosa*	8, 10, 12–14	•	•	•			
	Twinberry, *Lonicera involucrata*	1–8, 10–12	•	•		•		
	Osoberry, *Oemleria cerasiformis*	1–9 (also called Indian plum)	•	•				
	Devil's club, *Oplopanax horridum*	1–3, 5–9				•		
	Potentilla, *Potentilla fruticosa*	5–8 (higher elevations), 10, 13		•		•		
	Mock-orange, *Philadelphus lewisii*	1–5, 8–14	•	•				
	Ninebark, *Physocarpus capitatus*	1–13	•	•		•		
	Bitter cherry, *Prunus emarginata* var. *emarginata*	6–8 (east side), 10–14 (the west side tree form is *Prunus emarginata* var. *mollis*)	•	•				
	Chokecherry, *Prunus virginiana*	All (more common on east side)	•	•	•			
	Bitterbrush, *Purshia tridentata*	6–8 (east side), 10–14		•	•			
	Wild azalea, *Rhododendron occidentalis*	7–9	•	•				
	Skunkbush, *Rhus trilobata*	1, 3		•	•			
	Golden current, *Ribes aureum*	6–8 (east side), 10–14		•	•			
	Black gooseberry, *Ribes divaricatum*	1–5, 7, 8, 12	•	•		•		
	Swamp gooseberry, *Ribes lacustre*	1–9, 10, 13	•	•		•		
	Red-flowering currant, *Ribes sanguineum*	1–8	•	•				

✓	Common name, *Botanical name*	Regions/Notes	Woodlands	Waterways (Riparian)	Grasslands/Shrub-steppe	Shrub/Forested Wetlands	Fresh-water Marshes	Wet Meadows
	Deciduous Shrubs, cont.							
	Bald-hip rose, *Rosa gymnocarpa*	All	•	•				
	Nootka rose, *Rosa nutkana*	All	•	•	•	•		•
	Pea rose, *Rosa pisocarpa*	1-9 (coastal and west Cascades)	•	•				
	Wood's rose, *Rosa woodsii*	6-8 (east side), 10-14	•	•	•			
	Thimbleberry, *Rubus parviflorus*	1-3, 5-10, 13	•	•				
	Salmonberry, *Rubus spectabilis*	1-3, 4-9	•	•		•		
	Geyer willow, *Salix geyeriana*	1-9 (*Salix geyeriana* var. *geyeriana* east of mts.)		•		•	•	•
	Sitka willow, *Salix sitchensis*	1-12		•		•		
	Willow, *Salix* spp.	See "Deciduous Trees"						
	Blue elderberry, *Sambucus caerulea*	4, 6, 7, 10-14 (also called *S. mexicana*)	•	•		•		
	Red elderberry, *Sambucus racemosa*	4-7, 10-14 (also called *S. callicarpa*)	•	•		•		
	Black elderberry, *Sambucus racemosa* var. *melanocarpa*	6, 8 (east side), 10 (also called *S. melanocarpa*)	•	•		•		
	Buffaloberry, *Shepherdia canadensis*	2, 3, 10, 12			•			
	Cascade mountain-ash, *Sorbus scopulina*	5, 6-8 (mostly east side), 10, 11	•	•				
	Sitka mountain-ash, *Sorbus sitchensis*	1-8 (mostly mid to alpine elevations)	•	•				
	Hardhack, *Spiraea douglasii*	1-7, 8, 10-13		•		•	•	
	Birchleaf spirea, *Spiraea betulifolia* var. *lucida*	6-8, 10, 13, 14	•	•				
	Hybrid spirea, *Spiraea* x *pyramidata*	1-7, 8, 10-13		•				
	Snowberry, *Symphoricarpos albus*	1-4, 6-13	•	•		•		
	Alaska blueberry, *Vaccinium alaskaense*	1, 5-9	•					
	Black huckleberry, *Vaccinium membranaceum*	1-3, 5-9	•					
	Red huckleberry, *Vaccinium parvifolium*	1-5, 7, 8	•	•				
	Highbush-cranberry, *Viburnum edule*	1-3, 5, 7, 8	•	•				

✓	Common name, *Botanical name*	Regions/Notes	Woodlands	Waterways (Riparian)	Grasslands/Shrub-steppe	Shrub/Forested Wetlands	Fresh-water Marshes	Wet Meadows
	Evergreen Shrubs							
	Hairy manzanita, *Arctostaphylos columbiana*	1, 2, 8, 9	•		•			
	Green manzanita, *Arctostaphylos patula*	7-10, 13, 14	•					
	Sagebrush, *Artemisia tridentata*	8 (east side of mountains), 10-14			•			
	Hopsage, *Atriplex spinosa*	10, 12-14 (alkaline flats)			•			
	Snowbush, *Ceanothus velutinus* var. *laevigatus*	1, 2, 7-8 (west side of mountains), 9	•	•	•			
	Mountain balm, *C. velutinus* var. *velutinus*	6-8 (east side), 10-14	•		•			
	Rabbitbrush, *Chrysothamnus* spp.	8 (east side), 10-14			•			
	Silk-tasselbush, *Garrya elliptica*	1 (coastal areas)	•	•				
	Fremont's silk-tasselbush, *Garrya fremontii*	8, 9	•					
	Salal, *Gaultheria shallon*	1-9	•		•	•	•	
	Juniper, *Juniperus* spp.	See "Conifers"						
	Labrador tea, *Ledum groenlandicum*	1-3, 5, 6 (a bog plant in these areas)				•	•	
	Tall Oregon-grape, *Mahonia aquifolium*	1, 2, 4-8	•	•				
	Creeping Oregon-grape, *Mahonia* spp.	See "Ground Covers and Low Shrubs"						
	Wax-myrtle, *Myrica californica*	1, 5, 9 (a coastal plant)		•	•			
	Mountain-boxwood, *Pachistima myrsinites*	2, 6-11 (mostly found in forest understory)	•					
	Coffeeberry, *Rhamnus californica*	9 (forest and chaparral)	•	•				
	Pacific rhododendron, *Rhododendron macrophyllum*	1, 2, 5, 7-9	•	•				
	Evergreen huckleberry, *Vaccinium ovatum*	1, 2, 6-8 (west side)	•	•				
	Ground Covers and Low Shrubs							
	Kinnikinnik, *Arctostaphylos uva-ursi*	1, 2, 6-8, 10	•		•			
	Bunchberry, *Cornus unalaschkensis*	1, 2, 4-8 (also called *Cornus canadensis*)	•	•				
	Coast strawberry, *Fragaria chiloensis*	1, 3, 5 (coastal)	•	•	•			
	Woodland strawberry, *Fragaria vesca*	2, 7, 8, 10, 13	•	•				
	Mountain strawberry, *Fragaria virginiana*	2, 4, 6, 8, 10, 11, 13	•	•	•			
	Wintergreen, *Gaultheria ovatifolia*	1, 5-8 (middle to high elevations)	•	•				•
	Salal, *Gaultheria shallon*	See "Evergreen Shrubs"						
	Common juniper, *Juniperus communis*	1, 5-8 (subalpine to alpine), 10, 14			•			
	Twinflower, *Linnaea borealis*	2, 6-10	•	•				

✔	Common name, *Botanical name*	Regions/Notes	Woodlands	Waterways (Riparian)	Grasslands/Shrub-steppe	Shrub/Forested Wetlands	Fresh-water Marshes	Wet Meadows
	Ground Covers and Low Shrubs, cont.							
	Cascade Oregon-grape, *Mahonia nervosa*	1, 2, 4, 5, 6-8 (west side), 9	•	•				
	Creeping Oregon-grape, *Mahonia repens*	8 (east side), 9, 10, 12-14	•	•	•			
	Oxalis (Wood sorrel), *Oxalis oregana*	1, 2, 4, 5, 7-9	•	•		•		
	Trailing raspberry, *Rubus pedatus*	2, 5-8	•	•				
	Trailing blackberry, *Rubus ursinus*	1-10	•	•	•	•		
	Vines							
	Wild clematis, *Clematis ligusticifolia*	2, 6-8, 10, 12-14	•	•	•			
	Trumpet honeysuckle, *Lonicera ciliosa*	1, 2, 4, 6-9	•					
	Hairy honeysuckle, *Lonicera hispidula*	1, 2, 4, 6-9	•					
	Wild grape, *Vitus californica*	9	•	•				
	Perennials and Wildflowers							
	Yarrow, *Achillea millefolium*	All	•		•			
	American water-plantain, *Alisma plantago-aquatica*	All					•	•
	Pearly everlasting, *Anaphalis margaritacea*	All	•		•			
	Angelica, *Angelica lucida*	2		•			•	
	Columbine, *Aquilegia formosa*	2-8, 10-14	•	•				
	Seathrift, *Armeria maritima*	2, coastal			•			
	Goat's beard, *Aruncus sylvester*	1, 2, 4-8	•	•				
	Pacific aster, *Aster chilensis*	1-9 (mostly in coastal areas)		•	•			
	Douglas' aster, *Aster subspicatus*	1-9		•	•			•
	Deltoid balsamroot, *Balsamorhiza deltoidea*	2-3 (open prairie), 6 (east side), 11, 12			•			
	Arrowleaf balsamroot, *B. sagittata*	6-8 (east side), 10-14	•		•			
	Campanula, *Campanula rotundifolia*	1, 2, 5-8	•	•	•			
	Scarlet paintbrush, *Castilleja hispida*	1, 2, 5-14			•			
	Blue-eyed Mary, *Collinsia parviflora*	All			•			
	Varied-leaf collomia, *Collomia heterophylla*	2, 6, 7		•	•			
	Menzies' larkspur, *Delphinium menziesii*	4-8			•			
	Nutall's larkspur, *Delphinium nuttallii*	2, 6, 7			•			

✓ Common name, *Botanical name*	Regions/Notes	Woodlands	Waterways (Riparian)	Grasslands/Shrub-steppe	Shrub/Forested Wetlands	Fresh-water Marshes	Wet Meadows
Perennials and Wildflowers, cont.							
Bleeding heart, *Dicentra formosa*	1-9 (low to mid elevations, west side)	•	•				
Shooting-star, *Dodecatheon pulchellum*	1-3, 5-8, 10, 12-14		•	•			•
Fireweed, *Epilobium angustifolium*	All		•	•			
Showy fleabane, *Erigeron speciosus*	1-8, 10, 13			•			
Sulfur flower, *Eriogonum umbellatum*	6-8, 10-14			•			
Woolly-sunflower, *Eriophyllum lanatum*	1, 5-8, 10, 12-14			•			
California poppy, *Eschscholzia californica*	1 (Oregon south), 4, 8			•			
Blanket-flower, *Gaillardia aristata*	6-7 (east side), 10, 12			•			
Globe gilia, *Gilia capitata*	1, 2, 4, 6-8			•			
Coast gumplant, *Grindelia integrifolia*	1-4			•			
Cow-parsnip, *Heracleum lanatum*	1-8, 10-14		•		•	•	•
Alumroot, *Heuchera micrantha*	1-8 (low to subalpine elevations)	•	•				
Fern-leaved lomatium, *Lomatium dissectum*	1-8, 10, 12-14	•	•	•			
Barestem lomatium, *Lomatium nudicaule*	1-4, 5-8, 10-14			•			
Two-color lupine, *Lupinus bicolor*	1-4, 7, 8, 12						
Prairie lupine, *Lupinus lepidus*	7, 8, 10-14	•		•			
Beach lupine, *Lupinus littoralis*	1-3, 5 (coast only)						
Bigleaf lupine, *Lupinus polyphyllus*	2, 4, 6, 8, 10, 14	•	•		•		
Skunk cabbage, *Lysichiton americanus*	1-8 (mostly west of Cascade crest)	•	•		•		
Yellow monkey-flower, *Mimulus guttatus*	All		•		•	•	•
Cardwell's penstemon, *Penstemon cardwellii*	1, 2, 4-8 (mid-elevation)	•					
Davidson's penstemon, *Penstemon davidsonii*	1, 3, 5-8 (mid to high elevation)	•					
Small-flowered penstemon, *Penstemon procerus*	5, 6-8 (east side, mid to high elevation), 12	•		•			
Phlox, *Phlox diffusa*	1, 5-9			•			
Graceful cinquefoil, *Potentilla gracilis*	1-8, 10-14		•	•			
Self-heal, *Prunella vulgaris*	1-8, 10, 13, 14		•	•			
Wapato, *Sagittaria latifolia*	1-8 (low elevations)		•		•	•	•
Sedum, *Sedum spathulifolium*	1-9 (low to mid elevations)			•			

✓ Common name, *Botanical name*	Regions/Notes	Woodlands	Waterways (Riparian)	Grasslands/Shrub-steppe	Shrub/Forested Wetlands	Fresh-water Marshes	Wet Meadows
Perennials and Wildflowers, cont.							
Blue-eyed-grass, *Sisyrinchium angustifolium*	1-8, 10, 13		•	•			
Douglas' blue-eyed-grass, *S. douglasii*	1-8, 10, 12, 14			•			
Large Solomon's seal, *Smilacina racemosa*	All	•	•		•		
Star-flowered Solomon's seal, *S. stellata*	All	•	•				
Canada goldenrod, *Solidago canadensis*	All			•			
Hedge-nettle, *Stachys cooleyae*	1-9	•	•				
Piggy-back plant, *Tolmiea menziesii*	1, 2, 4-8	•	•				
Evergreen violet, *Viola sempervirens*	4, 6-8 (low to mid elevations)	•					
Grasses and Grass-like Plants							
Thickspike wheatgrass, *Agropyron dasystachyum*	8, 10, 14			•			
Bluebunch wheatgrass, *Agropyron spicatum*	8, 10-14	•		•			
Sloughgrass, *Beckmannia syzigachne*	1, 4, 8, 10, 12-14					•	•
Dewey's sedge, *Carex deweyana*	1-5, 7-14	•	•		•		
Henderson's sedge, *Carex hendersonii*	1, 3-9 (lower elevations)	•	•				
Kellog's sedge, *Carex lenticularis*	2, 3-5, 8, 10-14					•	•
Merten's sedge, *Carex mertensii*	1-3, 5, 6, 8, 9	•	•		•		•
Slough sedge, *Carex obnupta*	1-5, 8, 9		•		•	•	•
Inflated sedge, *Carex vesicaria* var. *major*	1-3, 5-14					•	•
Tufted hairgrass, *Deschampsia cespitosa*	1-5, 8, 10-14	•			•	•	•
Creeping spike-rush, *Eleocharis palustris*	1-4, 6-14		•		•	•	•
Basin wildrye, *Elymus cinereus*	1, 8, 10-14			•			
Blue wildrye, *Elymus glaucus*	1-4, 6-14	•		•			
Dune wildrye, *Elymus mollis*	1 (also called *Leymus mollis*)		•				
California fescue, *Festuca californica*	1, 4, 8, 9	•		•			
Idaho fescue, *Festuca idahoensis* var. *idahoensis*	8, 10-14	•		•			
Idaho fescue, *Festuca idahoensis* var. *roemeri*	1-9			•			
Red fescue, *Festuca rubra* spp. *littoralis*	1-3, 6-8			•			•
Tall mannagrass, *Glyceria elata*	1-3, 6-8, 10-14				•	•	•
Soft rush, *Juncus effusus*	All except high mountains (rare on east side)			•		•	•

✓ Common name, *Botanical name*	Regions/Notes	Woodlands	Waterways (Riparian)	Grasslands/Shrub-steppe	Shrub/Forested Wetlands	Fresh-water Marshes	Wet Meadows
Grasses and Grass-like Plants, cont.						•	
Hardstem bulrush, *Scirpus acutus*	All except high mountains						
Small-fruited bulrush, *Scirpus microcarpus*	All except high mountains		•		•	•	•
Cattail, *Typha latifolia*	All except high mountains					•	
Bear-grass, *Xerophyllum tenax*	1, 5-10, 13 (mid to high elevations)			•			
Bulbs							
Nodding onion, *Allium cernuum*	1, 2, 6, 7 (lower elevations)	•		•			
Harvest brodiaea, *Brodiaea coronaria*	1, 2, 4-6 (lower elevations)			•			
Common camas, *Camassia quamash*	2, 4, 6-8, 10-14			•			•
Glacier-lily, *Erythronium grandiflorum*	1, 3, 5-8, 10 (mid to high elevations)	•					
Fawn-lily, *Erythronium oregonum*	1, 2, 4-7, 10 (low elevations)	•		•			
Pink fawn-lily, *Erythronium revolutum*	1-3, 5 (coastal mountains)	•	•				
Tiger lily, *Lilium columbianum*	1, 2, 5-8	•					
Ferns							
Lady fern, *Anthyrium felix-femina*	1, 2, 4-10, 13	•	•		•		
Deer fern, *Blechnum spicant*	1, 2, 4-9	•	•		•		
Sword fern, *Polystichum munitum*	1-10	•	•				
Chain fern, *Woodwardia fimbriata*	1, 2, 5, 9 (coastal)	•	•				

PLANTS FOR PARTICULAR NEEDS AND CONDITIONS

The following chart is designed to help you match plants to specific site condition and landscape needs.
To create a list, place a pencil mark in the column to the left of plants that meet your requirements.
To locate your hardiness zone, see the "Plant Hardiness Zones" map, page 213.
For additional native species, see the "Native Plant Occurrence Chart," pages 214-223.
For an abbreviated list of plants, see "Short Plant Lists For Wildlife Landscapes," pages 234-238.
For additional landscape and wildlife information on many of these plants, see Appendix C.

Plants are listed alphabetically by their botanical name. The botanical name includes the *Genus* and the *species*.
When referencing more than one plant species in a particular genus, the abbreviation (spp.) is used.

✓ Common name, *Botanical name*	Hardiness zone	Native plants	Drought-tolerant plants	Plants for wet soil	Shade-tolerant plants	Fast-growing plants	Plants that form a thicket	Plants with thorns or spines	Plants with berries or fruits	Plants with seeds or nuts	Nectar plants for butterflies	Food plants for butterflies	Nectar plants for hummingbirds
Coniferous Trees													
Fir, *Abies* spp.	Varies	•								•		•	
Incense cedar, *Calocedrus decurrens*	All	•	•							•		•	
Lawson cypress, *Chamaecyparis lawsoniana*	4-7	•				•							
Hinoki false cypress, *Chamaecyparis obtusa*	3-7		•										
Juniper, *Juniperus* spp.	All	•	•						•			•	
Spruce, *Picea* spp.	All	•		•						•			
Pine, *Pinus* spp.	Varies	•	•			•				•		•	
Douglas-fir, *Pseudotsuga menziesii*	All	•	•			•				•		•	
Western red-cedar, *Thuja plicata*	All	•	•	•	•				•	•		•	
Western hemlock, *Tsuga heterophylla*	All	•		•	•	•				•			
Mountain hemlock, *Tsuga mertensiana*	All	•											
Deciduous Trees													
Hedge maple, *Acer campestre*	All		•			•					•		•
Vine maple, *Acer circinatum*	All	•		•	•	•					•		•
Douglas maple, *Acer glabrum* var. *douglasii*	All	•	•		•	•					•		•
Big-leaf maple, *Acer macrophyllum*	4-7	•		•	•	•					•		•

✓ Common name, *Botanical name*	Hardiness zone	Native plants	Drought-tolerant plants	Plants for wet soil	Shade-tolerant plants	Fast-growing plants	Plants that form a thicket	Plants with thorns or spines	Plants with berries or fruits	Plants with seeds or nuts	Nectar plants for butterflies	Food plants for butterflies	Nectar plants for hummingbirds
Deciduous Trees, cont.													
Alder, *Alnus* spp.	Varies	•		•	•	•				•		•	
Birch, *Betula* spp.	All	•		•	•	•				•		•	
Dogwood, *Cornus* spp.	Varies	•			•				•		•	•	•
Hawthorn, *Crataegus* spp.	All	•	•				•	•	•		•	•	•
Oregon ash, *Fraxinus latifolia*	4–7	•		•		•				•		•	
Walnut, *Juglans* spp.	All									•			
Pacific crabapple, *Malus fusca* (*Pyrus fusca*)	4–7	•		•	•		•		•		•	•	•
Flowering crabapple, *Malus* spp.	All								•		•	•	•
Black cottonwood, *Populus balsamifera* var. *trichocarpa* (*Populus trichocarpa*)	All	•		•		•					•	•	
Quaking aspen, *Populus tremuloides*	All	•		•		•	•					•	
Bitter cherry, *Prunus emarginata* var. *mollis* (also see "Deciduous Shrubs")	4–7	•	•		•	•	•		•		•	•	•
Oak, *Quercus* spp.	Varies	•	•							•		•	•
Cascara, *Rhamnus purshiana*	All	•		•		•	•		•			•	•
Sumac, *Rhus glabra*	All	•	•				•		•		•		
Willow, *Salix* spp.	All	•		•		•	•				•	•	•
Mountain-ash, *Sorbus* spp. (also see "Deciduous Shrubs")	All		•		•	•			•		•		•
Littleleaf linden, *Tilia cordata*	All					•					•	•	•
Broadleaf Evergreen Trees													
Madrona, *Arbutus menziesii*	3–7	•	•		•				•		•	•	•
Strawberry-tree, *Arbutus unedo*	4–7				•				•			•	
Chinquapin, *Chrysolepis chrysophylla*	4–7	•	•							•			
Tan-oak, *Lithocarpus densiflorus*	4–7	•	•							•			
Oak, *Quercus* spp.	Varies	•	•							•		•	
Bay tree (Oregon-myrtle), *Umbellularia californica*	4–7		•		•	•				•			

✓ Common name, *Botanical name*	Hardiness zone	Native plants	Drought-tolerant plants	Plants for wet soil	Shade-tolerant plants	Fast-growing plants	Plants that form a thicket	Plants with thorns or spines	Plants with berries or fruits	Plants with seeds or nuts	Nectar plants for butterflies	Food plants for butterflies	Nectar plants for hummingbirds
Deciduous Shrubs													
Serviceberry, *Amelanchier alnifolia*	All	•	•		•		•		•		•	•	•
Barberry, *Berberis* spp.	All		•		•			•					•
Bog birch, *Betula glandulosa*	All	•		•		•	•			•		•	
Butterfly bush, *Buddleia davidii*	All		•			•	•				•		•
Siberian peashrub, *Caragana arborescens*	1–3		•		•		•	•		•			
Flowering quince, *Chaenomeles speciosa*	All			•			•		•				•
Red-twig dogwood, *Cornus sericea (C. stolonifera)*	All	•		•	•	•		•		•	•		•
Hazelnut, *Corylus cornuta* var. *californica*	All	•	•		•	•			•				
Cotoneaster, *Cotoneaster* spp.	Varies		•		•			•					•
Hardy fuchsia, *Fuchsia magellanica*	3–7			•		•	•		•		•		•
Oceanspray, *Holodiscus* spp.	All	•	•		•		•			•		•	•
Twinberry, *Lonicera involucrata*	3–7	•		•	•	•	•		•		•		•
Tartarian honeysuckle, *Lonicera tatarica*	All		•		•	•	•		•		•		•
Osoberry (Indian plum), *Oemleria cerasiformis*	4–7	•		•	•		•		•				
Mock-orange, *Philadelphus lewisii*	All	•	•		•	•	•			•	•		•
Ninebark, *Physocarpus capitatus*	4–7	•	•	•	•		•			•			
Potentilla, *Potentilla fruticosa*	All	•	•			•					•	•	
Bitter cherry, *Prunus emarginata* var. *emarginata* (also see "Deciduous Trees")	1–3	•	•				•				•	•	•
Nanking cherry, *Prunus tomentosa*	All		•				•		•		•	•	•
Chokecherry, *Prunus virginiana*	All	•	•		•	•	•		•		•	•	•
Bitterbrush, *Purshia tridentata*	1–3	•	•							•			
Wild azalea, *Rhododendron occidentalis*	4–7	•			•						•	•	•
Skunkbush, *Rhus trilobata*	1–3	•	•			•	•		•				
Golden currant, *Ribes aureum*	All	•	•		•		•		•		•	•	•
Red-flowering currant, *Ribes sanguineum*	4–7	•	•		•	•	•		•		•	•	•
Wild gooseberry, *Ribes* spp.	4–7	•	•		•	•	•	•	•		•		•
Wild rose, *Rosa* spp.	All	•	•		•	•	•	•	•		•	•	•
Thimbleberry, *Rubus parviflorus*	4–7	•		•	•	•	•		•		•		•
Salmonberry, *Rubus spectabilis*	4–7	•		•	•	•	•	•	•		•		•

✓ Common name, *Botanical name*	Hardiness zone	Native plants	Drought-tolerant plants	Plants for wet soil	Shade-tolerant plants	Fast-growing plants	Plants that form a thicket	Plants with thorns or spines	Plants with berries or fruits	Plants with seeds or nuts	Nectar plants for butterflies	Food plants for butterflies	Nectar plants for hummingbirds
Deciduous Shrubs, cont.													
Elderberry, *Sambucus* spp.	Varies	•		•	•	•	•		•		•		•
Buffaloberry, *Shepherdia canadensis*	All	•	•				•		•				
Sitka mountain-ash, *Sorbus sitchensis*	All	•	•	•			•		•		•		•
Cascade mountain-ash, *Sorbus scopulina*	All	•	•	•			•		•		•		•
Hardhack (Spirea), *Spiraea douglasii*	All	•		•	•	•	•				•	•	•
Garden spirea, *Spiraea* spp.	All	•	•								•	•	•
Snowberry, *Symphoricarpos albus*	All	•	•		•	•	•		•			•	•
Red huckleberry, *Vaccinium parvifolium*	4–7	•			•				•			•	•
Viburnum, *Viburnum* spp.	All	•		•	•	•	•		•		•	•	•
Weigela, *Weigela florida*	All		•			•					•		
Evergreen Shrubs													
Strawberry-tree, *Arbutus unedo* 'Compacta'	4–7		•		•				•		•		•
Manzanita, *Arctostaphylos* spp.	Varies	•	•						•		•	•	•
Sagebrush, *Artemisia tridentata*	All	•	•		•							•	
Barberry, *Berberis* spp.	Varies		•					•	•	•			
Mountain balm, Snowbush, *Ceanothus* spp.	Varies	•	•			•	•				•	•	•
Rabbitbrush, *Chrysothamnus* spp.	1–3	•					•			•	•	•	•
Cotoneaster, *Cotoneaster* spp.	4–7		•			•	•	•	•				•
Silverberry, *Elaeagnus pungens*	4–7		•		•	•	•		•				
Escallonia, *Escallonia* spp.	4–7		•			•					•		•
Silk-tassel bush, *Garrya elliptica*	5–7	•	•		•				•				
Salal, *Gaultheria shallon*	4–7	•	•	•	•	•	•		•			•	•
Juniper, *Juniperus scopulorum*	All	•	•		•				•			•	
Tall Oregon-grape, *Mahonia aquifolium*	All	•	•		•	•	•	•	•		•		•
Wax-myrtle, *Myrica californica*	4–7	•	•		•	•	•			•			
Firethorn, *Pyracantha* spp.	Varies		•			•	•	•	•				•
Coffeeberry, *Rhamnus californica*	4–7	•	•		•							•	•
Rhododendron, *Rhododendron* spp.	2–7	•			•						•	•	•
Evergreen huckleberry, *Vaccinium ovatum*	4–7	•	•		•				•			•	•

✓ Common name, *Botanical name*	Hardiness zone	Native plants	Drought-tolerant plants	Plants for wet soil	Shade-tolerant plants	Fast-growing plants	Plants that form a thicket	Plants with thorns or spines	Plants with berries or fruits	Plants with seeds or nuts	Nectar plants for butterflies	Food plants for butterflies	Nectar plants for hummingbirds
Ground Covers and Low Shrubs													
Kinnikinnik, *Arctostaphylos uva-ursi*	All	•	•						•		•	•	•
Dwarf coyote bush, *Baccharis pilularis*	5–7		•			•					•		
Point Reyes creeper, *Ceanothus gloriosus*	4–7		•			•					•	•	
Bunchberry, *Cornus unalaschkensis (C. canadensis)*	All	•		•	•				•			•	
Cotoneaster, *Cotoneaster* spp.	All		•			•			•		•	•	•
Heather, *Erica* spp.	Varies										•		•
Wild strawberry, *Fragaria* spp.	Varies	•	•		•	•			•		•	•	•
Wintergreen, *Gaultheria* spp.	4–7	•			•				•				
Salal, *Gaultheria shallon*	4–7	•	•	•	•	•	•					•	•
Juniper, *Juniperus* spp.	All	•	•			•		•	•				
Dwarf Oregon-grape, *Mahonia aquifolium* 'Compacta'	All	•	•					•	•		•		
Cascade Oregon-grape, *Mahonia nervosa*	4–7	•	•		•		•	•	•		•		•
Creeping Oregon-grape, *Mahonia repens*	All	•	•				•	•	•		•		
Oxalis (Wood sorrel), *Oxalis oregana*	3–7	•		•	•	•							
Rubus, *Rubus calycinoides*	4–7		•		•	•			•				
Vines													
Trumpet vine, *Campsis radicans*	All		•			•	•				•		•
Clematis, *Clematis* spp.	Varies	•	•		•	•	•			•	•		
Honeysuckle, *Lonicera* spp.	Varies	•	•		•	•	•		•		•		
Virginia creeper, *Parthenocissus quinquefolia*	All		•		•	•			•				
Boston-ivy, *Parthenocissus tricuspidata*	All		•		•	•	•		•				
Scarlet runner bean, *Phaseolus* spp.	All					•					•	•	•
Grape, *Vitis* spp.	Varies	•	•		•	•		•	•				
Garden Perennials													
Yarrow, *Achillea* spp.	All	•	•			•					•	•	
Giant-hyssop, *Agastache foeniculum*	3–7					•						•	•
Columbine, *Aquilegia* spp.	All	•			•					•			•
Rockcress, *Arabis caucasica*	All		•								•	•	

✓ Common name, *Botanical name*	Hardiness zone	Native plants	Drought-tolerant plants	Plants for wet soil	Shade-tolerant plants	Fast-growing plants	Plants that form a thicket	Plants with thorns or spines	Plants with berries or fruits	Plants with seeds or nuts	Nectar plants for butterflies	Food plants for butterflies	Nectar plants for hummingbirds
Garden Perennials, cont.													
Seathrift, *Armeria maritima*	4–7	•	•			•					•		•
Goat's beard, *Aruncus sylvester*	4–7	•		•	•	•					•		•
Aster, *Aster* spp.	All	•	•			•				•	•	•	
Campanula, *Campanula* spp.	All	•	•		•	•					•		•
Daisy, *Chrysanthemum* spp.	All		•			•					•		
Coreopsis, *Coreopsis* spp.	All		•			•				•	•		
Delphinium, *Delphinium* spp.	All	•									•		•
Dianthus (Cottage pink), *Dianthus* spp.	All		•										
Bleeding heart, *Dicentra* spp.	All	•			•	•					•	•	•
Foxglove, *Digitalis* spp.	All		•		•	•					•		
Coneflower, *Echinacea purpurea*	All					•				•	•		•
Globe-thistle, *Echinops exaltatus*	All		•			•	•	•			•		
Fleabane, *Erigeron karvinskianus*	4–7	•	•			•	•			•	•		
Sea-holly, *Eryngium amethystinum*	All		•								•		
Wallflower, *Erysimum* spp.	Varies		•			•					•		•
Blanket flower, *Gaillardia* spp.	All	•	•			•				•	•		
Daylily, *Hemerocallis fulva*	All		•	•	•	•							•
Coral bells, *Heuchera sanguinea*	All				•								•
Hosta, *Hosta* spp.	All			•	•	•							•
Poker plant, *Kniphofia uvaria*	All		•			•							•
Gayfeather, *Liatris* spp.	All					•					•		•
Statice, *Limonium latifolium*	All		•								•		
Cardinal flower, *Lobelia cardinalis*	All			•	•								•
Lupine, *Lupinus* spp.	Varies	•	•			•				•	•	•	•
Beebalm, *Monarda didyma*	All			•		•					•		•
Catmint, *Nepeta* spp.	All		•			•					•		•
Penstemon, *Penstemon* spp.	Varies	•	•			•					•		•
Tall phlox, *Phlox* spp.	Varies		•			•					•		•
Cape-fuchsia, *Phygelius capensis*	4–7		•			•	•						•
Solomon's seal, *Polygonatum* spp.	All	•		•	•				•				•

✓ Common name, *Botanical name*	Hardiness zone	Native plants	Drought-tolerant plants	Plants for wet soil	Shade-tolerant plants	Fast-growing plants	Plants that form a thicket	Plants with thorns or spines	Plants with berries or fruits	Plants with seeds or nuts	Nectar plants for butterflies	Food plants for butterflies	Nectar plants for hummingbirds
Garden Perennials, cont.													
Black-eyed Susan, *Rudbeckia hirta*	All		•			•					•	•		•
Pincushion flower, *Scabiosa* spp.	All					•					•	•		•
Fall sedum, *Sedum spectabile*	All		•			•					•	•	•	
Dusty miller, *Senecio cineraria*	4–7		•			•					•	•	
Goldenrod, *Solidago* spp.	All	•	•			•					•	•		•
Verbena, *Verbena* spp.	Varies		•			•					•	•		•
California-fuchsia, *Zauschneria californica*	3–7		•			•								•
Garden Annuals													
Ageratum, *Ageratum houstonianum*	All										•		
Calendula, *Calendula officinalis*	All		•			•					•	•		•
Bachelor buttons, *Centaurea cyanus*	All		•			•					•	•		•
Spiderflower, *Cleome spinosa*	All					•	•	•			•		•
Cosmos, *Cosmos bipinnatus*	All		•			•					•	•	
Sweet William, *Dianthus barbatus*	All		•			•						•		•
Satin flower, *Godetia grandiflora*	All		•											•
Sunflower, *Helianthus annuus*	All					•					•	•	
Sweet pea, *Lathyrus* spp.	All					•					•	•	•	•
Sweet alyssum, *Lobularia maritima*	All		•		•	•						•	
Forget-me-not, *Myosotis* spp.	All			•	•	•					•	•	•
Flowering tobacco, *Nicotiana* spp.	All		•		•	•								•
Petunia, *Petunia* spp.	All		•			•								•
Sage, *Salvia* spp. (red-flower varieties)	All		•			•						•		•
French marigold, *Tagetes patula*	All					•					•	•	
Nasturtium, *Tropaeolum majus*	All				•	•					•	•	•	•
Verbena, *Verbena* spp.	All		•			•					•	•		•
Zinnia, *Zinnia elegans*	All					•					•	•		•

✓ Common name, *Botanical name*	Hardiness zone	Native plants	Drought-tolerant plants	Plants for wet soil	Shade-tolerant plants	Fast-growing plants	Plants that form a thicket	Plants with thorns or spines	Plants with berries or fruits	Plants with seeds or nuts	Nectar plants for butterflies	Food plants for butterflies	Nectar plants for hummingbirds
Wildflowers													
Yarrow, *Achillea millefolium*	All	•	•			•					•	•	
Nodding onion, *Allium cernuum*	All	•	•								•	•	•
Pearly everlasting, *Anaphalis margaritacea*	All	•	•			•					•	•	
Angelica, *Angelica lucida*	4–7	•		•							•	•	•
Columbine, *Aquilegia formosa*	All	•			•						•		•
Showy milkweed, *Asclepias speciosa*	All	•	•								•	•	•
Aster, *Aster* spp.	All	•	•			•					•	•	•
Balsamroot, *Balsamorhiza* spp.	All	•	•								•		
Campanula, *Campanula* spp.	All	•	•		•	•					•		•
Paintbrush, *Castilleja* spp.	All	•	•								•		•
Clarkia, *Clarkia* spp.	All	•	•								•		•
Delphinium, *Delphinium* spp.	Varies	•	•								•		•
Bleeding heart, *Dicentra formosa*	All	•			•	•					•	•	•
Fireweed, *Epilobium angustifolium*	All	•	•			•	•				•		•
Wild-buckwheat, *Eriogonum* spp.	Varies	•	•								•	•	•
California poppy, *Eschscholtzia californica*	All	•	•			•					•		
Blanket flower, *Gaillardia* spp.	All	•	•			•					•	•	•
Scarlet gilia, *Gilia aggregata*	All	•	•								•		•
Cow-parsnip, *Heracleum lanatum*	All	•		•	•	•	•				•	•	
Wild sweet pea, *Lathyrus* spp.	Varies	•	•			•					•	•	•
Desert-parsley, *Lomatium* spp.	Varies	•									•	•	
Lupine, *Lupinus* spp.	Varies	•	•			•					•	•	•
Monkey flower, *Mimulus* spp.	Varies	•		•	•						•		•
Checker mallow, *Sidalcea* spp.	Varies	•	•			•					•	•	•
Blue-eyed grass, *Sisyrinchium* spp.	Varies	•	•										
Goldenrod, *Solidago* spp.	All	•	•			•					•	•	•
Hedge-nettle, *Stachys cooleyae*	All	•		•	•	•							•
Stinging nettle, *Urtica dioica*	All	•		•	•	•			•			•	
Vetch, *Vicia* spp.	All		•			•	•				•	•	•

✓ Common name, *Botanical name*	Hardiness zone	Native plants	Drought-tolerant plants	Plants for wet soil	Shade-tolerant plants	Fast-growing plants	Plants that form a thicket	Plants with thorns or spines	Plants with berries or fruits	Plants with seeds or nuts	Nectar plants for butterflies	Food plants for butterflies	Nectar plants for hummingbirds
Grasses and Grass-like Plants													
Wheatgrass, *Agropyron* spp.	Varies	•	•							•			
Sloughgrass, *Beckmania syzigachne*	All	•		•		•				•			
Sedge, *Carex* spp.	All	•		•	•	•	•			•			
Tufted hairgrass, *Deschampsia cespitosa*	All	•		•	•	•				•			
Creeping spike-rush, *Eleocharis palustris*	All	•		•		•	•			•			
Wild ryegrass, *Elymus* spp.	Varies	•	•			•	•			•			
Fescue, *Festuca* spp.	Varies	•	•							•		•	
Tall mannagrass, *Glyceria elata*	4–7	•		•	•	•	•			•			
Rush, *Juncus* spp.	Varies	•		•	•	•	•			•			
Bulrush, *Scirpus* spp.	All	•		•		•	•			•			
Cattail, *Typha latifolia*	All	•		•	•	•	•			•			
Bear-grass, *Xerophyllum tenax*	All	•	•		•					•			•
Bulbs, Corms, Rhizomes, and Tubers													
Nodding onion, *Allium cernuum*	All	•	•							•	•		•
Brodiaea, *Brodiaea* spp.	All	•	•								•		•
Camas, *Camassia* spp.	All	•	•	•							•		•
Crocosmia, *Crocosmia* spp.	3–7		•	•		•	•						•
Fawn lily, *Erythronium* spp.	All	•		•	•								•
Iris, *Iris* spp.	All	•	•								•		•
Lily, *Lilium* spp.	All	•	•							•	•		•
Ferns													
Lady fern, *Anthyrium felix-femina*	All	•		•	•	•							
Deer fern, *Blechnum spicant*	All	•		•	•								
Sword fern, *Polystichum munitum*	All	•	•	•	•								
Chain fern, *Woodwardia fimbriata*	4–7	•		•	•								
Herbs													
Chives, *Allium schoenoprasum*	All					•					•	•	
Fennel, *Foeniculum vulgare*	All		•			•					•	•	•

✓ Common name, *Botanical name*	Hardiness zone	Native plants	Drought-tolerant plants	Plants for wet soil	Shade-tolerant plants	Fast-growing plants	Plants that form a thicket	Plants with thorns or spines	Plants with berries or fruits	Plants with seeds or nuts	Nectar plants for butterflies	Food plants for butterflies	Nectar plants for hummingbirds
Herbs, cont.													
Hyssop, *Hyssopsis officinalis*	All		•			•					•		•
Lavender, *Lavandula* spp.	All		•			•				•	•		•
Garden mint, *Mentha* spp.	All			•	•	•					•		•
Oregano, *Origanum vulgare*	All		•			•					•		•
Parsley, *Petroselinum crispum*	All				•	•				•	•	•	
Rosemary, *Rosmarinus officinalis*	3–7		•			•							•
Garden sage, *Salvia officinalis*	All		•			•					•		•
Thyme, *Thymus* spp.	All		•		•	•					•		

SHORT PLANT LISTS FOR WILDLIFE LANDSCAPES

Plants are listed alphabetically by their botanical name. The botanical name includes the *Genus* and the *species*. When referencing more than one plant species in a particular genus, the abbreviation (spp.) is used. Hardiness zones are to the left, see the "Plant Hardiness Zones" map on page 213. (N) indicates that the genus contains one or more native plants. For additional plant choices, see "Plants for Particular Needs and Conditions," pages 224–233. For landscape and wildlife information for many of the following plants, see Appendix C.

Food Plants for Butterfly Larvae

Highly recommended plants are **bold**. Also see "Some Common Pacific Northwest Butterflies" in Chapter 18.

Evergreen Trees

3–7	**Madrona**, *Arbutus menziesii* (N)
All	Incense-cedar, *Calocedrus* sp. (N)
Varies	**Pine**, *Pinus* spp. (N)
All	Douglas-fir, *Pseudotsuga* sp. (N)

Deciduous Trees

All	**Maple**, *Acer* spp. (N)
Varies	**Alder**, *Alnus* spp. (N)
All	Birch, *Betula* spp. (N)
Varies	Dogwood, *Cornus* spp. (N)
All	**Hawthorn**, *Crataegus* spp. (N)
All	Garden apple and crabapple, *Malus* spp. (N)
All	Cottonwood, *Populus* spp. (N)
All	**Cherry**, *Prunus* spp. (N)
All	Aspen, *Populus tremuloides* (N)
Varies	**Oak**, *Quercus* spp. (N)
All	Cascara, *Rhamnus purshiana* (N)
All	**Willow**, *Salix* spp. (N)

Evergreen Shrubs

Varies	Manzanita, *Arctostaphylos* spp. (N)
Varies	**Wild-lilac (Mountain balm)**, *Ceanothus* spp. (N)
Varies	**Buckbrush**, *Ceanothus* spp. (N)
4–7	Salal, *Gaultheria shallon* (N)
All	**Rhododendron**, *Rhododendron* spp. (N)

Deciduous Shrubs

All	**Serviceberry**, *Amelanchier alnifolia* (N)
All	Oceanspray, *Holodiscus* spp. (N)
All	**Chokecherry**, *Prunus virginiana* (N)
1–3	**Bitterbrush**, *Purshia tridentata* (N)
4–7	Coffeeberry, *Rhamnus californica* (N)
All	Wild rose, *Rosa* spp. (N)
All	Hardhack, *Spiraea douglasii* (N)
4–7	Garden blueberry, *Vaccinium* spp.
4–7	**Chaste tree**, *Vitex agnus-castus*

Garden Perennials and Wildflowers

All	**Pearly everlasting**, *Anaphalis margaritacea* (N)
4–7	Angelica, *Angelica lucida* (N)
All	**Butterfly weed**, *Asclepias* spp. (N)
All	**Rockcress**, *Arabis caucasica*
All	**Bleeding heart**, *Dicentra* spp. (N)
All	Cow-parsnip, *Heracleum lanatum* (N)
Varies	**Desert-parsley**, *Lomatium* spp. (N)
Varies	**Lupine**, *Lupinus* spp. (N)
All	Fall sedum, *Sedum spectabile*
4–7	**Dusty miller**, *Senecio cineraria*
Varies	Checker mallow, *Sidalcea oregana* (N)
All	Clover, *Trifolium* spp.
All	**Stinging nettle**, *Urtica dioica* (N)
All	Violet, *Viola* spp. (N)

Garden Annual Flowers

All	Borage, *Borago officinalis*
All	Sunflower, *Helianthus* spp.
All	**French marigold**, *Tagetes patula*
All	**Nasturtium**, *Tropaeolum majus*

Ground Covers

All	**Kinnikinnik**, *Arctostaphylos uva-ursi* (N)
All	Cotoneaster, *Cotoneaster* spp.
4–7	**Salal**, *Gaultheria shallon* (N)

Vines and Vine-like Plants

All	Hops, *Humulus lupulus*
All	**Trailing nasturtium**, *Tropaeolum majus*
All	Vetch, *Vicia* spp.

Garden Herbs and Vegetables

All	Dill, *Anethum graveolens*
All	Fennel, *Foeniculum vulgare*
All	Broccoli, carrot, kale, radish

Nectar Plants for Adult Butterflies

Highly recommended plants are **bold**. Also see "Some Common Pacific Northwest Butterflies" in Chapter 18.

Trees

3–7	Madrona, *Arbutus menziesii* (N)
Varies	Dogwood, *Cornus* spp. (N)
All	**Hawthorn**, *Crataegus* spp. (N)
All	**Crabapple**, *Malus* spp. (N)
All	**Cherry**, *Prunus* spp. (N)
All	**Willow**, *Salix* spp. (N)

Shrubs

All	**Butterfly bush**, *Buddleia* spp.
All	**Bluebeard**, *Caryopteris* spp.
Varies	**Wild-lilac**, *Ceanothus* spp. (N)
1–3	**Rabbitbrush**, *Chrysothamnus* spp. (N)
All	Red-twig dogwood, *Cornus sericea (stolonifera)* (N)
Varies	Daphne, *Daphne* spp.
4–7	Escallonia, *Escallonia* spp.
All	Tall Oregon-grape, *Mahonia aquifolium* (N)
All	**Mock-orange**, *Philadelphus lewisii* (N)
All	**Chokecherry**, *Prunus virginiana* (N)
1–3	**Bitterbrush**, *Purshia tridentata* (N)
4–7	Wild azalea, *Rhododendron occidentale* (N)
Varies	Rhododendron, *Rhododendron* spp. (N)
All	Wild rose, *Rosa* spp. (N)
Varies	Elderberry, *Sambucus* spp. (N)
All	**Spirea**, *Spiraea* spp. (N)
All	Lilac, *Syringa* spp.
All	**Germander**, *Teucrium chamaedrys*

Large and Small-scale Ground Covers

4–7	**Seathrift**, *Armeria maritima* (N)
All	Cotoneaster, *Cotoneaster* spp.
Varies	Heather, *Erica* spp.
Varies	Wild strawberry, *Fragaria* spp. (N)
All	**Candytuft**, *Iberis* spp.

Vines and Vine-like Plants

Varies	Clematis, *Clematis* spp. (N)
Varies	**Honeysuckle**, *Lonicera* spp. (N)
All	**Trailing nasturtium**, *Tropaeolum majus*

Garden Herbs and Vegetables

All	Fennel, *Foeniculum vulgare*
All	**Hyssop**, *Hyssopsis officinalis*
All	**Lavender**, *Lavandula* spp.
All	**Garden mint**, *Mentha* spp.
All	**Oregano**, *Origanum vulgare*
All	Garden sage, *Salvia* spp.
All	**Thyme**, *Thymus* spp.
All	Broccoli, carrot, kale, radish

Garden Perennials and Wildflowers

All	**Yarrow**, *Achillea* spp. (N)
All	**Pearly everlasting**, *Anaphalis margaritacea* (N)
All	**Butterfly weed** (milkweed), *Asclepias* spp. (N)
All	**Aster**, *Aster* spp. (N)
All	**Yellow alyssum**, *Aurinia saxatilis*
Varies	Campanula, *Campanula* spp. (N)
All	**Daisy**, *Chrysanthemum* spp.
All	Coreopsis, *Coreopsis* spp.
All	Clove (Cottage) pink, *Dianthus* spp.
All	**Coneflower**, *Echinacea purpurea*
All	Globe-thistle, *Echinops* spp.
All	**Wild-buckwheat**, *Erigonium* spp. (N)
All	**Sea-holly**, *Eryngium amethystinum*
Varies	Wallflower, *Erysimum* spp.
All	**Blanket flower**, *Gaillardia* spp. (N)
All	**Gilia**, *Gilia* spp. (N)
All	Heliotrope, *Heliotropium* spp.
All	Gayfeather, *Liatris* spp.
All	**Statice**, *Limonium latifolium*
All	**Desert-parsley**, *Lomatium* spp. (N)
Varies	**Lupine**, *Lupinus* spp. (N)
Varies	**Monkey flower**, *Mimulus* spp. (N)
All	**Bee balm**, *Monarda didyma*
All	Catmint, *Nepeta* spp.
Varies	Penstemon, *Penstemon* spp. (N)
Varies	Phlox, *Phlox* spp. (N)
All	Black-eyed Susan, *Rudbeckia hirta*
All	**Pincushion flower**, *Scabiosa* spp.
All	**Fall sedum**, *Sedum spectabile*
4–7	Dusty miller, *Senecio cineraria*
All	**Goldenrod**, *Solidago* spp. (N)
All	**Tall verbena**, *Verbena bonariensis*

Garden Annual Flowers

All	**Ageratum**, *Ageratum houstonianum*
All	**Alyssum**, *Alyssum maritima*
All	**Calendula**, *Calendula officinalis*
All	**Clarkia**, *Clarkia* spp. (N)
All	Spiderflower, *Cleome spinosa*
All	Cosmos, *Cosmos bipinnatus*
All	**Sweet William**, *Dianthus barbatus*
All	**Globe gilia**, *Gilia capitata* (N)
All	Forget-me-not, *Myosotis* spp.
All	**French marigold**, *Tagetes patula*
All	Nasturtium, *Tropaeolum majus*
All	Low verbena, *Verbena* spp.
All	**Zinnia**, *Zinnia elegans*

Nectar Plants for Hummingbirds

Recommended plants are **bold**

Trees

3-7	Madrone, *Arbutus menziesii* (N)
4-7	Strawberry-tree, *Arbutus unedo*
Varies	**Dogwood**, *Cornus* spp. (N)
All	**Crabapple**, *Malus* spp. (N)
All	Cascara, *Rhamnus purshiana* (N)

Shrubs and Brambles

5-7	Glossy abelia, *Abelia grandiflora*
Varies	Manzanita, *Arctostaphylos* spp. (N)
All	**Butterfly bush**, *Buddleia davidii*
1-3	**Siberian pea-shrub**, *Caragana arborescens*
All	Flowering quince, *Chaenomeles japonica*
All	Red-twig dogwood, *Cornus sericea (stolonifera)* (N)
3-7	**Hardy fuchsia**, *Fuchsia magellanica*
4-7	Salal, *Gaultheria shallon* (N)
All	Oceanspray, *Holodiscus* spp. (N)
4-7	Rose-of-Sharon, *Hibiscus syriacus*
All	Bog-laurel, *Kalmia polifolia* (N)
All	Tatarian honeysuckle, *Lonicera tatarica*
4-7	**Wild azalea**, *Rhododendron occidentale* (N)
Varies	Rhododendron, *Rhododendron* spp. (N)
4-7	**Red-flowering current**, *Ribes sanguineum* (N)
4-7	Thimbleberry, *Rubus parviflora* (N)
4-7	**Salmonberry**, *Rubus spectabilis* (N)
Varies	Elderberry, *Sambucus* spp. (N)
All	Snowberry, *Symphoricarpos* spp. (N)
All	Lilac, *Syringa* spp.
All	**Weigela**, *Weigela florida*

Garden Perennials and Wildflowers

All	**Columbine**, *Aquilegia formosa* (N)
All	Hollyhock, *Althaea rosea*
4-7	Goat's beard, *Aruncus sylvester* (N)
Varies	Campanula, *Campanula* spp. (N)
All	Canna, *Canna* spp.
Varies	**Paintbrush**, *Castilleja* spp. (N)
All	**Clarkia**, *Clarkia* spp. (N)
All	**Delphinium**, *Delphinium* spp. (N)
All	**Bleeding heart**, *Dicentra* spp. (N)
All	Foxglove, *Digitalis* spp.
All	Fireweed, *Epilobium angustifolium* (N)
All	**Gilia**, *Gilia* spp. (N)
All	**Coral bells**, *Heuchera* spp. (N)
All	Hosta, *Hosta* spp.
All	Poker plant, *Kniphofia uvaria*
All	**Lavender**, *Lavandula* spp.

Garden Perennials and Wildflowers, cont.

All	**Cardinal flower**, *Lobelia cardinalis*
Varies	Lupine, *Lupinus* spp. (N)
Varies	Monkey flower, *Mimulus* spp. (N)
All	**Beebalm**, *Monarda didyma*
All	Catmint, *Nepeta* spp.
Varies	**Penstemon**, *Penstemon* spp. (N)
4-7	**Cape-fuchsia**, *Phygelius capensis*
All	Obedience plant, *Physostegia virginiana*
All	Balloon flower, *Platycodon grandiflorus*
All	Lungwort, *Pulmonaria*, spp.
All	**Scabiosa**, *Scabiosa* spp.
All	Goldenrod, *Solidago* spp. (N)
4-7	**Hedge-nettle**, *Stachys cooleyae* (N)
All	Lamb's ears, *Stachys lantana*
All	Veronica, *Veronica* spp.
3-7	**California-fuchsia**, *Zauschneria californica*

Annual Garden Flowers

All	Snapdragon, *Antirrhinum* spp.
All	Begonia, *Begonia* spp.
All	**Clarkia**, *Clarkia* spp. (N)
All	Spider flower, *Cleome spinosa*
All	**Sweet William**, *Dianthus barbatus*
All	Jewelweed, *Impatiens capensis*
All	Flowering tobacco, *Nicotiana* spp.
All	Geranium, *Pelargonium* hybrids
All	Petunia, *Petunia* spp.
All	**Sage**, *Salvia* spp. (Red-flower varieties)
All	Nasturtium, *Tropaeolum majus*
All	Zinnia, *Zinnia elegans*

Vines and Vine-like Plants

All	**Trumpet vine**, *Campsis radicans*
Varies	**Honeysuckle**, *Lonicera* spp. (N)
All	**Scarlet runner bean**, *Phaseolus* spp.
All	Trailing nasturtium, *Tropaeolum speciosum*
All	**Vetch**, *Vicia* spp.

Bulbs, Corms, Rhizomes, and Tubers

All	Nodding onion, *Allium cernuum* (N)
All	**Firecracker plant**, *Brodiaea ida-maia*
All	Dahlia, *Dahlia merkii*
All	**Fritillary**, *Fritillaria* spp.
All	**Gladiolus**, *Gladiolus* spp.
All	Iris, *Iris* spp. (N)
All	**Lily**, *Lilium* spp. (N)

Deer-resistant (or close to it) Plant

Many factors influence the degree of plant usage by deer: the number of deer in the area, the availability of alternative food sources, winter weather conditions, and plant preferences. In addition, deer in certain areas have different tastes and young plants may be eaten and older plants left alone.

Ornamental plants (including fruit trees), which may be expected to receive heavy use by deer and are often permanently damaged, should not be planted unless they are carefully protected.

The following list includes "best bets" in deer country. The list should be considered a guide rather than the final word. Plants are listed alphabetically by their botanical name. Plants may need to be protected when young.

Trees

Varies	Fir, *Abies* spp. (N)
Varies	Maple, *Acer* spp. (N)
All	Birch, *Betula* spp. (N)
3-7	False cypress, *Chamaecyparis* spp.
4-7	Fig, *Ficus carica*
4-7	Oregon ash, *Fraxinus latifolia* (N)
All	Juniper, *Juniperus* spp. (N)
All	Spruce, *Picea* spp. (N)
Varies	Pine, *Pinus* spp. (N)
All	Douglas-fir, *Pseudotsuga menziesii* (N)
Varies	Oak, *Quercus* spp. (N)
All	Sumac, *Rhus* spp. (N)
All	Willow, *Salix* spp. (N)
All	Western red-cedar, *Thuja plicata* (N)
All	Little-leaf linden, *Tilia cordata*
All	Hemlock, *Tsuga* spp. (N)
4-7	Bay (Oregon-myrtle), *Umbellularia californica* (N)

Deciduous Shrubs

All	Serviceberry, *Amelanchier alnifolia* (N)
All	Barberry, *Berberis* spp.
All	Butterfly bush, *Buddleia* spp.
All	Red-twig dogwood, *Cornus sericea (stolonifera)* (N)
All	Hazelnut (Filbert), *Corylus* spp. (N)
1-3	Winterfat, *Eurotia lanata* (N)
3-7	Winter jasmine, *Jasminum nudiflorum*
All	Potentilla, *Potentilla fruticosa* (N)
All	Chokecherry, *Prunus virginiana* (N)
All	Golden currant, *Ribes aureum* (N)
4-7	Red-flowered currant, *Ribes sanguineum* (N)
4-7	Wild gooseberry, *Ribes* spp. (N)
All	Wild rose, *Rosa* spp. (N)
Varies	Elderberry, *Sambucus* spp. (N)
All	Spirea, *Spiraea* spp. (N)
All	Snowberry, *Symphoricarpos* spp. (N)
All	Lilac, *Syringa* spp.

Evergreen Shrubs

Varies	Manzanita, *Arctostaphylos* spp. (N)
All	Sagebrush, *Artemisia tridentata* (N)
All	Barberry, *Berberis* spp.
6-7	Bottlebrush, *Callistemon* spp.
1-3	Rabbitbrush, *Chrysothamnus* spp. (N)
All	Mock-orange, *Choisya* spp. (N)
4-7	Silverberry, *Elaeagnus pungens*
5-7	Silk-tassel bush, *Garrya elliptica* (N)
4-7	Salal, *Gaultheria shallon* (N)
All	Juniper, *Juniperus* spp. (N)
All	Mountain-laurel, *Kalmia latifolia* (N)
All	Oregon-grape, *Mahonia aquifolium* (N)
4-7	Wax-myrtle, *Myrica californica* (N)
All	Oregon-boxwood, *Pachystima myrsinites* (N)
All	Mugho pine, *Pinus mugo*
4-7	Coffeeberry, *Rhamnus californica* (N)
Varies	Rhododendron, *Rhododendron* spp. (N)
4-7	Evergreen huckleberry, *Vaccinium ovatum* (N)

Perennial Flowers

All	Yarrow, *Achillea* spp. (N)
All	Rockcress, *Arabis* spp.
4-7	Seathrift, *Armeria maritima* (N)
All	Snow-in-summer, *Cerastium tomentosum*
All	Daisy, *Chrysanthemum maximum*
All	Coreopsis, *Coreopsis* spp.
All	Bleeding heart, *Dicentra* spp. (N)
All	Foxglove, *Digitalis* spp.
All	Coneflower, *Echinacea purpurea*
All	Globe-thistle, *Echinops exaltus*
All	Wild buckwheat, *Eriogonum* spp. (N)
All	Sea-holly, *Eryngium amethystinum*
Varies	Wallflower, *Erysimum* spp.
All	Blanket flower, *Gaillardia aristata* (N)
All	Baby's breath, *Gypsophila paniculata*
Varies	Hellebore, *Helleborus* spp.

Deer-resistant (or close to it) Plants, cont.

<u>Perennials, cont.</u>

All	Daylily, *Hemerocallis fulva*
All	Hosta, *Hosta* spp.
All	Iris, *Iris* spp. (N)
All	Poker plant, *Kniphofia* spp.
All	Gayfeather, *Liatris spicata*
All	Blue flax, *Linum* spp.
All	Lobelia, *Lobelia cardinalis*
Varies	Lupine, *Lupinus* spp. (N)
All	Bee balm, *Monarda didyma*
All	Catmint, *Nepeta* spp.
All	Poppy, *Papaver* spp.
All	Russian-sage, *Perouskia atriplicifolia*
All	Solomon's-seal, *Polygonatum* spp. (N)
All	Lungwort, *Pulmonaria* spp.
All	Black-eyed Susan, *Rudbeckia* spp.
All	Fall sedum, *Sedum spectabile*
All	Blue-eyed grass, *Sisyrinchium* spp. (N)
3-7	California fuchsia, *Zauschneria* spp.

<u>Annual Flowers</u>

All	Ageratum, *Ageratum houstonianum*
All	Calendula, *Calendula officinalis*
All	Bachelor buttons, *Centaurea cyanus*
All	Clarkia, *Clarkia* spp. (N)
All	Larkspur, *Consolida ambigua*
All	Cosmos, *Cosmos bipinnatus*
All	California poppy, *Eschscholtzia californica* (N)
All	Geranium, *Pelargonium* spp.
All	Sunflower, *Helianthus annuus*
All	Sweet alyssum, *Lobularia maritima*
All	Zinnia, *Zinnia* spp.

<u>Bulbs, Corms, and Tubers</u>

3-7	Crocosmia, *Crocosmia crocosmiiflora*
All	Crocus, *Crocus* spp.
All	Fritillary, *Fritillaria* spp.
All	Garden corn-lily, *Ixia* spp.
All	Trillium, *Trillium* spp. (N)

<u>Ground Covers and Low Shrubs</u>

All	Kinnikinnik, *Arctostaphylos uva-ursi* (N)
5-7	Dwarf coyote brush, *Baccharis pilularis* (N)
All	Bunchberry, *Cornus unalaschkensis (canadensis)* (N)
All	Cotoneaster, *Cotoneaster* spp.
All	Heather, *Erica* spp.
Varies	Wild strawberry, *Fragaria* spp. (N)
2-7	Wintergreen, *Gaultheria procumbens*
3-7	Salal, *Gaultheria shallon* (N)
All	Sunrose, *Helianthemum* spp.
All	Juniper, *Juniperus* spp. (N)
5-7	Lithodora, *Lithodora diffusa*
Varies	Oregon-grape, *Mahonia* spp. (N)
4-7	Oxalis (Wood sorrel), *Oxalis oregona* (N)
4-7	Trailing rosemary, *Rosmarinus officinalis*
3-7	Trailing raspberry, *Rubus pedatus*
3-7	Trailing blackberry, *Rubus ursinus*

<u>Garden Herbs</u>

All	Garden chive, *Allium schoenoprasum*
All	Garlic chive, *Allium tuberosum*
All	Hyssop, *Hyssopsis officinalis*
All	Garden mint, *Mentha* spp.
4-7	Rosemary, *Rosmarinus officinalis*
All	Lavender, *Lavandula* spp.
All	Thyme, *Thymus* spp.
All	Sweet marjoram, *Origanum majorana*
All	Oregano, *Origanum vulgare*
All	Rue, *Ruta graveolens*
All	Santolina, *Santolina* spp.

<u>Vines</u>

Varies	Clematis, *Clematis* spp. (N)
Varies	Honeysuckle, *Lonicera* spp. (N)
All	Wisteria, *Wisteria* spp.

LANDSCAPE AND WILDLIFE INFORMATION FOR SPECIFIC PLANTS

Appendix C Key

The plants listed in Appendix C attract wildlife. They grow well, given reasonable plant selection and maintenance, and are generally available from nurseries in the Pacific Northwest (see Appendix E).

Appendix C is divided into the following categories:

Coniferous Trees
Deciduous Trees
Broadleaf Evergreen Trees
Deciduous Shrubs
Evergreen Shrubs
Ground Covers and Low Shrubs
Vines
Garden Perennials (and Some Wildflowers)
Grasses (and Some Grass-like Plants)

KEY

Hardiness zones represent the zones of adaptability in the Pacific Northwest. See the "Plant Hardiness Zone Map" in Appendix B to determine your zone. For information on where Pacific Northwest native plants occur naturally, see the "Native Plant Occurrence Map" in Appendix B.

Plant size is expressed in feet and represents estimated sizes in typical landscape situations at "landscape maturity."

For most plants, landscape maturity means:
2–3 years for most perennials, wildflowers, and grasses.
3–5 years for most shrubs and ground covers.
10–15 years for most trees.

(The size estimates given are for planning and spacing purposes and do not necessarily reflect the ultimate size of the plant.)

Light needs are as follows:
F-sun: Plants grow in full sun during all, or most of the day.
P-shade: Plants grow in light shade, dappled shade, or afternoon shade.
F-shade: Plants grow in little or no direct sunlight.

Soil needs are as follows:
Wet: Plants grow in soil that may be saturated in winter and moist during the growing season.
Moist: Plants grow in soil that contains constant moisture during the growing season.
Dry: Plants grow in soil that may become quite dry during the growing season.

Landscape information includes a brief description of the plant, suggested landscape settings, and other helpful information.

Wildlife information provides information on what wildlife species are known to use the plant. This information has been obtained from literature, stomach analysis, and field observations. These lists of wildlife plants are suggestive, rather than definitive. Weather conditions, time of year, location of plants, and other factors will influence wildlife usage.

CONIFEROUS TREES

Plants are listed alphabetically by their botanical name.
The botanical name first includes the *Genus* and then the *species*.
When referencing more than one species in a genus, the abbreviation (spp.) is used.
(N) = Plants that are native to Oregon, Washington, or southern British Columbia.
For definitions of headings, see the "Key" at the beginning of Appendix C.
For information on where native plants occur naturally, see the "Native Plant Occurrence Chart" in Appendix B.

Common name *Botanical name*	Hardiness Zones	Plant Size	Light Needs	Soil Needs
Native fir (N) *Abies* spp.	All	75-150' tall 30' wide	F-sun	Moist, good drainage

LANDSCAPE INFORMATION: Most native firs are high-mountain trees that grow best near their natural habitats (see Appendix B). For hot, dry, windy areas, it may be better to choose a more appropriate native conifer or a non-native fir. **White fir** (*Abies concolor*) is a moderate or slow grower when young, but eventually requires plenty of space. The needles are bluish-green and curved. It's somewhat tolerant of poor soils but not air pollution. **Grand fir** (*Abies grandis*) is a tall stately tree with a somewhat narrow crown. It grows best in cool coastal areas, is a moderate to rapid grower, and gets very large. Grand fir can grow in more shade than can other firs. **Noble fir** (*Abies procera*) has a beautiful conical shape, short stiff branches, and blue-green needles. The immense new cones are striking in summer. It likes cool, moist soil. In lowland urban settings it tends to become ratty-looking with age.

WILDLIFE INFORMATION: Firs provide shelter for many species of birds and mammals including squirrels, porcupines, and deer. Birds that eat the seeds include grouse, nuthatches, chickadees, grosbeaks, finches, and crossbills. Sapsuckers, woodpeckers, and many other insect-eating birds forage on both live and dead standing firs. Cavity-nesting birds and other wildlife nest and roost in tree cavities. The foliage is used by pine white butterfly larvae.

Fir *Abies* spp.	Varies	20-50' tall	F-sun	Moist to dry, good drainage

LANDSCAPE INFORMATION: **Korean fir** (*Abies koreana*) (Zones 3-7) is a slow-growing, compact, pyramidal tree with blunt, shiny needles. **Norman fir** (*Abies nordmanniana*) (Zones All) is another vigorous, densely foliaged fir with commercially available compact forms. Both are good alternatives to most other native conifers where eventual height is of concern.

WILDLIFE INFORMATION: Mid-size conifers provide shelter for many species of birds and mammals, including squirrels. Seed production of non-native species may not be reliable, but when present seed may be eaten by nuthatches, chickadees, grosbeaks, finches, and crossbills. Insect-eating birds will forage on firs.

Incense-cedar (N) *Calocedrus decurrens*	All	75-90' tall	F-sun	Moist to dry

LANDSCAPE INFORMATION: A distinctive columnar to narrowly pyramidal tree with rich-green foliage and reddish-brown bark. It is very adaptable to landscape conditions and difficult sites including hot,

Common name *Botanical name*	Hardiness Zones	Plant Size	Light Needs	Soil Needs

dry roadsides. Its narrow growth habit is ideal for screens, shelterbelts, informal groupings, or single plantings. Often a better choice than western red-cedar in hot, dry areas. The tree occurs naturally in southern Oregon and in scattered areas farther north in the state.

WILDLIFE INFORMATION: Incense cedar provides shelter for many species of birds and mammals, including squirrels, porcupines, and deer. Chipping sparrows, hermit thrushes, siskins, flickers, and nuthatches eat the seeds; sapsuckers, woodpeckers, and other insect-eating birds forage on live and dead standing trees. Cavity-nesting birds and other wildlife nest and roost in tree cavities.

Common name *Botanical name*	Hardiness Zones	Plant Size	Light Needs	Soil Needs
Lawson cypress (N) (Port Orford cedar) *Chamaecyparis lawsoniana*	4-7	60-150' tall	F-sun P-shade	Moist to dry, good drainage

LANDSCAPE INFORMATION: A pyramidal tree with lacy, drooping foliage. Many different size, shape, and color selections used to be available from nurseries but now few are because of susceptibility to root rot. The columnar forms are still often used as screens and windbreaks; other selections are useful where size is a consideration. Where space isn't such a consideration, the tree form found in the wild is the most desirable.

WILDLIFE INFORMATION: Its dense foliage creates important shelter and nest sites for birds and small mammals, including chickadees, juncos, jays, warblers, and tree squirrels. Cavity-nesting birds and other wildlife nest and roost in cavities in mature trees.

Common name *Botanical name*	Hardiness Zones	Plant Size	Light Needs	Soil Needs
Hinoki false cypress *Chamaecyparis obtusa*	4-7	Varies	F-sun P-shade	Moist to dry, good drainage

LANDSCAPE INFORMATION: This is a common and generally a small ornamental conifer. Many variations in size, growth rate, form, and color are available from nurseries. The frequently planted "slender hinoki cypress" (*Chamaecyparis obtusa* 'Gracilis') grows to 20 feet and is a good conifer for small landscapes.

WILDLIFE INFORMATION: Its thick form creates important shelter and nest sites for songbirds.

Common name *Botanical name*	Hardiness Zones	Plant Size	Light Needs	Soil Needs
Juniper (N) *Juniperus* spp.	All	Varies	F-sun	Dry, good drainage

LANDSCAPE INFORMATION: **Rocky Mountain juniper** (*Juniperus scopulorum*) is a small tree (to 20 feet) or large shrub with scale-like foliage, shredding bark, and berry-like cones that are pea-sized and bluish-purple. This is an excellent plant for dry, sunny sites, needing no summer water. Plant as a windbreak, screen, or train it into a small tree. Several different cultivars are available from nurseries. **Western juniper** (*Juniperus occidentalis*) grows to 40-50 feet, has an open, irregular branching pattern, and is more appropriate for landscapes east of the Cascade mountains, where it occurs naturally.

WILDLIFE INFORMATION: Jays, red-winged blackbirds, waxwings, and solitaires are common consumers of the berries. Berries are also eaten by grouse, fox, deer, quail, coyotes, and chipmunks.

Common name *Botanical name*	Hardiness Zones	Plant Size	Light Needs	Soil Needs
Native spruce (N) *Picea* spp.	Varies	90-150' tall	Varies	Moist, cool

LANDSCAPE INFORMATION: Most native spruces planted in landscapes at lower altitudes eventually succumb to a combination of aphids and air pollution. Because of this and their eventual size, most are not

Common name *Botanical name*	Hardiness Zones	Plant Size	Light Needs	Soil Needs

appropriate for small landscapes or areas far from their natural habitats. **Brewer's spruce** (*Picea breweriana*) (Zones 4-7) has drooping branches and is the most ornamental. Its slow growth also makes it the most appropriate for small landscapes. **Engelmann spruce** (*Picea engelmannii*) (Zones All) is a narrow grower with fine-textured, slender needles. The lower branches remain dense and it resembles a droopy, less stiff Colorado blue spruce (*Picea pungens*). It grows in full sun and in deep, moist soil. **Sitka spruce** (*Picea sitchensis*) (Zones 4-7) is a fast-growing, broad tree with stiff needles that are prickly to touch. It is best suited for moist areas in full sun or partial shade. It is often found in coastal and lowland wetlands, and is a good conifer to plant in an area dominated by reed-canary grass.

WILDLIFE INFORMATION: Birds that eat the seeds include nuthatches, grosbeaks, finches, chickadees, siskins, goldfinches, crossbills, and sparrows. Grouse eat the needles, sapsuckers harvest insects from the sap, and woodpeckers forage on bark beetles. Many birds and mammals including deer, squirrels, chipmunks, and porcupines benefit from the shelter spruces provide, especially in winter. Large birds of prey may nest in mature trees. Cavity-nesting birds and other wildlife nest and roost in tree cavities.

Common name *Botanical name*	Hardiness Zones	Plant Size	Light Needs	Soil Needs
Spruce *Picea* spp.	All	60' tall	F-sun	Moist to dry, good drainage

LANDSCAPE INFORMATION: **Norway spruce** (*Picea abies*) is deep-green and pyramidal in shape when young. Older trees tend to grow out horizontally with strongly drooping branches. A good conifer for windbreaks in cold, dry areas. The **Serbian spruce** (*Picea omorika*) is a narrow, short-branched, slender-trunked tree that grows a bit taller, but narrower than the Norway spruce. Both are very hardy, wind-resistant, and fast-growing when they receive ample water.

WILDLIFE INFORMATION: The thick form creates important shelter (especially in winter) and nest sites for birds.

Common name *Botanical name*	Hardiness Zones	Plant Size	Light Needs	Soil Needs
Native pine (N) *Pinus* spp.	Varies	Varies	F-sun	Moist to dry, good drainage

LANDSCAPE INFORMATION: All native pines eventually get quite tall and there are several non-native species that may be more appropriate in landscapes where eventual height is of concern. **Shore pine** (*Pinus contorta* var. *contorta*) (Zones 4-7) is generally mid-sized. It may grow as a straight tree to 90 feet tall, or with a crooked trunk and an irregular dense crown on the coast or in poor soil. It grows in sand dunes, wetland edges, and rocky hilltops in coastal areas and lower elevations. **Lodgepole pine** (*Pinus contorta* var. *latifolia*) (Zones All) grows rather narrow, has short rich-green needles and squat, prickly cones. It grows rapidly to 60-100 feet tall and makes an effective screen when grown in informal groups. **White pine** (*Pinus monticola*) (Zones All) is a tall, symmetrical grower that grows very fast when young. The dense foliage is dark bluish-green; the cones look like large bananas in size and shape. **Ponderosa pine** (*Pinus ponderosa*) (Zones 1-3) has a tall, narrow, conical crown and grows to over 100 feet tall. It's best suited for large landscapes and windbreaks in hot, dry areas east of the Cascade mountains and in southern Oregon where it is more common.

WILDLIFE INFORMATION: The seeds are eaten by grouse, crossbills, grosbeaks, chickadees, band-tailed pigeons, quail, mourning doves, jays, nuthatches, finches, siskins, squirrels, and chipmunks. Bushtits, kinglets, chickadees, and woodpeckers glean pine beetles and other insects from the branches and cones. Many animals benefit from the evergreen cover and many songbirds nest in pines. The foliage is used by pine white butterfly larvae. Cavity-nesting birds and other wildlife nest and roost in cavities in mature trees.

Common name *Botanical name*	Hardiness Zones	Plant Size	Light Needs	Soil Needs
Pine *Pinus* spp.	All	Varies	F-sun	Moist to dry, good drainage

LANDSCAPE INFORMATION: **Mugho pine** (*Pinus mugo*) is a common landscape tree or tall shrub to 10 feet tall and wide. (There are several varieties that grow much smaller.) Give it space to grow equally wide as tall and don't plant where it will need to be pruned. **Austrian black pine** (*Pinus nigra*) is a broad, flat-topped tree to 75 feet tall. The attractive needles are dark-green and very stiff. It's very hardy, tolerant of a wide range of urban conditions, and makes a good specimen, screen, or windbreak. **Scots pine** (*Pinus sylvestris*) is pyramidal in youth, becoming irregular and drooping with age. It has medium-size, stiff, blue-green needles, and attractive reddish bark. It grows to about 60 feet tall and makes a good specimen, screen, or windbreak. Other equally good small pines include the **Japanese red pine** (*Pinus densiflora*) and the **limber pine** (*Pinus flexus*).

WILDLIFE INFORMATION: When available, the seeds are eaten by seed-eating birds. Bushtits, kinglets, chickadees, and woodpeckers glean pine beetles and other insects from the branches and cones. Many animals benefit from the evergreen cover and many songbirds nest in pines. The foliage is used by pine white butterfly larvae.

Douglas-fir (N) *Pseudotsuga menziesii*	All	80-150' tall	F-sun P-shade	Moist to dry, good drainage

LANDSCAPE INFORMATION: A fast-growing conifer that occurs in almost all regions of the Pacific Northwest that can support trees. It even appears from seedlings in urban settings. Douglas-fir forms a dense screen of soft, medium-green needles and is an excellent conifer for large landscapes or a quick temporary screen in smaller ones. At maturity it may shed big limbs during storms or blow over completely. Not a "true fir."

WILDLIFE INFORMATION: Birds that eat the seeds include grouse, crossbills, siskins, and many others. Squirrels and chipmunks also eat the seeds. Chickadees, nuthatches, brown creepers, and woodpeckers glean insects from the trunk, branches, and twigs. The foliage and twigs are browsed by beavers, porcupines, deer, and elk. The foliage is also eaten by pine white butterfly larvae, silver-spotted tiger moth larvae, and numerous other moths. Cavity-nesting birds and other wildlife, including flying squirrels, nest and roost in cavities in mature trees.

Western red-cedar (N) *Thuja plicata*	All	150' tall	F-sun to F-shade	Moist to dry

LANDSCAPE INFORMATION: The provincial tree of British Columbia. A large ornamental conifer with drooping branchlets that form graceful sprays of lacy foliage. When the foliage is crushed it has an unusual, sweet fragrance. The bark is attractive, gray-to-reddish. It is wind-firm and rarely sheds limbs. Red-cedar does poorly as seedlings in full sun and is usually found growing in the shade of other trees. In landscapes east of the mountains it will require summer water. Red-cedar forms a very dense tall screen and is highly recommended for buffer plantings, planting in with existing woodland trees and shrubs, or placed anywhere in a landscape where it can attain its natural form and size. This conifer can take intense pruning, however.

WILDLIFE INFORMATION: Birds that eat the winged seeds include grosbeaks, sparrows, waxwings, nuthatches, and siskins. Its dense foliage creates important shelter and nest sites for birds such as juncos, jays, and warblers, and small mammals including tree squirrels. The bark tears off in fibrous strips and is used by tree squirrels and porcupines for nest material. Deer and elk may browse the twigs and foliage. Cavity-nesting birds and other wildlife nest and roost in cavities created in mature trees.

Common name *Botanical name*	Hardiness Zones	Plant Size	Light Needs	Soil Needs
Western hemlock (N) *Tsuga heterophylla*	All	50-100' tall	F-sun to F-shade	Moist to dry

LANDSCAPE INFORMATION: The state tree of Washington. A large, fast-growing ornamental tree with down-sweeping branches, delicate feathery foliage, and many small cones. Although found in forested wetlands, it also grows in drier soil and is often found with red-cedar and Douglas-fir. Western hemlock does poorly as seedlings in full sun and is usually found growing in the shade of other trees. Western hemlock creates an excellent screen where there is room for it to attain its natural form and size. It does best in woodland conditions west of the Cascade mountains, and in dry urban areas it is often short-lived.

WILDLIFE INFORMATION: The seeds are available in September through January and are eaten by juncos, siskins, chickadees, grouse, finches, crossbills, chipmunks, and squirrels. The wood is used by porcupines, mountain beavers, and beavers (cut for building material, not food). Pileated woodpeckers eat carpenter ants on snags. Deer and elk browse the twigs and needles. Cavity-nesting birds and other wildlife nest and roost in cavities created in mature trees.

Common name *Botanical name*	Hardiness Zones	Plant Size	Light Needs	Soil Needs
Mountain hemlock (N) *Tsuga mertensiana*	All	25' tall	F-sun	Moist

LANDSCAPE INFORMATION: A slow-grower with dense, gray-green, short needles. It is a perfect conifer for small landscapes, particularly when planted in groups of three or five with differing heights. Trees blend well with salal, kinnikinnik, and evergreen huckleberry. Mountain hemlock commonly occurs west of the Cascade crest at elevations over 4,000 feet, but grows well in landscapes down to sea level. In nature it can reach 100 feet tall but in gardens it tends to stay smaller. To help prevent the destruction of their native habitat, don't buy plants that have been dug from the wild.

WILDLIFE INFORMATION: The seeds are eaten by juncos, siskins, chickadees, finches, crossbills, chipmunks, and squirrels. In older trees, the dense foliage creates important winter cover and potential nest sites for small birds.

For other conifers, see the tables in Appendix B.

DECIDUOUS TREES

Plants are listed alphabetically by their botanical name.

The botanical name first includes the *Genus* and then the *species*.

When referencing more than one species in a genus, the abbreviation (spp.) is used.

(N) = Plants that are native to Oregon, Washington, or southern British Columbia.

For definitions of headings, see the "Key" at the beginning of Appendix C.

For information on where native plants occur naturally, see the "Native Plant Occurrence Chart" in Appendix B.

Common name *Botanical name*	Hardiness Zones	Plant Size	Light Needs	Soil Needs
Vine maple (N) *Acer circinatum*	All	15-30' tall and wide	P-shade F-shade	Moist to dry

LANDSCAPE INFORMATION: A reliable ornamental that grows naturally as a multi-stemmed large understory shrub or small tree. Plant singularly or in groups in a woodland landscape, along a stream, or on a moist hillside. It has excellent soil-binding characteristics and is tolerant of shade. Trees grown in shade will often perform poorly if suddenly moved to sunny sites.

WILDLIFE INFORMATION: Birds that eat the seeds include grosbeaks, woodpeckers, nuthatches, finches, quail and grouse. Deer, mountain beavers, and beavers eat the wood and twigs. A larvae plant for the brown tissue moth and polyphemus moth, and a good nectar source for bees.

Douglas maple (N) *Acer glabrum* var. *douglassii*	All	15-25' tall	F-sun P-shade	Moist to dry

LANDSCAPE INFORMATION: This small maple is more common in intermountain and mountain regions than the vine maple. The twigs of new growth are bright red. It has interesting-looking seeds and excellent fall color. Generally, it is similar to vine maple in form but is more tolerant of dry soils and low humidity.

WILDLIFE INFORMATION: Birds that eat the seeds include grosbeaks, woodpeckers, nuthatches, finches, quail, and grouse. Deer, mountain beavers, and beavers eat the wood and twigs. A larvae plant for the brown tissue moth and polyphemus moth, and a good nectar source for bees.

Big-leaf maple (N) *Acer macrophyllum*	4-7	70' tall and 50' wide	F-sun P-shade	Moist to dry

LANDSCAPE INFORMATION: A fast-growing maple that is only suited for large landscapes. Plant in woodland edges along streams, or in the open where deep shade is desired. Underplant with salal, Oregon-grape, and sword fern. Bigleaf maple ranks only after red alder as the most abundant deciduous tree in Northwest coastal forests. The nooks and crannies in its bark are the favored habitats of mosses, lichens, and licorice fern.

WILDLIFE INFORMATION: Limbs of bigleaf maple break off and create sites for rot which make it easy for cavity-nesting birds to excavate. Rot also attracts insects, creating a food source for insectivorous birds. Birds associated with bigleaf maple include grouse, grosbeaks, kinglets, siskins, vireos, warblers, sapsuckers, woodpeckers, nuthatches, song sparrows, finches, and quail. Deer, muskrats, and beavers eat the wood and twigs. Good nectar source for bees and the foliage may be eaten by swallowtail butterfly larvae.

Common name *Botanical name*	Hardiness Zones	Plant Size	Light Needs	Soil Needs
Alder (N) *Alnus* spp.	Varies	Varies	F-sun P-shade	Wet to moist

LANDSCAPE INFORMATION: **Red alder** (*Alnus rubra*) (Zones 4-6) is a common tree in lowland areas, often forming pure stands. The bark is gray, smooth, often with white patches of lichens. Clusters of dark cones (jellybean-sized) containing winged nutlets remain on the tree through winter. It tolerates brackish soil and lives 70 to 100 years. **Mountain alder** (*Alnus incana*) (Zones 1-3) and **White alder** (*Alnus rhombifolia*) (Zones 1-3) are fast-growing and more tolerant of heat and wind. Like the red alder, they can reach 70' tall. **Sitka alder** (*Alnus viridis* ssp. *sinuata*) (Zones All) is a miniature looking alder that forms a shrub or small tree 12- 20' tall. It is generally thicket-forming and grows better in full sun. Plant alders along streams, moist slopes, wet clearings, and moist bottomlands. Excellent for revegetating disturbed sites and creating quick cover or a screen in poor, moist soil.

WILDLIFE INFORMATION: Birds that use alder for cover and nesting include warblers, bushtits, and sparrows. Birds that eat seeds, buds, and insects associated with alder include mallards, widgeons, grouse, bushtits, kinglets, siskins, vireos, warblers, and chickadees. Mammals that eat the leaves, twigs, or wood include snowshoe hares, beavers, porcupines, deer, and elk. Alders create organic debris for soil organisms and are good cover for fish. The leaves are eaten by swallowtail and other butterfly larvae, and tent caterpillars. Cavity-nesting birds and other wildlife nest and roost in tree cavities.

Birch (N) *Betula* spp.	All	30'-50' tall	F-sun	Wet to moist

LANDSCAPE INFORMATION: **Water birch** (*Betula occidentalis*) is a very hardy clump-grower with coppery-brown, shiny bark and deep-green leaves that turn pale yellow in fall. It's not quite as tall as the paper birch and is a better tree for wet areas that dry out some in summer. **Paper birch** *(Betula papyrifera)* is often a multi-stemmed tree with papery or smooth, attractive white bark occurring on older trees. It has yellowish catkins in spring and nice yellow leaves in fall. Don't plant birches near drain pipes or foundations.

WILDLIFE INFORMATION: Seed is eaten on and under the trees by juncos, siskins, finches, sparrows, and grouse. Birches supply many insects to birds including kinglets, sapsuckers, woodpeckers, warblers, nuthatches, and chickadees. Bark is used as nesting material by birds and small mammals. Mammals that eat the leaves, twigs, or wood include beaver, hare, chipmunk, deer, and elk. Leaves are eaten by mourning cloak and swallowtail butterfly larvae. Cavity-nesting birds and other wildlife nest and roost in tree cavities.

Pacific dogwood (N) *Cornus nuttallii*	2-7	50' tall 25' wide	F-sun P-shade	Moist to dry, good drainage

LANDSCAPE INFORMATION: An attractive, round-to-conical–formed tree with large creamy-white, showy flower bracts. The flowers appear in spring (and often again in fall), and are followed by half-inch, orange-red button-like fruits. Plant in open to dense woodland gardens, in forest edges, and along waterways, not in lawns. Trees are intolerant of watering and fertilizing, susceptible to anthracnose disease, and difficult to transplant. Protect the trunk from hot sun. The hybrid 'Eddie's White Wonder' is a better choice for ornamental areas receiving summer water.

WILDLIFE INFORMATION: Birds that eat the fruit include sapsuckers, woodpeckers, bluebirds, tree swallows, vireos, thrushes, evening grosbeaks, white-crowned sparrows, song sparrows, towhees, grouse, jays, and house finches. Flower parts may be eaten by spring azure butterfly larvae.

Common name *Botanical name*	Hardiness Zones	Plant Size	Light Needs	Soil Needs
Dogwood *Cornus* spp.	Varies	20–30' tall 15–25' wide	F-sun P-shade	Moist to dry

LANDSCAPE INFORMATION: **Kousa dogwood** (*Cornus kousa*) (Zones 3–7) is a full, bushy, horizontal-spreading tree with profuse white blooms in June. Interesting, edible, strawberry-like red fruit remains on tree in fall. Fall color is red. It's a nice ornamental specimen tree and preferable to pink dogwood in disease resistance, bud hardiness, sun-and-drought tolerance. It looks better with summer water and can be grown in a lawn. The **Cornelian-cherry** (*Cornus mas*) (Zones All) is a multi-stemmed and bushy large shrub or tree with small yellow blossoms that occur on bare twigs in very early spring. The berries are bright red, then dark red, and edible by fall. It's an easy-care tree planted for its interesting, early flower habit and fruit.

WILDLIFE INFORMATION: The flowers attract a variety of flying insects, and different urban bird species eat fruit on and under the tree.

Common name *Botanical name*	Hardiness Zones	Plant Size	Light Needs	Soil Needs
Native hawthorn (N) *Crataegus* spp.	Varies	15–30' tall and wide	F-sun	Moist to dry

LANDSCAPE INFORMATION: **Black hawthorn** (*Crataegus douglasii*) (Zones All) and **Columbia hawthorn** (*Crataegus columbiana*) (Zones 1–3) grow in moist and dry areas as multi-stemmed, slow-growing shrubs or sturdy small trees. The **Columbia hawthorn** occurs mainly east of the Cascade mountains and the **black hawthorn** occurs on both the east and west sides. The branches are thorned and the bark is light-brown, rough, and scaly. The flower clusters are white, about 3 inches in diameter, and are followed by berries in August. Fruits on the black hawthorn are blackish, and on the Columbia hawthorn they are reddish. Both have nice, dark fall color. Plant in open areas, forest edges, hedges, sunny streamsides, roadsides, and coastal bluffs. They adapt to disturbed sites, have good soil stability value, and can eventually form an impenetrable thicket.

WILDLIFE INFORMATION: Berries are eaten by solitaires, robins, waxwings, grosbeaks, thrushes, woodpeckers, band-tailed pigeons, wood ducks, grouse, pheasants, and turkeys, also black bear, coyotes, and foxes. Rabbits and deer browse leaves and twigs. The plants are resistant to beaver damage. Leaves are food for swallowtail butterfly larvae.

Common name *Botanical name*	Hardiness Zones	Plant Size	Light Needs	Soil Needs
Hawthorn *Crataegus* spp.	All	15–30' tall	F-sun	Moist to dry, good drainage

LANDSCAPE INFORMATION: **Lavalle hawthorn** (*Crataegus* x *lavallei*) has lustrous dark-green foliage in summer that turns to orange-red in late fall. The attractive white flowers are followed by orange-red fruit that persists into winter. The thorns are 2 inches long. **Washington hawthorn** (*Crataegus phaenopyrum*) is a rounded, dense tree with lustrous, green summer foliage that turns red-orange in fall. White flowers occur in May and are followed by tiny, orange-red fruit that persists into winter. These two hawthorns are more disease-resistant than other non-native hawthorns and both have 2- to 3-inch thorns.

WILDLIFE INFORMATION: The dense, twiggy crown provides good escape cover for songbirds. Fruit hangs on trees well into winter and provides food for birds and other wildlife on and under the tree. Birds may not eat the fruit until snow covers other food sources. Leaves provide food for swallowtail butterfly larvae.

Common name *Botanical name*	Hardiness Zones	Plant Size	Light Needs	Soil Needs
Oregon ash (N) *Fraxinus latifolia*	4-7	85' tall 60' wide	F-sun	Moist to wet

LANDSCAPE INFORMATION: A large, straight-trunked tree that grows in wet soils. Female specimens produce dry, winged seeds which look like miniature canoe paddles. Fall color is yellow. Plant near waterways, on bottomlands, around the margins of wetlands, or as a large shade tree. Good for large-scale landscapes and areas with standing water all winter.

WILDLIFE INFORMATION: Seeds are eaten by grosbeaks, wood ducks, finches, grouse, and other birds. Sapsuckers often use trunks as drill sites. Wood is used by beavers; twigs and leaves are eaten by deer and elk. Leaves are eaten by butterfly larvae and other types of insects. Cavity-nesting birds and other wildlife nest and roost in tree cavities.

Walnut *Juglans* spp.	All	50' tall and wide	F-sun	Moist to dry

LANDSCAPE INFORMATION: Walnuts eventually become big, hardy shade trees. **Black walnut** (*Juglans nigra*) is a rounded tree with furrowed, black-brown bark. The nuts are thick-shelled, very hard, and strongly flavored. Grafted varieties are available from nurseries. The **English walnut** (*Juglans regia*) has smooth gray-white bark and large branches. Both prefer hot summers and different varieties may be necessary for optimum nut production.

WILDLIFE INFORMATION: Nutmeats are often eaten by birds after the nuts have fallen on a hard surface, been run over by a car, or been opened by squirrels, chipmunks or other mammals. Birds that eat the nutmeat include woodpeckers, jays, orioles, crows, chickadees, nuthatches, finches, and sparrows.

Pacific crabapple (N) *Malus (Pyrus) fusca*	4-7	20-30' tall and wide	F-sun P-shade	Moist

LANDSCAPE INFORMATION: A thornless tree or large shrub that can eventually grow into a dense thicket. The ½-inch white flowers are followed by small, elongated apple-like fruits. The small tart fruits vary in shape, start green, and turn to yellow or red. Plant near waterways, wetlands, or other moist sites. Very salt-tolerant.

WILDLIFE INFORMATION: Birds that eat the fruits include evening grosbeaks, towhees, sapsuckers, woodpeckers, waxwings, and grouse. Fruits are also eaten by coyotes, foxes, and other mammals. Butterflies associated with Pacific crabapple include the spring azure. Cavity-nesting birds and other wildlife may nest and roost in tree cavities of large trees.

Flowering crabapple *Malus* spp.	All	Varies	F-sun	Moist

LANDSCAPE INFORMATION: The many different varieties of flowering crabapple vary in size, flower color, and disease susceptibility. All are spectacular, small ornamental trees when in bloom. The fruits vary in size from ¼-inch to 2 inches in diameter and are dark-red to yellow or green. Some varieties retain fruit on the tree throughout the winter. Contact your local nursery for recommendations; beware of sterile varieties that don't produce fruit.

WILDLIFE INFORMATION: Birds that eat the fruit on and under the tree include grouse, geese,

Common name *Botanical name*	Hardiness Zones	Plant Size	Light Needs	Soil Needs

pheasants, sapsuckers, woodpeckers, jays, robins, waxwings, starlings, orioles, and towhees. Finches, sparrows, and waxwings eat the flowers. Nectar is used by hummingbirds and some adult butterflies. Leaves are food for butterfly larvae.

Common name *Botanical name*	Hardiness Zones	Plant Size	Light Needs	Soil Needs
Quaking aspen (N) *Populus tremuloides*	All	40-70' tall	F-sun	Wet to moist or slightly dry

LANDSCAPE INFORMATION: A fast-growing tree with smooth, pale gray-green to whitish trunk and limbs. The dainty, light-green, round leaves flutter and "quake" in the slightest breeze. Beautiful gold fall color. Plant in groups along large waterways and other moist areas. Aspen is naturally thicket-forming and is not suited for a small garden or areas near drain lines or foundations.

WILDLIFE INFORMATION: The catkins are eaten by grouse, pheasants, and siskins. Mature trees and snags are used by cavity-nesting birds including sapsuckers, woodpeckers, small owls, and chickadees. Winter buds are eaten by grouse, possibly orioles and purple finches. Mammals that use leaves, twigs, or wood include deer, mountain beavers, hares, rabbits, and porcupines. Leaves are eaten by Lorquin's admiral and swallowtail butterfly larvae.

Common name *Botanical name*	Hardiness Zones	Plant Size	Light Needs	Soil Needs
Black cottonwood (N) *Populus balsamifera* var. *trichocarpa*	All	70-150' tall	F-sun	Moist to wet

LANDSCAPE INFORMATION: A very large, fast-growing straight-trunked tree. The sticky leaf-buds emit balsamic odor, and in late spring seeds from female trees fill the air with cottony down. Plant in areas where quick cover and fast growth is needed and in forested wetlands where it is allowed plenty of space. It's an important tree for streambank protection. It can be propagated by placing 6-foot dormant cuttings in the ground, even in areas where reed-canary grass dominates. It will also sprout from the stump.

WILDLIFE INFORMATION: Provides roosting and nesting sites for many species of birds, including bald eagles, hawks, owls, and great blue herons. Mature trees and snags are used by cavity-nesting birds including wood ducks, woodpeckers, and chickadees. Other birds associated with this tree include vireos, grosbeaks, sapsuckers, finches, towhees, grouse, and quail. Also a food source for mountain beaver, beaver, deer, and elk. Leaves are eaten by Lorquin's admiral, mourning cloak, and swallowtail butterfly larvae; also dagger tussock and hornet moth larvae. Cavity-nesting birds and other wildlife nest and roost in tree cavities.

Common name *Botanical name*	Hardiness Zones	Plant Size	Light Needs	Soil Needs
Bitter cherry (N) *Prunus emarginata* var. *mollis*	4-7	70' tall 30' wide	F-sun P-shade	Moist to dry

LANDSCAPE INFORMATION: A thicket-forming tall shrub or single-stem tree with reddish-brown, shiny bark. The flowers are small, white, and in 2- to 5-inch clusters. The small, bright-red bitter cherries occur in mid to late summer and the leaves turn a bright yellow in fall. Plant in sunny, dry clearings and along margins of woods and woodland landscapes.

WILDLIFE INFORMATION: Many birds eat fruits, including grouse, band-tailed pigeons, flickers, jays, robins, bluebirds, waxwings, tanagers, orioles, grosbeaks, finches, mourning doves, and towhees. Fruits are also eaten by squirrels, foxes, black bears, coyotes, chipmunks, and raccoons. Deer and elk will browse leaves and twigs. Butterflies associated with this species includes Sara orangetip, silvery blue, swallowtail, Lorquin's admiral, and spring azure.

Common name *Botanical name*	Hardiness Zones	Plant Size	Light Needs	Soil Needs
Native oak (N) *Quercus* spp.	Varies	65' tall 50' wide	F-sun	Dry, good drainage

LANDSCAPE INFORMATION: **Oregon white oak** (*Quercus garryana*) (Zones 4–7), also known as **Garry oak,** is a thick-limbed tree with a broad crown; it's often short and crooked in rocky soils. It is the only oak native to Washington and British Columbia. A long-lived tree, some specimens indicate a life span of over 300 years. Plant in dry to moist, well-drained gravely or sandy soils. **California black oak** (*Quercus kelloggii*) (Zones 3–4) has a handsome branch pattern, attractive leaves, and nice fall color. It is slightly larger than the white oak. For other oaks, see "Broadleaf Evergreen Trees."

WILDLIFE INFORMATION: Many birds eat the acorns, including wood ducks, mallards, turkeys, band-tailed pigeons, quails, grouse, woodpeckers, nuthatches, thrushes, towhees, jays, and Clark's nutcrackers. Mammals that eat acorns include black bears, deer, muskrats, raccoons, tree squirrels, gophers, ground squirrels, and mice. Many insects are associated with oaks and the leaves are eaten by certain butterfly larvae. Cavity-nesting birds and other wildlife nest and roost in tree cavities.

Common name *Botanical name*	Hardiness Zones	Plant Size	Light Needs	Soil Needs
Oak *Quercus* spp.	All	Varies	F-sun	Moist to dry, good drainage

LANDSCAPE INFORMATION: **Red oak** (*Quercus rubra*) and **scarlet oak** (*Quercus coccinea*) are moderate- to fast-growing trees when planted in deep, rich soil. They have an open-branching habit, beautiful fall color, and are easy to garden under. They also make good street, lawn, or shade trees. They need plenty of space and can grow to 80 feet tall. **Pin oak** (*Quercus palustris*) is smaller and grows better in drier zones.

WILDLIFE INFORMATION: See information on native oak.

Common name *Botanical name*	Hardiness Zones	Plant Size	Light Needs	Soil Needs
Cascara (N) *Rhamnus purshiana*	All	To 30' tall and wide	F-sun P-shade	Moist to dry

LANDSCAPE INFORMATION: An erect, tall shrub or small tree. The flowers are inconspicuous; the berries are up to ½-inch long, yellow or red. They ripen to black in late summer and hang on into fall. Plant with conifers, vine maple, and red alder in wet clearings. It has a good soil-binding characteristics, grows well on disturbed sites, but is sensitive to air pollution.

WILDLIFE INFORMATION: Berries are eaten by grosbeaks, woodpeckers, grouse, band-tailed pigeons, mourning doves, jays, robins, and tanagers. It also attracts many insectivorous birds, including bushtits, kinglets, chickadees, flycatchers, and nuthatches. Mammals that eat fruit include black bears, foxes, coyotes, and raccoons. Leaves and other plant parts are used by swallowtail, gray hairstreak, and other butterfly larvae.

Common name *Botanical name*	Hardiness Zones	Plant Size	Light Needs	Soil Needs
Sumac (N) *Rhus glabra*	All	6–15' tall and wide	F-sun P-shade	Moist to dry

LANDSCAPE INFORMATION: A large shrub or small tree with a striking growth form, brilliant fall color, and showy fruit clusters. It is adaptable to poor soil conditions and hot climates. It can spread vigorously by shallow underground roots and form thickets. Plant with sagebrush and bunch grasses in dry landscapes, also along a sunny woodland edge. Plant in clumps or as a single specimen. **Staghorn sumac** (*Rhus typhina*) grows larger, has velvety branches, and is from the eastern U.S.

WILDLIFE INFORMATION: Its leggy growth offers little winter protection, and few birds nest in the

Common name _Botanical name_	Hardiness Zones	Plant Size	Light Needs	Soil Needs

branches. However, birds find cover in sumac thickets and nest at the base among stems. Birds that may use the fruits include turkeys, grouse, pheasants, hermit thrushes, crows, finches, pine grosbeaks, vireos, jays, and chickadees. Foliage may be used by spring azure butterfly larvae.

Common name _Botanical name_	Hardiness Zones	Plant Size	Light Needs	Soil Needs
Willow (N) _Salix_ spp.	Varies	20-50' tall and wide	F-sun	Moist to dry

LANDSCAPE INFORMATION: More than a dozen species of willow are native to the Pacific Northwest, many with subtle differences. The more common species include **Hooker willow** (_Salix hookeriana_, includes _S. piperi_) (Zones 4-7), **Pacific willow** (_Salix lucida_ var. _lasiandra_) (Zones All), **Scouler willow** (_Salix scouleriana_) (Zones All), and **Sitka willow** (_Salix sitchensis_) (Zones All). All have excellent soil-binding abilities and help to control erosion along waterways, stream and creek banks, and wetlands. Use as streambank restoration and stabilization, general wildlife habitat enhancement, water quality improvement, and shelterbelts.

WILDLIFE INFORMATION: Excellent forage and cover for wildlife, including fish when planted next to water. Insectivorous birds associated with willow include bushtits, kinglets, warblers, and sapsuckers. Buds and small portions of the twigs are eaten by grouse and grosbeaks. Rabbits, elk, and deer also eat twigs, foliage, and bark. Beavers feed on bark, buds, and wood. Hare, deer, elk, and porcupines eat the wood. Nectar is used by many insects, including brown elfin, Sara orangetip, mourning cloak, Milbert's tortoiseshell, and other adult butterflies. Leaves are used by mourning cloak, tiger swallowtail, Lorquin's admiral, and satyr anglewing butterfly larvae.

Common name _Botanical name_	Hardiness Zones	Plant Size	Light Needs	Soil Needs
Mountain-ash _Sorbus_ spp.	All	50' tall 40' wide	F-sun P-shade	Moist to dry, good drainage

LANDSCAPE INFORMATION: Mountain-ash are ornamental trees with dense, oval to round crowns and fern-like foliage. The flat clusters of small, creamy-white flowers appear in April or May followed by ¼-inch berries. Nice fall color. Berries may be messy on patios or busy walkways. Different varieties and species of mountain-ash are available from nurseries. **European mountain-ash** (_Sorbus aucuparia_) and **Chinese mountain-ash** (_Sorbus hupehensis_) are very popular. Species with pink or white fruit seem less favored by birds. For information on the native mountain ash, see "Deciduous Shrubs."

WILDLIFE INFORMATION: The berry-like fruits persist into winter and are a favorite of robins. Other birds that eat the berries include ruffed grouse, Swainson's thrush, waxwings, orioles, grosbeaks, and finches. Deer will browse leaves and twigs.

For other deciduous trees, see the tables in Appendix B.

BROADLEAF EVERGREEN TREES

Plants are listed alphabetically by their botanical name.
The botanical name first includes the *Genus* and then the *species*.
When referencing more than one species in a genus, the abbreviation (spp.) is used.
(N) = Plants that are native to Oregon, Washington, or southern British Columbia.
For definitions of headings, see the "Key" at the beginning of Appendix C.
For information on where native plants occur naturally, see the "Native Plant Occurrence Chart" in Appendix B.

Common name *Botanical name*	Hardiness Zones	Plant Size	Light Needs	Soil Needs
Madrona (N) *Arbutus menziesii*	3-7	30-75' tall, broad, round	F-sun P-shade	Dry, good drainage

LANDSCAPE INFORMATION: This is a bold and massive, yet elegant broadleaf tree. It has large, oval, shiny-green, rhododendron-like leaves and brown-red bark that peels in thin flakes. The fragrant flowers are small, white bells that occur from April to June. The fruit is an orange-red berry ½-inch in diameter that sometimes persists on the tree until Christmas. Plants are tolerant of infertile soil and grow well with salal, Oregon-grape, mountain-boxwood, and snowberry. The root system should not be disturbed and plants should not be irrigated once they're established.

WILDLIFE INFORMATION: Fruit is eaten by band-tailed pigeons, quail, flickers, varied thrushes, waxwings, evening grosbeaks, mourning doves, and robins. Fruit is also eaten by raccoons and other mammals. The flowers are visited by spring azure butterflies and bees. It is a larval food plant for the ceanothus silk moth and the brown elfin butterfly.

Strawberry tree *Arbutus unedo*	4-7	20-25' tall and broad	F-sun P-shade	Moist to dry, good drainage

LANDSCAPE INFORMATION: A slow to moderately fast-growing tree with foliage that looks nice all year. Delicate ivory flowers appear in October along with previous spring's fruit. The berries are ¾-inch, round and somewhat strawberry-like in appearance, not in taste. The strawberry tree makes a wonderful screen when mass-planted and left unpruned, and an open branched tree when pruned. Plants may need protection in cold winters. Also see *Arbutus unedo* 'Compacta' under "Evergreen Shrubs."

WILDLIFE INFORMATION: The flowers attract bees and moths. The fruit is eaten by birds, on and under the plant.

Chinquapin (N) *Chrysolepis chrysophylla*	4-7	To 50'	F-sun to F-shade	Dry, good drainage

LANDSCAPE INFORMATION: A massive, slow-growing tree or large shrub depending on its location. It has deep-green leaves that have yellow fuzz underneath. The showy summer blooms have fluffy, upright spikes with strongly scented, cream-white flowers. It is best grown as a specimen tree and it shouldn't be watered after it is established.

WILDLIFE INFORMATION: The small, sweet, chestnut-like nuts occur in spiny burs and are consumed by squirrels, chipmunks, ground squirrels, and birds.

Common name *Botanical name*	Hardiness Zones	Plant Size	Light Needs	Soil Needs
Tan-oak (N) *Lithocarpus densiflorus*	4-7	25-60' tall	F-sun P-shade	Dry, good drainage

LANDSCAPE INFORMATION: A fine ornamental with stiff, leathery leaves and a dense crown. Male catkins are erect and creamy-white in bloom; acorns are about 1 inch long. It is tolerant of shade and poor soil. It can be planted as a specimen or in widely spaced groups in large landscapes. It makes an excellent screen and can be used as a street tree.

WILDLIFE INFORMATION: Acorns are eaten by wood ducks, mallards, band-tailed pigeons, woodpeckers, pheasants, nuthatches, thrushes, towhees, grouse, quail, jays, and Clark's nutcrackers, also bear, deer, muskrats, raccoons, tree squirrels, gophers, ground squirrels, and mice. Older trees may be used by cavity-nesting species.

Canyon live oak (N) *Quercus chrysolepis*	5-7	20-60'	F-sun	Dry, good drainage

LANDSCAPE INFORMATION: A handsome tree with a dense, rounded form that tolerates infertile, rocky soil. The leaves are dark-green and glossy, and the bark is white and smooth. The acorn cups are covered with a golden fuzz. Plant alone, or in widely spaced groups.

WILDLIFE INFORMATION: See information for tan oak (*Lithocarpus densiflorus*).

Holly oak *Quercus ilex*	4-7	20-50' tall	F-sun P-shade	Dry, good drainage

LANDSCAPE INFORMATION: A clean, formal-looking oak native to the Mediterranean region. It tolerates wind and salt air, makes a good street tree, and it can be pruned if necessary. **Southern live oak** (*Quercus virginiana*) is another good evergreen oak for suburban landscapes and it has all the attributes of the holly oak.

WILDLIFE INFORMATION: Acorns are eaten by mallards, band-tailed pigeons, woodpeckers, nuthatches, towhees, quail, and jays, also deer, raccoons, tree squirrels, gophers, and mice.

Bay tree (N) (Oregon-myrtle) *Umbellularia californica*	4-7	Varies	F-sun to F-shade	Moist to dry

LANDSCAPE INFORMATION: This is the Oregon "myrtlewood" tree. It is a neat-looking large shrub or dense tall tree that casts deep shade. Growth rate is variable depending on exposure, soil, and moisture. Trees can reach 60 feet or more when located out of the wind. Crushed leaves release a memorable, spicy-sweet fragrance. Leaves are used in the same way as "bay leaves" purchased as a spice at the store. Plant on moist streambanks, floodplains, and lower mountain slopes with alder, maple, and conifers. Very shade-tolerant and a good choice for a large, shady, screen planting. Recommended for large landscapes or a temporary planting in smaller areas.

WILDLIFE INFORMATION: Fruit is green, olive-like, and contains one large seed. Seeds may be eaten by squirrels. Mature trees provide good thermal cover.

For other evergreen trees, see the tables in Appendix B.

DECIDUOUS SHRUBS

Plants are listed alphabetically by their botanical name.
The botanical name first includes the *Genus* and then the *species*.
When referencing more than one species in a genus, the abbreviation (spp.) is used.
(N) = Plants that are native to Oregon, Washington, or southern British Columbia.
For definitions of headings, see the "Key" at the beginning of Appendix C.
For information on where native plants occur naturally, see the "Native Plant Occurrence Chart" in Appendix B

Common name *Botanical name*	Hardiness Zones	Plant Size	Light Needs	Soil Needs.
Serviceberry (N) (Saskatoon) *Amelanchier alnifolia*	All	8-30' tall and wide	F- sun P-shade	Moist to dry, good drainage

LANDSCAPE INFORMATION: A broad, multi-branched, large shrub or small tree that tends to grow larger east of the mountains. The white flowers from early spring are followed by purple ¼-inch edible berries in June or July. The fall foliage can be a beautiful red and yellow. Plant in an open woodland, shrub border, hedgerow, or on a sunny bank. Plants are slow to establish and are sensitive to competition around the roots.

WILDLIFE INFORMATION: Berries are eaten by woodpeckers, crows, chickadees, thrushes, towhees, bluebirds, waxwings, orioles, tanagers, grosbeaks, goldfinches, juncos, grouse, and pheasants; also chipmunks, marmots, skunks, foxes, ground squirrels, raccoons, and bear. Deer and elk browse the leaves and twigs. Nectar is used by spring azure butterflies and foliage is eaten by swallowtail and other butterfly larvae.

Barberry *Berberis* spp.	All	5'-8' tall 5' wide	F–sun P-shade	Moist to dry, good drainage

LANDSCAPE INFORMATION: Barberries are popular landscape shrubs and have a dense vase shape, small rounded leaves, small yellow flowers, and many thorns. The bright-red, egg-shaped berries hang on into winter, and shrubs have brilliant fall color when grown in full sun. Many species are available from nurseries; two common ones are **Japanese barberry** (*Berberis thunbergii*) and **Korean barberry** (*Berberis koreana*). A purple form of Japanese barberry is common. Dwarf varieties lack the shelter quality of taller forms. Plant in a mixed hedge, large shrub border, and as a foundation plant around a building.

WILDLIFE INFORMATION: Older plants provide good shelter and nesting habitat for finches and other urban songbirds. Berries may be eaten by robins, waxwings, and juncos.

Bog birch (N) *Betula glandulosa*	All	6-10' tall 6' wide	F–sun	Wet to moist

LANDSCAPE INFORMATION: A graceful shrub or small tree with small, rounded leaves that turn orange and coppery in fall. Plant in a wet place in the yard, the back of a border, at the edge of a pond, or mass-plant with other shrubs and trees in large wetland landscapes.

WILDLIFE INFORMATION: Birds that eat the seeds on and under the shrub include grosbeaks, crossbills, juncos, siskins, chickadees, and grouse. Insectivorous birds including warblers and vireos glean insects off the shrub. Beavers, hare, chipmunks, elk, and deer eat woody parts. Leaves are eaten by swallowtail, mourning cloak, and other butterfly larvae.

Common name *Botanical name*	Hardiness Zones	Plant Size	Light Needs	Soil Needs
Butterfly bush *Buddleia davidii*	All	Varies	F-sun	Moist to dry, good drainage

LANDSCAPE INFORMATION: A vigorous shrub that can grow 6 feet or more in a year. The species can grow 15 feet tall; dwarf varieties grow to about 8 feet. It's highly valued for a long show of colorful, fragrant summer flowers and its adaptability to poor soils. It's easily propagated from hardwood cuttings stuck in the ground in winter. Varieties available from nurseries include those with white, pink, blue, or purple flowers. Only dwarf or 'Petite' selections are recommended for small spaces. The flowers are produced on new growth and plants can be pruned heavily (nearly to the ground) each year prior to spring growth for maximum flower production and to keep plants compact. Deadhead blossoms to prolong bloom. Butterfly bush is considered an invasive plant in some areas; contact your county noxious weed board or county extension service for current information in your area.

WILDLIFE INFORMATION: Flowers attract swallowtail, pine white, red admiral, painted lady, and mourning cloak butterflies. Flowers are also visited by moths and hummingbirds.

Siberian peashrub *Caragana arborescens*	1-3	10-15' tall and wide	F-sun P-shade	Moist to dry

LANDSCAPE INFORMATION: A fast-growing, multi-stemmed, upright shrub or small tree with small thorny spines. Bright-yellow flowers occur in early May. It tolerates harsh conditions and alkaline soils and is a nearly indestructible hedge, windbreak, or small specimen tree in hot, dry areas.

WILDLIFE INFORMATION: Flowers attract hummingbirds and butterflies. When unpruned it provides good cover for songbirds.

Deer brush (N) (Mountain lilac) *Ceanothus integerrimus*	All	6-12' tall 6' wide	F-sun P-shade	Dry, good drainage

LANDSCAPE INFORMATION: An openly branched shrub that is sometimes partially evergreen. The tiny flowers occur in large, open clusters in late spring and are white to deep blue-lilac. A good plant for open woodlands, woodland edges, roadsides, and large landscapes.

WILDLIFE INFORMATION: Nectar is used by spring azure and other adult butterflies, also many other flying insects. The leaves may be used by pale swallowtail butterfly and ceanothus moth larvae. Leaves and twigs are browsed by deer and elk.

Red-twig dogwood (N) *Cornus sericea (C. stolonifera)*	All	4-12' tall	F-sun P-shade	Wet to moist

LANDSCAPE INFORMATION: A fast-growing multi-stemmed shrub that eventually spreads into a thicket from root sprouts. White flower clusters appear in May to June followed by white or bluish berries in late summer. Red or purple twigs and branches are attractive, especially in winter, and fall color may be a deep purplish-red. Plant in wet areas along creeks, lakes or large ponds, meadows, or any wet or irrigated area where the plant can spread. A good barrier plant that makes a dense thicket. **Yellow-twig dogwood** (*C. sericea* 'Flaviramea') has yellow twigs and branches; the variety 'Isanti' is compact; the variety 'Eligantissima' is variegated.

WILDLIFE INFORMATION: Berries are eaten by vireos, warblers, kingbirds, robins, flickers, flycatch-

Common name *Botanical name*	Hardiness Zones	Plant Size	Light Needs	Soil Needs

ers, wood ducks, grouse, band-tailed pigeons, and quail, also bear, foxes, skunks, and chipmunks. The wood is eaten by porcupines and the twigs are browsed by deer, elk, and rabbits. Beavers and muskrats use twigs to repair dams or build dens. Nectar is used by orange sulphur and other adult butterflies. The leaves are used by spring azure and other butterfly larvae.

Common name *Botanical name*	Hardiness Zones	Plant Size	Light Needs	Soil Needs
Hazelnut (N) *Corylus cornuta* var. *californica*	All	10-30' tall and wide	F-sun P-shade	Moist to dry

LANDSCAPE INFORMATION: A tall, multi-branched, spreading shrub with sucker growth from the base. East of the mountains plants stay under 10 feet in the wild; west of the mountains they easily grow twice that size. The velvety-textured leaves are pale yellow in fall and the long male (pollen) tassels are striking in late winter. An attractive landscape plant when used as a tall background shrub in a woodland setting. It's also suitable for streamsides, shelterbelts, and large informal hedges when allowed to sucker freely. European, commercial hazelnuts, also known as **filberts**, are also good choices for the landscape; two or more varieties are required for good nut production.

WILDLIFE INFORMATION: Edible nuts are much sought-after by squirrels and Steller's jays, even before they are ripe. Chipmunks, raccoons, and red foxes also eat the nuts. Rabbits and beavers eat the leaves and wood. Larger plants provide good tangled cover and nesting habitat for birds close to the ground.

Common name *Botanical name*	Hardiness Zones	Plant Size	Light Needs	Soil Needs
Cotoneaster *Cotoneaster* spp.	Varies	Varies	F-sun P-shade	Moist to dry, good drainage

LANDSCAPE INFORMATION: **Peking cotoneaster** (*Cotoneaster acutifolia*) (Zones 1–3) is an attractive shrub with glossy dark-green leaves, small pink flowers, and small purple berries. It grows to 10 feet tall and wide and has orange-red fall color. Plant to create a screen, or as a tall hedge with other shrubs in hot areas. **Rock cotoneaster** (*Cotoneaster horizontalis*) (Zones All) is a heavily textured shrub with a herringbone branch pattern. It has tiny, glossy dark-green leaves with orange, red, or purplish fall color. Numerous small red berries hold on the plant well into the winter. It works well as a tall ground cover or as an accent plant in a rockery; also attractive when cascading over a sunny bank. Don't plant too close to a walkway and prune only minimally to retain shape. For other cotoneasters, see "Evergreen Shrubs."

WILDLIFE INFORMATION: Berries are eaten by finches, robins, and other fruit-eating birds. Bees and butterflies feed on the nectar.

Common name *Botanical name*	Hardiness Zones	Plant Size	Light Needs	Soil Needs
Hardy fuchsia *Fuchsia magellanica*	3-7	3-6' tall and wide	F-sun P-shade	Moist, good drainage

LANDSCAPE INFORMATION: A fast-growing shrub with many 1½ inch long, pendent flowers that are crimson with rich purple petticoats. Flowers occur July to frost. This is a choice late-flowering shrub for a protected area in the landscape. In severe winters it may be killed to the ground, but established plants and those with mulched roots will rebound in the spring. In cold areas treat the plant like a perennial.

WILDLIFE INFORMATION: Hummingbirds are fond of the flowers. The fruit may be eaten by robins and other fruit-eating birds.

Common name *Botanical name*	Hardiness Zones	Plant Size	Light Needs	Soil Needs
Oceanspray (N) *Holodiscus* spp.	All	4-15' tall and wide	F-sun P-shade	Moist to dry, good drainage

LANDSCAPE INFORMATION: **Oceanspray** (*Holodiscus discolor*) is a tall, erect shrub with slender, arching branches. The tiny, cream-white flowers are profuse, fluffy clusters. The seeds are in clusters of tiny, dry pods which persist through winter. The foliage is medium-scale with a pleasing texture. It grows well on disturbed and dry sites, has good soil-binding abilities, tolerates salt air, and is attractive combined with other native trees and shrubs. **Desert oceanspray** (*Holodiscus dumosus*) is a smaller, intricately branched shrub that is best suited for hot, dry landscapes east of the Cascade mountains.

WILDLIFE INFORMATION: Twiggy growth provides songbirds with good cover. In winter, insectivorous birds such as chickadees and bushtits forage for insects in the shrub. Foliage is browsed by elk and deer, and eaten by swallowtail, brown elfin, Lorquin's admiral, and spring azure butterfly larvae. Nectar may be obtained by swallowtail butterflies.

Common name *Botanical name*	Hardiness Zones	Plant Size	Light Needs	Soil Needs
Twinberry (N) *Lonicera involucrata*	3-7	6-8' tall and wide	F-sun to F-shade	Wet to moist

LANDSCAPE INFORMATION: A large-leafed shrub that suckers and eventually forms a thicket. The flowers are mostly tubular with pale-yellow petals that turn slightly purplish when mature. The fruit is glossy-black and about ¼-inch in diameter. It has good soil-binding properties and is a good choice for wet, large-scale projects. Its coarse, large growth may not be suited for small gardens.

WILDLIFE INFORMATION: Berries are eaten by grouse, grosbeaks, juncos, waxwings, thrushes, flickers, finches, and quail. Deer browse the twigs and leaves. Hummingbirds obtain nectar from the flowers.

Common name *Botanical name*	Hardiness Zones	Plant Size	Light Needs	Soil Needs
Tartarian honeysuckle *Lonicera tatarica*	All	8-10' tall and wide	F-sun P-shade	Moist to dry

LANDSCAPE INFORMATION: An upright shrub that forms a dense mass of twiggy branches. The small pink or rose flowers occur in late spring or early summer, followed by bright red fruit. It grows in poor, dry soil and makes a good screen, hedge, windbreak, or background plant. It is recommended for hot dry landscapes east of the Cascade mountains.

WILDLIFE INFORMATION: Berries are eaten by ruffed grouse, pheasants, flickers, waxwings, finches, robins, and juncos. Hummingbirds feed on the nectar and the hollow stems of older branches may attract cavity-nesting bees. The bushy growth of this shrub offers protection for small mammals and ideal nesting habitat for songbirds.

Common name *Botanical name*	Hardiness Zones	Plant Size	Light Needs	Soil Needs
Osoberry (N) (Indian-plum) *Oemleria cerasiformis*	4-7	10- 20' tall 10' wide	F-sun to F-shade	Moist to dry

LANDSCAPE INFORMATION: An open shrub with arched branches that often sucker from the base. When grown in sun it's more compact. One of the first native plants to flower and leaf out in the spring. The white flowers occur in February and early March; the purple-blue "plums" or berries ripen between June and August. Male and female flowers occur on different plants so plant both to assure fruit production. The crushed leaves smell like cucumbers or pea pods and drop in late summer or early fall. Plant along streambanks,

Common name *Botanical name*	Hardiness Zones	Plant Size	Light Needs	Soil Needs

roadsides, large woodlands, and in the back of smaller gardens. It grows in clay soils and is often planted by birds.

WILDLIFE INFORMATION: Berries are eaten by waxwings, robins, and other birds, also foxes, coyotes, deer, and bear. Nectar may be used by Anna's hummingbirds and early bumblebees in lowland areas.

Mock-orange (N) *Philadelphus lewisii*	All	5-10' tall and wide	F-sun P-shade	Moist to dry

LANDSCAPE INFORMATION: A multi-stemmed shrub with snow-white, fragrant blooms in June and July. A useful tall, informal shrub for large-scale plantings. It's also useful in informal hedges, the back of large shrub borders, or as a specimen plant when lightly shaped. It is fast-growing and a good erosion-control plant.

WILDLIFE INFORMATION: Dry seed capsules disperse seed beginning in September. Birds known to eat seeds include catbirds, grosbeaks, juncos, thrushes, bluebirds, chickadees, flickers, finches, quail, and grouse. Deer and elk browse the shrub. Swallowtail, common wood nymph, and other butterflies harvest the nectar.

Ninebark (N) *Physocarpus capitatus*	4-7	10-15' tall and wide	F-sun P-shade	Moist to dry

LANDSCAPE INFORMATION: An erect or spreading, tall multi-stemmed shrub or small tree that eventually develops brown, shredding bark. The creamy-white flowers appear in clusters much like the snowball bush. Plant in moist sites, open woods, wooded edges, and along the water, also in dry brushy areas. It is suited to larger landscape projects or as a plant of interest in the smaller landscape. It has excellent soil-binding characteristics.

WILDLIFE INFORMATION: Fruits occur in reddish bunches of seed pods and may be eaten by birds and small mammals. Deer browse on twigs and leaves.

Potentilla (Buttercup-bush) *Potentilla fruticosa*	All	1-3' tall and wide	F-sun	Dry, good drainage

LANDSCAPE INFORMATION: Many buttercup-bush cultivars exist in the nursery trade. They are valued for their dependable white, orange, or red flowers, which occur in spring and summer. The leaves are bright-green or grayish and divided into small leaflets. The eventual size depends on the variety. They adapt well to landscape settings and are useful in a shrub border, mass-planted in a sunny hot area, or as a low hedge. Cut back oldest stems to make room for new growth. A native buttercup bush exists but is not often available in nurseries.

WILDLIFE INFORMATION: Nectar is used by many flying insects including adult purplish copper and spring azure butterflies. Finches and other birds and small mammals eat the seeds.

Bitter cherry (N) *Prunus emarginata* var. *emarginata*	1-3	To 12' tall 6' wide	F-sun	Moist to dry

LANDSCAPE INFORMATION: A twiggy, upright shrub that grows naturally east of the Cascade mountains and is much smaller than its westside relative (see "Deciduous Trees"). The flowers are small, white, and in 2- to 5-inch clusters. Small, bright-red, bitter cherries ripen in mid to late summer and the leaves turn a bright yellow in fall. **Nanking cherry** (*Prunus tomentosa*) is a tall shrub or small ornamental tree with soft-

Common name *Botanical name*	Hardiness Zones	Plant Size	Light Needs	Soil Needs

textured leaves and very early, fragrant flowers. It's more ornamental and more readily available than the native east side bitter cherry. It is also tolerant of poor soil and very hardy in Zones 1–3. Plant either of these small cherries with other plants in a hedgerow, in a woodland edge, any grassland landscape, or in the back of the flower border.

WILDLIFE INFORMATION: Berries are eaten by grouse, band-tailed pigeons, flickers, jays, robins, waxwings, tanagers, orioles, mourning doves, and magpies. Fruit is also eaten by squirrels, foxes, black bear, coyotes, chipmunks, and raccoons. Deer and elk will browse leaves and twigs. Nectar is used by bees and sara orangetip, silvery blue, and swallowtail butterflies.

Common name *Botanical name*	Hardiness Zones	Plant Size	Light Needs	Soil Needs
Chokecherry (N) *Prunus virginiana*	All	8–20' tall and wide	F-sun P-shade	Moist to dry, good drainage

LANDSCAPE INFORMATION: An openly branched, suckering shrub or small tree. Different varie-ties occur on either side of the Cascade mountains; the smaller (8 to 12 feet tall) variety is more common on the west side. The tiny white flowers occur after leaves unfold and are followed by an abundance of black, red, or yellow fruit. Plants have nice fall color. The taller east-side variety is a superb fast-growing plant for hot, dry areas, including disturbed sites and open areas along roads, fields, and burns. The fruit drop can be messy on walks and patios.

WILDLIFE INFORMATION: Berries are eaten by woodpeckers, pheasants, grouse, jays, robins, thrushes, waxwings, vireos, orioles, and grosbeaks. Deer and elk browse leaves and twigs. Nectar is consumed by silvery blue, swallowtail, and other butterflies; the foliage is eaten by Lorquin's admiral butterfly larvae.

Common name *Botanical name*	Hardiness Zones	Plant Size	Light Needs	Soil Needs
Bitterbrush (N) *Purshia tridentata*	1–3	3–6' tall	F-sun	Moist to dry, good drainage

LANDSCAPE INFORMATION: A twiggy, sometimes semi-evergreen shrub that resembles big sage in leaf and form. The yellow flowers are neat, dainty, and star-shaped. They are also abundant, making the shrub very attractive when it flowers in late spring. Plant in hot, dry areas including roadsides, grasslands, and any areas recently disturbed. Combine with sage, rabbitbrush, serviceberry, and juniper.

WILDLIFE INFORMATION: The leaves, buds, and small twigs are eaten by deer and other browsers, also ground squirrels and other smaller mammals. Plants are used as cover by songbirds. Butterflies harvest the nectar.

Common name *Botanical name*	Hardiness Zones	Plant Size	Light Needs	Soil Needs
Wild azalea (N) *Rhododendron occidentalis*	4–7	3–8' tall and wide	F-sun P-shade	Moist, good drainage

LANDSCAPE INFORMATION: A highly prized, multi-stemmed shrub with upright, twiggy growth. It grows naturally in thickets along streams and creeks. The flowers are white to pink; the lower lobe has a deep yellow blotch. It grows best in soil that is deep and rich in humus and looks nice mass-planted or interspersed with evergreen shrubs.

WILDLIFE INFORMATION: Fragrant flowers attract many insects, including swallowtail butterflies. The dense, twiggy growth is used by birds for nesting and shelter. Mountain beavers will sometimes gnaw on the lower branches.

Common name *Botanical name*	Hardiness Zones	Plant Size	Light Needs	Soil Needs
Skunk bush (N) *Rhus trilobata*	1–3	3–5' tall, spreads wider	F-sun	Moist to dry

LANDSCAPE INFORMATION: A fast-growing and spreading shrub with twiggy branches that eventually form a dense thicket. The oak-like leaves provide rich red fall color. A very useful shrub for dry, sunny locations.

WILDLIFE INFORMATION: The small red berries are attractive to many birds including pheasants, grouse, jays, robins, thrushes, and orioles.

Golden currant (N) *Ribes aureum*	All	3–6' tall and wide	F-sun to P-shade	Moist to dry, good drainage

LANDSCAPE INFORMATION: An attractive, erect shrub with light-green lobed leaves and twiggy branches that are attractive all year. Clusters of bright yellow, spicy-fragrant flowers adorn the shrub in May and are followed by gold-colored edible fruit. It's very useful for dry, sunny spots and performs well mixed with other plants in an informal hedge or thicket. West of the mountains it prefers the hottest area in the landscape. Another useful currant for winter cold/summer hot areas is the **squaw current** (*Ribes cereum*). It has attractive orange-red fruit.

WILDLIFE INFORMATION: Berries are eaten by pheasants, grouse, jays, robins, thrushes, and orioles. Deer and elk browse the twigs and leaves.

Red-flowering currant (N) *Ribes sanguineum*	4–7	6–10' tall 6' wide	F-sun P-shade	Moist to dry, good drainage

LANDSCAPE INFORMATION: An upright shrub with very attractive red-to-pink flowers in pendant 4-inch clusters in March and April. The berries are small, powder blue-black, and ripen from June to August. It is highly recommended for both large and small landscapes where the striking flowers can be seen. Plants tend to be leggier and flower less in shade. This currant grows best in soil that is on the dry side.

WILDLIFE INFORMATION: Berries are eaten by grouse, pheasants, robins, towhees, thrushes, waxwings, sparrows, jays, and woodpeckers. Fruit is also eaten by coyotes, foxes, mountain beavers, raccoons, skunks, squirrels, and chipmunks. Twigs and foliage are browsed by deer and elk. Nectar is consumed by hummingbirds and some adult butterflies; foliage is eaten by zephyr and other butterfly larvae.

Wild gooseberry (N) *Ribes* spp.	4–7	3–5' tall 5' wide	F-sun P-shade	Moist

LANDSCAPE INFORMATION: Native gooseberries are useful plants for streamsides, forested areas, and sunny edges. All are thorny and make excellent hedgerow plants or individual wildlife plants. Two that are becoming more common in nurseries are **swamp gooseberry** (*Ribes lacustre*) and **gummy gooseberry** (*Ribes lobbii*), which has small flowers like tiny white and rose-purple fuchsias.

WILDLIFE INFORMATION: Flowers are often visited by hummingbirds. Fruit on and off the shrub is eaten by towhees, waxwings, woodpeckers, and grouse; also coyote, foxes, mountain beavers, raccoons, skunks, chipmunks, squirrels, porcupines, deer, and elk.

Common name *Botanical name*	Hardiness Zones	Plant Size	Light Needs	Soil Needs
Wild rose (N) *Rosa* spp.	All	5–10' tall and wide	F-sun P-shade	Moist to dry

LANDSCAPE INFORMATION: All native roses grow in a variety of habitats, from open to wooded, dry to moist. All eventually grow into thickets. Roses require no tending and are suited for large informal wildlife gardens, hedgerows, fencelines, and large restoration projects. With summer water and fertilizer all grow much taller (to 10 feet) and will quickly invade other plantings. **Bald-hip rose** (*Rosa gymnocarpa*) is less thorny than other native roses and the flowers and hips are orange to scarlet, and small. **Nootka rose** (*Rosa nutkana*) has large, solitary, pink-to-purplish flowers, which are followed by red pear-shaped hips. **Pea rose** (*Rosa pisocarpa*) is the most dainty and grows into thickets on stream banks and along wetland borders. **Woods' rose** (*Rosa woodsii*) has small, light-pink to deep-rose flowers that occur in clusters.

WILDLIFE INFORMATION: Birds that eat the hips include grouse, bluebirds, juncos, grosbeaks, quail, pheasants, and thrushes. Seeds are used by birds as a source of grit. Mammals that eat the hips include chipmunks, rabbits, hares, porcupines, coyotes, deer, elk, and bear. Rose thickets are important cover for pheasants, grouse, and other birds, also small mammals. Leaves may be eaten by mourning cloak butterfly larvae and are a favorite building material used by the leaf-cutter bee, which cuts neat circles to cart away to its nest. Young shoots are popular with aphids, which in turn provide food for a wide range of predators, including ladybugs and songbirds.

Thimbleberry (N) *Rubus parviflorus*	4–7	2–6' tall	F-sun P-shade	Moist

LANDSCAPE INFORMATION: A thornless, thicket-forming bramble with large, maple-shaped, fuzzy leaves. The flowers are white and bloom when most other natives are done flowering. The raspberry-like fruit is domed, red, and edible. It has good soil-binding characteristics and is well-adapted to eroded and disturbed sites such as ditches. It's a vigorous grower and suited for revegetating moist areas in sun or shade.

WILDLIFE INFORMATION: Berries are eaten by finches, wrens, jays, bushtits, and quail. Coyote, foxes, and bear also eat the berries.

Salmonberry (N) *Rubus spectabilis*	4–7	5–8' tall, spreads wide	F-sun P-shade	Moist to dry

LANDSCAPE INFORMATION: An erect, thicket-forming grower with sparsely thorned, woody stems. Hot-pink to red flowers appear before the foliage in early spring and bright-red or orange, edible berries occur in late June. It has good soil-binding characteristics and is well-adapted to eroded or disturbed sites, where it tends to a form a monoculture. Plant on moist slopes or bottoms of streambanks, ravines, and open areas at edges of marshes or lakes. The native **blackcapped raspberry** (*R. leucodermis*) is less aggressive, has tall bluish-gray stems, and delicious fruits.

WILDLIFE INFORMATION: Berries are eaten by finches, wrens, bushtits, thrushes, robins, towhees, grouse, pheasants, and quail. Coyotes, bears, raccoons, chipmunks, and squirrels eat the berries, and leaves are browsed by deer and rabbits. It is one of the first blooming plants visited by hummingbirds and it also attracts bumblebees.

Willow (*Salix* spp.), see "Deciduous Trees"

Common name *Botanical name*	Hardiness Zones	Plant Size	Light Needs	Soil Needs
Elderberry (N) *Sambucus* spp.	Varies	8-20' tall and wide	F-sun P-shade	Moist to dry

LANDSCAPE INFORMATION: **Blue elderberry** (*Sambucus caerulea*) (Zones All) is a tall, upright shrub or small tree that is very attractive when adorned with the gray-blue berries. West of the Cascade mountains it blooms and fruits later than the red elderberry. It has long, lance-shaped leaves and white, flat-topped flower clusters. It's finicky about transplanting, but once established is tough, and has good soil-binding characteristics. **Red elderberry** (*Sambucus racemosa*) (Zones 4-6) is a tall, somewhat lanky, upright bush that is attractive in flower and fruit. The tiny, creamy-white flowers are concentrated in dense, conical clusters, and are followed by bitter, bright-red berries in summer. **Black elderberry** (*Sambucus racemosa* var. *melanocarpa*) (Zones All) looks like the red elderberry, but has black or purple fruits. Plant elderberries in woodland landscapes, in informal screens, and as specimen plants in the back of the border.

WILDLIFE INFORMATION: Fruits are eaten by sparrows, thrushes, warblers, bluebirds, jays, tanagers, grosbeaks, sapsuckers, woodpeckers, and band-tailed pigeons. Small mammals also eat the berries. Deer and elk browse the twigs and leaves. Nectar is used by hummingbirds, bumblebees, and butterflies. Broken branches are nest sites for native cavity-nesting bees.

Common name *Botanical name*	Hardiness Zones	Plant Size	Light Needs	Soil Needs
Buffaloberry (N) *Shepherdia canadensis*	All	6-8' tall and wide	F-sun	Moist to dry

LANDSCAPE INFORMATION: A suckering shrub that tolerates poor (including alkaline) soil conditions. The currant-sized fruits are yellowish-red and extremely bitter. It's a rugged shrub that will grow in moist or dry open woods. It forms a thicket and is a good choice for the informal mixed hedge. Male and female flowers occur on different plants so plant both to assure fruit.

WILDLIFE INFORMATION: Bear, deer, porcupines, chipmunks, grouse, and many other birds are known to eat the fruit.

Common name *Botanical name*	Hardiness Zones	Plant Size	Light Needs	Soil Needs
Mountain-ash (N) *Sorbus* spp.	All	6-10' tall and wide	F-sun	Moist to dry

LANDSCAPE INFORMATION: Unlike the well-known **European mountain-ash** (*S. aucuparia*), the native mountain ash **Sitka mountain-ash** (*Sorbus sitchensis*) and **Cascade mountain-ash** (*Sorbus scopulina*) are not trees, but large slow-growing, suckering shrubs. The flowers are small, white, flat-topped clusters. The orange-red fruits are large and remain on the plant into fall and winter. Although they will grow in a sunny spot at lower elevations, they usually produce more flowers and fruit at higher elevations.

WILDLIFE INFORMATION: Berries are eaten by grouse, thrushes, waxwings, orioles, grosbeaks, robins, finches, and a variety of mammals. Deer will browse leaves and twigs.

Common name *Botanical name*	Hardiness Zones	Plant Size	Light Needs	Soil Needs
Hardhack (N) *Spiraea douglasii*	All	6-8' tall, spreads wide	F-sun	Wet to moist

LANDSCAPE INFORMATION: A common, multi-stemmed shrub with pink, 3-inch, spike-like flower clusters at the tips of the branches. Its aggressive, spreading habit and good soil-binding characteristics make it effective along streamside buffers, ditches, and wet waste areas in suburbia. Due to its ability to spread widely, this is not a plant for small sites.

| Common name | Hardiness | Plant | Light | Soil |
Botanical name	Zones	Size	Needs	Needs

WILDLIFE INFORMATION: The thicket-like growth creates shelter and the stems are used by beavers for dam construction. Foliage may be eaten by pale swallowtail, spring azure, Lorquin's admiral, and other butterfly larvae. Flowers attract butterfly species and other flying insects.

| **Spirea** (N) | All | 2-4' tall | F-sun | Moist to dry |
| *Spiraea* spp. | | and wide | P-shade | |

LANDSCAPE INFORMATION: Nice ornamental shrubs that produces tiny cream-white, pink, or red flowers in large, flat-topped clusters. **Birchleaf spirea** (*Spiraea betulifolia* var. *lucida*) is a good understory plant for a hot, dryish woodland garden or mixed with ornamental shrubs and perennials in a drought-tolerant flower bed. *Spiraea* x *pyramidata* has pink flowers and can grow in moist or wet soil. The many non-native spireas are all useful plants.

WILDLIFE INFORMATION: Grouse and deer eat the leaves. The nectar and leaves are used by certain butterflies. The flat-topped flowers seem to attract a variety of predatory insects including syrphid flies, which eat aphids.

| **Snowberry** (N) | All | 3-5' tall | F-sun to | Moist to dry |
| *Symphoricarpos albus* | | and wide | F-shade | |

LANDSCAPE INFORMATION: A suckering, many-branched and twiggy shrub. The flowers are small, bell-shaped, pinkish or white; the berries are white, marble-size, and grow in tight clusters. The berries persist into winter and are striking when mixed with plants bearing native rose hips. Plant in shady areas under large trees, mix with other plants in informal hedges to form thickets; also plant on roadsides and open slopes to help control errosion.

WILDLIFE INFORMATION: Plants provide low shelter and nesting cover for small animals. Berries are not a first choice by birds, but are eaten by grosbeaks, waxwings, robins, thrushes, towhees, grouse, pheasants, and quail when other wild foods are scarce. Known to provide nesting habitat for gadwall ducks. Leaves and twigs browsed by deer and pheasants. Leaves are eaten by the sphinx moth larvae. Hummingbirds and bumblebees feed on the nectar.

| **Huckleberry** (N) | 4-7 | 4-10' tall | F-sun | Moist to dry |
| *Vaccinium* spp. | | and wide | P-shade | |

LANDSCAPE INFORMATION: **Red huckleberry** (*Vaccinium parvifolium*) is an open-spreading shrub with thin, pretty, light-green branches. The inconspicuous pink flowers are followed by bright-red, showy, edible berries. Plant at the edge of a woodland and in a woodland landscape. It does best in soils with decaying wood, such as areas next to an old stump or log. Large plants can provide a sturdy framework for the native honeysuckles. The **Black huckleberry** (*Vaccinium membranaceum*) grows in a similar setting and has very tasty berries.

WILDLIFE INFORMATION: Berries are eaten by pheasants, mourning doves, flickers, jays, robins, waxwings, orioles, tanagers, towhees, sparrows, and chickadees. Many insectivorous birds glean spiders and other insects from the leaves and branches. Nectar is harvested by hummingbirds and bumblebees.

Common name *Botanical name*	Hardiness Zones	Plant Size	Light Needs	Soil Needs
Viburnum *Viburnum* spp.	All	6–12' tall and wide	F–sun P–shade	Moist to dry

LANDSCAPE INFORMATION: Viburnums are often inconspicuous shrubs that blend well with trees and other shrubs. The leaves are maple-like and the fall foliage is often brilliant. The early, sometimes fragrant white flowers occur in clusters followed by showy clusters of fruit that stay on plant well into winter. They are very hardy, easy to grow, develop quickly to form screens, and all are suited for the woodland garden or larger wildlife plantings. Three species available from nurseries include **Korean spice viburnum** (*Viburnum carlesii*), **Nannyberry** (*Viburnum lentago*), and **Arrowwood** (*Viburnum dentatum*).

WILDLIFE INFORMATION: Many birds eat the fruit, including thrushes, jays, bluebirds, flickers, grouse, pheasants, flycatchers, robins, woodpeckers, finches, and waxwings. Fruit is also eaten by coyote, foxes, and smaller mammals. Butterflies associated with these species include the spring azure.

Highbush-cranberry (N) *Viburnum edule*	All	6–10' tall	F–sun P–shade	Moist

LANDSCAPE INFORMATION: An inconspicuous shrub that blends its rather straggly branches in with other trees and shrubs. The early white flowers occur in clusters followed by showy clusters of red-orange berries that stay on the plant well into winter. Overripe berries have a musky odor. The leaves are maple-like and the fall foliage is often brilliant. It is easy to grow, and grows quickly into a screen. Plant in the woodland landscape, moist forest, forest edge, margin of a wetland, and on a streambank.

WILDLIFE INFORMATION: The fruit is eaten by thrushes, jays, bluebirds, flickers, grouse, pheasants, robins, woodpeckers, finches, and waxwings. Fruit is also eaten by coyotes, foxes, and smaller mammals. Butterflies associated with these species include the spring azure.

For other deciduous shrubs, see the tables in Appendix B.

EVERGREEN SHRUBS

Plants are listed alphabetically by their botanical name.
The botanical name first includes the *Genus* and then the *species*.
(N) = Plants that are native to Oregon, Washington, or southern British Columbia
When referencing more than one species in a genus, the abbreviation (spp.) is used.
For additional explanation of the headings, see the "Key" at the beginning of Appendix C.
For information on where native plants occur naturally, see "Native Plant Occurrence Chart" in Appendix B.

Common name *Botanical name*	Hardiness Zones	Plant Size	Light Needs	Soil Needs
Strawberry-tree *Arbutus unedo* 'Compacta'	4-7	8' tall and wide	F-sun P-shade	Moist to dry, good drainage

LANDSCAPE INFORMATION: A dense, neat-looking shrub that is slow-growing and eventually equally tall as broad. The ivory-colored, bell-shaped flowers appear in early fall and are attractive seen with the previous spring's fruit. The red-and-yellow, ¾-inch fruit is somewhat strawberry-like in appearance but bland and gritty. The handsome, dark-green leaves eventually make an effective screen when mass-planted; single plants fit nicely into a small area. Plants may need some protection in colder areas.

WILDLIFE INFORMATION: Fruit is eaten on the bush by robins and other fruit-eating birds. Fruit and flowers are also commonly eaten on the ground. A good late-flowering bee plant and a good shelter plant in windy, coastal areas.

Common name *Botanical name*	Hardiness Zones	Plant Size	Light Needs	Soil Needs
Manzanita (N) *Arctostaphylos* spp.	Varies	Varies	F-sun	Dry, good drainage

LANDSCAPE INFORMATION: Manzanitas are spectacular, slow-growing shrubs that prefer a sunny, hot, well-drained site and resent summer watering. Salvaging plants is nearly impossible. **Hairy manzanita** (*Arctostaphylos columbiana*) (Zones 4-6) is a broad, uniform grower with reddish-brown bark that flakes and peels. It can reach 3-8 feet tall and wide. The leaves are gray-green, the flowers are white and bell-like, and the fruit is red and ripens in summer. Another nice mid-size manzanita for Zones 6-7 is the non-native **Vine Hill manzanita** (*Arctostaphylos densiflora*). **Green manzanita** (*Arctostaphylos patula*) (Zones All) has tough oval-shaped leaves, bright-pink flowers, and can reach 3-6 feet tall and wide.

WILDLIFE INFORMATION: Fruit is eaten by grouse, evening grosbeaks, quail, and other birds, also deer, foxes, raccoons, skunks, coyotes, chipmunks, and ground squirrels. The nectar is eaten by humming-birds and bees. The flowers are eaten by brown elfin and spring azure butterfly larvae. Ceanothus silkmoth larvae may eat the leaves.

Common name *Botanical name*	Hardiness Zones	Plant Size	Light Needs	Soil Needs
Sagebrush (N) *Artemisia tridentata*	All	3-6' tall and wide	F-sun	Dry, good drainage

LANDSCAPE INFORMATION: This well-known plant from western dryland regions has many branches, silver aromatic foliage, and minute flowers in small yellow plume-like clusters in late summer. It's a quick grower when watered and can be clipped to control height. A nice plant for large and small-scale landscapes that feature other native dryland plants, or in an unclipped hedge. The silver foliage provides

Common name *Botanical name*	Hardiness Zones	Plant Size	Light Needs	Soil Needs

contrast to green-leaved trees and shrubs. **Hopsage** (*Atriplex spinosa*) is similar to sagebrush in size and form. The leaves are narrow and gray, the flowers are minute, and the seed bracts on female plants can be whitish, greenish, or shades of red.

WILDLIFE INFORMATION: Sagebrush is particularly valuable in wild areas because it is both abundant and evergreen. Wintering herds of elk and deer rely heavily on it, particularly during winters when other forage is scarce. Sage grouse use the shrub for both food and cover and may travel long distances in winter to find stands uncovered by snow. Small animals such as rabbits and quail depend on plants for thermal cover in hot and cold weather.

Common name *Botanical name*	Hardiness Zones	Plant Size	Light Needs	Soil Needs
Evergreen barberry *Berberis* spp.	Varies	Varies	F-sun P-shade	Moist to dry, good drainage

LANDSCAPE INFORMATION: Barberries are easy shrubs to grow, make good thorny hedges, and look nice all year. All have small dark-blue berries. Shearing tends to reduce flowering and fruiting so it is best to let the plants develop a natural form. This also makes plants more accessible to nesting birds. **Darwin barberry** (*Berberis darwinii*) (Zones 5-6) has small holly-like leaves, profuse showy, orange flowers, and upright, fountain-like growth to 10 feet. **Wintergreen barberry** (*Berberis julianae*) (Zones 2-7) is a dense, upright grower to 6 feet with leathery, spined dark-green leaves and yellow flowers. **Warty barberry** (*Berberis verruculosa*) (Zones 2-7) is a smaller-scaled, more tailored shrub with spiny, glossy dark-green leaves and yellow flowers.

WILDLIFE INFORMATION: Berries are eaten in late winter by robins, waxwings, and other fruit-eating birds. Large barberry plants are often one of the best shelter shrubs in traditional suburban landscapes.

Common name *Botanical name*	Hardiness Zones	Plant Size	Light Needs	Soil Needs
Mountain-lilac *Ceanothus* spp.	5-7	6-10' tall and wide	F-sun	Dry, good drainage

LANDSCAPE INFORMATION: Mountain-lilacs are fast-growing shrubs that feature an abundance of beautiful flowers. They prefer a hot, sunny site with plenty of space. They are not long-lived; 8-12 years is typical, and they don't transplant well once established. *Ceanothus* **'Julia Phelps'** forms a densely branched mass of small foliage with sprays of dark indigo-blue flowers in mid-spring. *Ceanothus* **'Dark Star'** has even tinier leaves and striking cobalt-blue flowers.

WILDLIFE INFORMATION: Flowers are visited by many flying insects, especially honeybees. Butterflies associated with the plant include the pale swallowtail and the brown elfin. Deer browse the leaves and twigs.

Common name *Botanical name*	Hardiness Zones	Plant Size	Light Needs	Soil Needs
Snowbush (N) *Ceanothus velutinus* var. *laevigatus*	4-7	8-15' tall and wide	F-sun	Dry, good drainage

LANDSCAPE INFORMATION: An open-branched shrub with a mass of glossy, aromatic foliage and sprays of cream-white flowers in mid-spring. It prefers a sunny, hot, well-drained site and resents summer water. It is best suited for large areas.

WILDLIFE INFORMATION: Flowers are visited by many flying insects, especially bees. Ceanothus silkmoth larvae eat the leaves. Butterflies associated with the plant include the pale swallowtail, brown elfin, and the hedgerow hairstreak. Deer and elk browse the twigs and leaves.

Common name *Botanical name*	Hardiness Zones	Plant Size	Light Needs	Soil Needs
Mountain balm (N) *Ceanothus velutinus* var. *velutinus*	1-3	2-8' tall and wide	F-sun	Dry, good drainage

LANDSCAPE INFORMATION: Mountain balm forms a twiggy, open mound and makes a good large-scale ground cover in sunny, hot landscapes. The branches are covered with glossy, aromatic foliage and sprays of cream-white fragrant flowers in mid-spring. It is attractive planted together with Oregon-grape and other natives that don't require much summer water once established.

WILDLIFE INFORMATION: See information for snowbush (*Ceanothus velutinus* var. *laevigatus*).

Rabbitbrush (N) *Chrysothamnus* spp.	1-3	2-3' tall and wide	F-sun	Dry, good drainage

LANDSCAPE INFORMATION: **Gray rabbitbrush** (*Chrysothamnus nauseosus*) has short white or gray, woolly hairs that densely mat the stems and leaves. **Green rabbitbrush** (*Chrysothamnus viscidiflorus*) has green stems. Both have bright-yellow flowers in showy clusters in late summer. Plant in large and small-scale projects in hot, dry areas. Both species often naturally occur with sagebrush. These are particularly good plants for butterfly gardens east of the Cascade mountains.

WILDLIFE INFORMATION: Plants are important shelter for pheasants, grouse, quail and songbirds. The leaves, buds, and small twigs are eaten by deer and elk, especially in winter when not covered with snow. Ground squirrels, mice, voles, and rabbits may use the bark, stems, or seeds for food and nesting. Nectar is used by painted lady and other butterflies.

Cotoneaster *Cotoneaster* spp.	4-7	Varies	F-sun	Dry, good drainage

LANDSCAPE INFORMATION: **Willowleaf cotoneaster** (*Cotoneaster salicifolia*) is a large, spreading shrub or small tree to 15 feet tall and wide. It has long, narrow, dark-green leaves and bright red berries. It creates a graceful and effective screen or background plant when given plenty of space. **Red clusterberry** (*Cotoneaster lacteus*) is a graceful, arching shrub to about 8 feet tall and wide. It has light-green leaves, clustered white flowers, and heavy crops of red berries. It makes a nice informal hedge or screen. All cotoneasters produce more fruit and provide better shelter for wildlife when they are not pruned or only lightly pruned. For other cotoneasters, see "Deciduous shrubs."

WILDLIFE INFORMATION: The berries are generally eaten in late winter by robins, waxwings, and other birds needing food at this time. Honeybees, spring azure butterflies and other butterflies feed on flower nectar.

Silverberry *Elaeagnus pungens*	4-7	6-15' tall and wide	F-sun P-shade	Moist to dry

LANDSCAPE INFORMATION: A dense, grayish-green shrub that grows quite fast when young. The ½-inch fruits are oval, red with silver dots. The spiny branches and dense foliage make silverberry an effective barrier plant and an excellent screen, especially in coastal landscapes. Different varieties are available from nurseries.

WILDLIFE INFORMATION: The dense evergreen growth creates shelter and nest places for songbirds. The fruit is eaten by robins and other birds.

Common name *Botanical name*	Hardiness Zones	Plant Size	Light Needs	Soil Needs
Escallonia *Escallonia* spp.	5-7	3-10' tall and wide	F-sun	Moist to dry, good drainage

LANDSCAPE INFORMATION: Escallonias are prized for their glossy, dark-green leaves and summer blooms that vary from white to pink to crimson, depending on the species or variety. Some may freeze badly at 10-15 degrees F, but will generally recover in spring. Taller varieties make good fast-growing screens; smaller ones make good mini-hedges and are nice evergreen accents in areas planted with perennials. Plants produce more fruit and provide better shelter for wildlife when they are not pruned or only lightly pruned.

WILDLIFE INFORMATION: The flowers attract bees and hummingbirds and are visited by spring azure and other butterflies.

Common name *Botanical name*	Hardiness Zones	Plant Size	Light Needs	Soil Needs
Silk-tasselbush (N) *Garrya elliptica*	5-7	8-10' tall and wide	F-sun P-shade	Dry, good drainage

LANDSCAPE INFORMATION: A dense shrub clad with wavy, oblong leaves that are dark green above, gray and woolly beneath. The attractive male flower-tassels appear in late winter. These are yellowish, slender, graceful, and 3-8 inches long. The variety 'James Roof' was selected for its extra-long catkins. Because male and female catkins occur on separate plants, both must be planted to produce the grape-like clusters of purplish fruit on female plants. This is a good screen, informal hedge, or display shrub with late winter interest. **Fremont's silk-tasselbush** (*Garrya fremontii*) is hardier in heat and cold than the coast silk-tasselbush, but less ornamental. It grows 4-6 feet tall and wide and is for landscapes in Zones 1-4.

WILDLIFE INFORMATION: Although not a favorite food, the fruit is eaten by robins and other fruit-eating birds. Plants serve as shelter for birds during the cold windy months.

Common name *Botanical name*	Hardiness Zones	Plant Size	Light Needs	Soil Needs
Salal (N) *Gaultheria shallon*	4-7	2-6' tall and wide	F-sun to F-shade	Moist to dry

LANDSCAPE INFORMATION: A versatile plant that can take wind, poor soil, snow, and little water. In dry soil and full sun it can be a 2-foot groundcover; in shade with rich soil and water it can reach 8 feet. The leathery leaves are generally dark green; the flowers are white-to pink, urn-shaped, and occur on long racemes. The purplish-black berries are ripe by midsummer and remain on shrubs until fall. A good choice for growing under trees in woodland areas, on slopes and roadsides, and in informal hedges. It can be slow to establish and difficult to remove.

WILDLIFE INFORMATION: The berries are eaten by grouse, band-tailed pigeons, towhees, and other ground-feeding birds. Bear, foxes, coyotes, and smaller mammals also eat the berries. The twigs are eaten by deer. Leaves are eaten by brown elfin butterfly larvae.

Common name *Botanical name*	Hardiness Zones	Plant Size	Light Needs	Soil Needs
Tall Oregon-grape (N) *Mahonia (Berberis) aquifolium*	All	6-8' tall 6' wide	F-sun P-shade	Moist to dry

LANDSCAPE INFORMATION: A common native shrub and Oregon's state flower. It has bright yellow flowers, new bronze growth in spring, and attractive blue-black fruit in summer. The foliage looks good all year. The plant's erect form makes it an excellent barrier and informal screen plant when mass-planted. It does not transplant easily and is best planted from a well-rooted container. The dwarf selection, **dwarf Oregon-grape** (*Mahonia aquifolium* 'Compacta') grows only 2 feet high (see "Ground Covers and Low Shrubs").

Common name *Botanical name*	Hardiness Zones	Plant Size	Light Needs	Soil Needs

WILDLIFE INFORMATION: The berries are eaten by many birds including grouse, pheasants, robins, waxwings, juncos, sparrows, and towhees. Foxes, raccoons, and coyotes also eat the berries. Deer and elk will occasionally browse the leaves and flowers. Orchard mason bees and painted lady butterflies use the nectar.

Common name *Botanical name*	Hardiness Zones	Plant Size	Light Needs	Soil Needs
Cascade Oregon-grape (N) *Mahonia (Berberis) nervosa*	4-7	12-24" tall and wide	P-shade	Moist or dry, good drainage

LANDSCAPE INFORMATION: A slow-spreading evergreen shrub or ground cover that grows well in the filtered shade of trees and shrubs. The clusters of blue fruit, attractive leaves, and early bright yellow flowers make this a very desirable landscape plant. Plant in large groups with other natives such as sword fern and salal or at a smaller scale with other woodland plants.

WILDLIFE INFORMATION: The berries are ripe in late summer and may be eaten by grouse, waxwings, thrushes, towhees, grouse, pheasants, and other birds. Deer, elk, and rabbits may browse the foliage. Nectar may be used by certain adult butterflies and other flying pollinators.

Common name *Botanical name*	Hardiness Zones	Plant Size	Light Needs	Soil Needs
Creeping Oregon-grape (N) *Mahonia (Berberis) repens*	All	18-36" tall and wide	F-sun P-shade	Moderate to dry, good drainage

LANDSCAPE INFORMATION: An evergreen plant that spreads fairly slowly from underground stems. The yellow flowers occur in early spring and are followed by blue berries in fall and purplish foliage in winter. The spiny foliage looks good all year. A good plant for grouping under large shrubs and trees. Not easy to transplant and best planted out of a well-rooted container.

WILDLIFE INFORMATION: See information on Cascade Oregon-grape, *Mahonia (Berberis) nervosa*.

Common name *Botanical name*	Hardiness Zones	Plant Size	Light Needs	Soil Needs
Wax-myrtle (N) *Myrica californica*	4-7	10-15' tall 8' wide	F-sun P-shade	Moist to dry

LANDSCAPE INFORMATION: A fast-growing, upright shrub that looks good all year. The branches are densely clad with glossy leaves that are fragrant when crushed. The flowers are inconspicuous and are followed by small purplish fruits. It can be used as a screen or a hedge and can be clipped if needed. This is an excellent shrub for coastal landscapes.

WILDLIFE INFORMATION: The fruits are eaten by quail, waxwings, and other birds, and the twigs and leaves are browsed by deer. Plants provide important shelter in windy, coastal areas.

Common name *Botanical name*	Hardiness Zones	Plant Size	Light Needs	Soil Needs
Firethorn *Pyracantha coccinea*	2-7	8-10' tall and wide	F-sun P-shade	Moist to dry

LANDSCAPE INFORMATION: A much-branched, erect and broad-spreading bush with many thorns. It has lustrous green leaves and profuse white flower-clusters in spring or early summer. The orange or reddish berries persist into winter. The foliage can burn in a harsh winter, may even die back to ground, but will recover. A good barrier or informal hedge. The variety 'Lelandei' is hardy and widely grown.

WILDLIFE INFORMATION: The berries are popular with woodpeckers, robins, thrushes, waxwings, finches, and sparrows. Many small flying insects visit flowers. Shearing creates dense, impenetrable shrubs and fewer flowers. It is best to let the plants develop an open form which is more accessible to feeding and nesting birds.

Common name *Botanical name*	Hardiness Zones	Plant Size	Light Needs	Soil Needs
Pacific rhododendron (N) *Rhododendron macrophyllum*	4-7	7-12' tall and wide	P-shade or F-sun	Moist to dry

LANDSCAPE INFORMATION: A large shrub with bold, leathery leaves. The large flower trusses are rose-purple (rarely white) and appear in May through June. It grows under conifers in woodlands but flowers more in the open. It needs plenty of space. One of the most drought-tolerant rhododendrons and resistant to root weevils. The Washington state flower.

WILDLIFE INFORMATION: Bumblebees and swallowtail butterflies are attracted to the flowers. Deer will eat the flowers and flower buds. Songbirds nest in large specimens.

Common name *Botanical name*	Hardiness Zones	Plant Size	Light Needs	Soil Needs
Evergreen huckleberry (N) *Vaccinium ovatum*	4-7	3-8' tall and wide	F-sun F-shade	Moist to dry

LANDSCAPE INFORMATION: A slow-grower with attractive, bronzy new foliage. The flowers are small, pinkish, bell-shaped, and occur from April to July. The edible fruit is black or blue and occurs in small clusters in midsummer to fall. A fairly compact grower in sunny spots and an erect and much taller shrub in shade. Berry crops are better in some sun. Plant in coniferous forests, especially in edges and openings. Can be used in hedges, screens, and foundation plantings.

WILDLIFE INFORMATION: The fruits are an important food source in late summer and fall for wildlife. Birds that eat berries include grouse, pheasants, band-tailed pigeons, flickers, chickadees, robins, bluebirds, waxwings, orioles, towhees, and sparrows. Fruits are also eaten by foxes, raccoons, black bears. Twigs are browsed by deer.

For other evergreen shrubs, see tables in Appendix B.

GROUND COVERS AND LOW SHRUBS

Plants are listed alphabetically by their botanical name.

The botanical name first includes the *Genus* and then the *species*.

When referencing more than one species in a genus, the abbreviation (spp.) is used.

(N) = Plants that are native to Oregon, Washington, or southern British Columbia.

For definitions of headings, see the "Key" at the beginning of Appendix C.

For information on where native plants occur naturally, see the "Native Plant Occurrence Chart" in Appendix B.

Common name *Botanical name*	Hardiness Zones	Plant Size	Light Needs	Soil Needs
Kinnikinnik (N) *Arctostaphylos uva-ursi*	All	8" tall 5' wide	F-sun P-shade	Dry, good drainage

LANDSCAPE INFORMATION: Kinnikinnik is a versatile ground cover with red-brown bark, small leathery leaves, and attractive flowers and fruit. In mid-spring, tight racemes of white or pink blossoms nod from the ends of branches. These are followed by attractive red berries which may persist into winter. The evergreen foliage looks fresh all year. Plant in sunny woodland edges, on dry slopes, and as a trailing mat over rock walls. Plants start off slow, so mulch around them to prevent weeds from establishing.

WILDLIFE INFORMATION: The berries are eaten by ruffed grouse, band-tailed pigeons, evening grosbeaks, sparrows, and other ground-feeding birds, also bear, coyotes, and foxes. Deer will browse the twigs. The flowers attract bees and brown elfin butterflies.

Dwarf coyote bush *Baccharis pilularis*	5-7	6-18" tall 4' wide	F-sun	Dry, good drainage

LANDSCAPE INFORMATION: A dense, billowy, mat-like ground cover with small, bright-green leaves. A good evergreen plant for slopes in coastal areas. Very drought-tolerant.

WILDLIFE INFORMATION: The small flowers attract a variety of flying insects including the painted lady, west coast lady, orange sulphur, and skipper butterflies. The dense growth creates good cover for insects and other invertebrates.

Point Reyes creeper *Ceanothus gloriosus*	4-7	4-18" tall 6' wide	F-sun	Moist to dry, good drainage

LANDSCAPE INFORMATION: An evergreen ground cover with dark-green leaves and light-blue flowers in 1-inch clusters. It's fast-growing and tolerant of infertile soil. Plant on sunny, dry slopes or large open areas where plants can spread freely. Not a good choice for small areas and places next to walkways where annual pruning will become necessary.

WILDLIFE INFORMATION: The flowers are nectar sources for different flying insects, especially bumblebees.

Common name *Botanical name*	Hardiness Zones	Plant Size	Light Needs	Soil Needs
Bunchberry (N) *Cornus unalaschkensis (canadensis)*	All	6" tall 24" wide	P-shade	Moist, acid, gritty

LANDSCAPE INFORMATION: A deciduous carpet plant that grows from creeping rootstocks. The dogwood-type flowers occur in early summer and are followed by dense clusters of red berries in August. The leaves turn yellow in fall and die back in winter. It can be difficult to grow, but worth the effort. It grows best in woodland settings in soil that has plenty of rotting organic matter.

WILDLIFE INFORMATION: The fruit is eaten by sparrows, thrushes, vireos, grouse (will also eat buds), and pheasants.

Common name *Botanical name*	Hardiness Zones	Plant Size	Light Needs	Soil Needs
Cotoneaster *Cotoneaster* spp.	All	18" tall 6' wide	F-sun P-shade	Moist to dry, good drainage

LANDSCAPE INFORMATION: **Lowfast cotoneaster** (*Cotoneaster* 'Lowfast') is evergreen and fast-growing. **Creeping cotoneaster** (*Cotoneaster adpressus*) is deciduous and slower-growing. **Cranberry cotoneaster** (*Cotoneaster apiculatus*) grows a bit denser and taller and can take some shade. All are common landscape plants with attractive small, dark-green leaves and red berries. Plant on hot, dry slopes, next to large rocks, over a wall, and in large level areas. Many other cotoneaster species are available in nurseries.

WILDLIFE INFORMATION: Many ground-feeding birds including robins eat the berries in winter. The flowers attract bees and spring azure butterflies.

Common name *Botanical name*	Hardiness Zones	Plant Size	Light Needs	Soil Needs
Heather *Erica* spp.	Varies	12" tall	F-sun	Moist to dry, good drainage

LANDSCAPE INFORMATION: Heathers are low-growing plants with needle-like foliage and small urn-shaped flowers of purple, rose, pink, or white. They are effective planted singly or as a mass color accent. To attract the largest number of flying insects, plant varieties that flower in the spring and summer.

WILDLIFE INFORMATION: The spring and summer-flowering varieties attract bumblebees and purplish copper butterflies.

Common name *Botanical name*	Hardiness Zones	Plant Size	Light Needs	Soil Needs
Wild strawberry (N) *Fragaria* spp.	All	4" tall	F-sun P-shade	Moist to dry, good drainage

LANDSCAPE INFORMATION: **Coast strawberry** (*Fragaria chiloensis*) is a fast-growing plant with leathery, glossy leaves and dime-size white flowers throughout the spring. The thick evergreen growth can eventually prevent weed establishment and can take light foot-traffic. Plants need room to spread and can quickly dominate a small area. It naturally occurs by the sea. **Woodland strawberry** (*Fragaria vesca*) is daintier and less aggressive, but hardier, and has delicious fruit. It grows in moist woodlands. **Mountain strawberry** (*Fragaria virginiana*) grows in drier, more open areas. Only the coast strawberry is truly evergreen.

WILDLIFE INFORMATION: Selections for landscaping vary in fruit production. Birds known to eat the fruit include robins, towhees, pine grosbeaks, waxwings, and grouse. Fruit is also eaten by mice and other small mammals. The flowers attract bees and sara orangetip butterflies.

Common name *Botanical name*	Hardiness Zones	Plant Size	Light Needs	Soil Needs
Wintergreen (N) *Gaultheria* spp.	4-7	6" tall 12" wide	F-shade P-shade	Moist

LANDSCAPE INFORMATION: A fresh-looking ground cover with glossy, oval, two-inch evergreen leaves. The small white flowers are followed by scarlet berries. The leaves and fruits have the flavor of wintergreen. It is best planted in small areas and in soil well-prepared with organic matter. The non-native *Gaultheria procumbens* is the most available species; the native *Gaultheria ovatifolia* is also nice and makes a fine ground cover in the woodland.

WILDLIFE INFORMATION: The berries remain on the plant into winter and may be eaten by ground-feeding and forest-edge birds, and small mammals. Because the fruits are low to the ground, they are often eaten by slugs.

Common juniper (N) *Juniperus communis*	All	1' tall 6' wide	F-sun P-shade	Dry, good drainage

LANDSCAPE INFORMATION: This native juniper has many different forms. The dwarf varieties form mats or large clumps and are well-suited for ground cover. The fruits are berry-like cones, pale green at first, and bluish-black when mature. To secure berries, do not plant sterile varieties. Plant in sunny areas, open rocky slopes, and other harsh locations.

WILDLIFE INFORMATION: The fruit is eaten by a variety of birds, generally in winter. Large plants can provide year-round shelter and protect the burrows made by small mammals.

Lavender (and others) *Lavendula* spp.	All	18-24" high	F-sun	Moist to dry

LANDSCAPE INFORMATION: **Lavender** is prized for its fragrant lavender or purple flowers and grayish or gray-green aromatic foliage. Plant as a hedge or edging, in herb gardens, or in borders with plants needing similar conditions. Plants need loose, fast-draining soil and are very drought-tolerant. **Germander** (*Teucrium chamaedrys*) is another tough plant that can endure sun, heat and poor rocky soils. Many upright stems carry toothed dark green leaves and red-purple, ¾-inch flowers in loose spikes all summer. Useful as edging, foreground, or low hedge. A somewhat similar plant that takes wetter soil better than lavender is **hyssop** (*Hyssopus officinalis*). It's a compact sub-shrub with narrow, dark green, pungent leaves and a profusion of dark blue flower spikes from July to frost. All the above mentioned plants can be pruned immediately after bloom to keep plants compact and neat.

WILDLIFE INFORMATION: The flowers attract hummingbirds and butterflies, including tiger swallowtails, painted ladies, and woodland skippers.

Dwarf Oregon-grape (N) *Mahonia (Berberis) aquifolium* 'Compacta'	All	3' tall 2' wide	F-sun P-shade	Moist or dry, good drainage

LANDSCAPE INFORMATION: A tough, low evergreen shrub that looks good all year. It's a selection of the native tall Oregon-grape. The bright-yellow flowers and new bronze growth are followed by attractive blue-black fruit. It spreads slowly from underground runners and is a good choice for tight spaces such as narrow planting strips. Also plant in masses in the woodland or on slopes.

Common name *Botanical name*	Hardiness Zones	Plant Size	Light Needs	Soil Needs

WILDLIFE INFORMATION: The berries are eaten by grouse, pheasants, robins, waxwings, juncos, sparrows, towhees, and other birds; foxes and coyotes also eat the berries. Deer and elk occasionally browse the leaves and flowers. Orchard mason bees and painted lady butterflies use the nectar.

| **Rubus**
Rubus calycinoides | 4-7 | 2" high | F-sun | Moist to dry,
good drainage |

LANDSCAPE INFORMATION: An attractive low evergreen ground cover that spreads underground from creeping stems. It forms a thick mat that spreads about 1 foot a year. The small white flowers resemble strawberry flowers; the fruits are salmon colored. It's a tough plant that is perfect for small confined areas. It looks best with a little watering.

WILDLIFE INFORMATION: The berries are eaten by robins and other ground-feeding birds.

| **Trailing raspberry** (N)
Rubus pedatus | 4-7 | 2' high
6' wide | P-shade | Moist |

LANDSCAPE INFORMATION: An attractive evergreen ground cover that spreads underground from creeping stems. The fruits occur in small clusters of bright-red drupelets and are juicy and flavorful. It is easy to grow and has no thorns or prickles. Plant in moist woodlands, wet banks and other moist areas, also in small confined planting strips. Our only native blackberry, the **trailing blackberry** (*Rubus ursinus*) grows in full sun and can quickly form a 1-foot-high impenetrable thicket. The fruits are delicious but, because male and female plants are on separate plants and individual plants spread over large areas, it's not uncommon to find large patches of male plants and no fruit.

WILDLIFE INFORMATION: The berries are eaten by white-crowned sparrows, chickadees, pheasants, grouse, quail, band-tailed pigeons, and robins, also slugs, chipmunks, and other mammals because the fruit is low to the ground. The twigs and leaves are browsed by rabbits and deer.

For other ground covers and low shrubs, see the tables in Appendix B.

VINES

Plants are listed alphabetically by their botanical name.
The botanical name first includes the *Genus* and then the *species*.
(N) = Plants that are native to Oregon, Washington, or southern British Columbia
When referencing more than one species in a genus, the abbreviation (spp.) is used.
For additional explanation of the headings, see the "Key" at the beginning of Appendix C.
For information on where native plants occur naturally, see the "Native Plant Occurrence Chart" in Appendix B.

Common name *Botanical name*	Hardiness Zones	Plant Size	Light Needs	Soil Needs
Trumpet vine *Campsis radicans*	All	30-40'	F-sun	Moist to dry

LANDSCAPE INFORMATION: A fast-growing deciduous vine with large, orange, trumpet-shaped flowers in summer. The variety 'Flava' has apricot-yellow flowers. It clings to structures with aerial roots and may eventually develop a woody trunk. It will require annual pruning to keep it within bounds in a small area. Trumpet vine is a good plant for covering a cyclone fence, arbor, or a large sturdy trellis. It is especially suited for hot inland areas where it may freeze to the ground but will grow new stems in the spring.
WILDLIFE INFORMATION: The bright tubular flowers attract hummingbirds.

Clematis (N) *Clematis* spp.	Varies	Varies	F-sun P- shade	Moist to dry

LANDSCAPE INFORMATION: **Evergreen clematis** (*Clematis armandii*) (Zones 4–7) is a vigorous grower with dark-green, leathery leaves and showy clusters of white, fragrant flowers in March. This bold evergreen vine will climb 20 or 30 feet and requires little pruning. **Small-flowered clematis** (*Clematis flammula*) (Zones All) is more typical of the ornamental deciduous clematis. It has multiple fragrant, white blooms in August and September. The stems reach 10 feet or more and tend to become bare below. **Wild clematis** (N) (*Clematis ligusticifolia*) (Zones 1–3) is an extremely vigorous, drought-tolerant vine. It has abundant clusters of showy, white flowers and ornamental seed heads. Wild clematis creates quick cover for wildlife and is also good for erosion control. Its aggressive growth makes it suitable only for large, exposed landscapes, including chain-link fences. Although similar in appearance, wild clematis is not the aggressive non-native **Traveler's joy** (*Clematis vitalba*) common in urban greenbelts west of the Cascade mountains.
WILDLIFE INFORMATION: The plants are valued for their cover, nectar source, and seed sources when available.

Honeysuckle (N) *Lonicera* spp.	Varies	10-20 feet	F-sun P-shade	Moist to dry

LANDSCAPE INFORMATION: **Brown's honeysuckle** (*Lonicera x brownii* 'Dropmore Scarlet') (Zones All) blooms for a long period, is the hardiest of the non-native garden honeysuckles, but is not fragrant. It is a popular landscaping plant that will easily grow up into a tree. **Trumpet honeysuckle** (N) (*Lonicera ciliosa*) (Zones All) is a deciduous vine valued for its reddish-orange flowers that occur in clusters in May and June. It has red berries and the flowers are not fragrant. It will spread over the ground or climb 10-20 feet.

Common name *Botanical name*	Hardiness Zones	Plant Size	Light Needs	Soil Needs

Hairy honeysuckle (N) (*Lonicera hispidula*) (Zones 4-7) has tough, hairy leaves, and small pink or pink-tinged yellow blossoms followed by red berries. It grows 10-20 feet and will crawl or climb. Plant on dry banks, over fences (will need some help), and mix in with hedgerow plants. A fragrant honeysuckle is the **gold flame honeysuckle** (*Lonicera heckrottii*). It blooms from spring until frost.

WILDLIFE INFORMATION: The flowers of most honeysuckles yield abundant nectar for humming-birds and some larger butterflies. The fruit is eaten by grouse, pheasants, flickers, robins, thrushes, bluebirds, waxwings, grosbeaks, finches, and juncos. Its twining habit creates nest sites for small birds.

Common name *Botanical name*	Hardiness Zones	Plant Size	Light Needs	Soil Needs
Virginia creeper *Parthenocissus quinquefolia*	All	60'	F-sun P-shade	Moist to dry

LANDSCAPE INFORMATION: This woody vine has large, attractive leaves that turn bright red in fall. It can be invasive and may self-seed. It's an effective cover for a chain-link fence or arbor, but not the side of a wood or shingle house where it can do damage to the structure. A slightly less-aggressive plant, **Boston-ivy** (*Parthenocissus tricuspidata*), is a self-supporting vine with large leaves and bright-red fall color. It grows fast and clings tightly to concrete and brick walls.

WILDLIFE INFORMATION: The fruit is eaten by flickers, crow, chickadees, nuthatches, robins, thrushes, starlings, house sparrows, tanagers and finches.

Common name *Botanical name*	Hardiness Zones	Plant Size	Light Needs	Soil Needs
Grape (N) *Vitis* spp.	Varies	Varies	F-sun	Moist to dry

LANDSCAPE INFORMATION: The **garden grape** (*Vitis* spp.) (Zones All) is a popular fruiting vine with nice fall color. To get quality fruit, choose a variety that fits your climate, train carefully, and prune year-ly. It can also be allowed to run freely on the ground to form a loose thicket or used over an archway, trellis, or arbor. The native **wild grape** (*Vitis californica*) (Zones 4-7) grows along streambanks and in wet deciduous woodlands of southern Oregon. It is aggressive and only suited for large projects where it can grow freely to 60 feet or more.

WILDLIFE INFORMATION: All berry-eating birds and mammals will eat the fruit, especially on varie-ties that hold the fruits on the vines into fall. Opossums will eat the grape and leave the peel.

For other vines, see the tables in Appendix B.

GARDEN PERENNIALS (AND SOME WILDFLOWERS)

Plants are listed alphabetically by their botanical name.
The botanical name first includes the *Genus* and then the *species*.
When referencing more than one species in a genus, the abbreviation (spp.) is used.
(N) = Plants that are native to Oregon, Washington, or southern British Columbia.
For definitions of headings, see the "Key" at the beginning of Appendix C.
For information on where native plants occur naturally, see the "Native Plant Occurrence Chart" in Appendix B.

Common name *Botanical name*	Hardiness Zones	Plant Size	Light Needs	Soil Needs
Yarrow (N) *Achillea* spp.	All	Varies	F-sun	Moist to dry

LANDSCAPE INFORMATION: Yarrows are carefree perennials that bloom from summer to early fall. There are low growing and taller species. The leaves are gray or green, aromatic, and usually finely divided. The flowers are red, pink, yellow, or white depending on the species and variety. Tall species can spread quickly and are easily divided when clumps get crowded. The **native yarrow** (*Achillea millefolium*) grows to about 2 feet and has white flowers. It can thrive in a lawn and will grow in any sunny, well-drained landscape area.

WILDLIFE INFORMATION: The flower heads are flattish clusters and excellent platforms for pollinating insects, including butterflies and syrphid flies, to land on. Because of the abundance of individual flowers in each cluster, insects often spend a long time on each flower head.

Pearly everlasting (N) *Anaphalis margaritacea*	All	24" high	F-sun	Dry

LANDSCAPE INFORMATION: Each flower head is surrounded by several rows of pearly colored, translucent, papery bracts which are already dry in texture when in bloom. The little flower heads are scarcely ⅓ inch across but a hundred or more may be grouped in a cluster at the top of each stem, making quite a showy inflorescence in late summer. A great plant for dry forest openings, roadcuts, and disturbed sites. Plants spread from rhizomes and once established they can compete with aggressive grasses.

WILDLIFE INFORMATION: The flowers attract pollinators including syrphid flies and small wasps, also woodland skipper and mylitta crescent butterflies. The larvae of painted lady butterflies feed on the foliage.

Columbine (N) *Aquilegia* spp.	All	Varies	F-sun P-shade	Moist to dry

LANDSCAPE INFORMATION: Columbines have lacy, fern-like foliage and beautiful nodding flowers. The flowers occur on erect flower stocks that can vary from 6 to 24 inches. Both flower color and form vary according to species and variety. They bloom in spring and early summer. Older, less hybridized varieties will have qualities that appeal to their pollinators. Cut back old stems for a second crop of flowers; leave some to

Common name *Botanical name*	Hardiness Zones	Plant Size	Light Needs	Soil Needs

seed if you want plants to self-sow and attract seed-eating birds. The native **red columbine** (*Aquilegia formosa*) is found in meadows, rocky outcrops, and sunny streamsides throughout the Pacific Northwest. It grows up to 3 feet high, has red and yellow flower spurs, and is attractive in woodland gardens.

WILDLIFE INFORMATION: The flowers are often visited by hummingbirds and the seeds are eaten by sparrows, juncos, and finches.

Common name *Botanical name*	Hardiness Zones	Plant Size	Light Needs	Soil Needs
Seathrift (N) *Armeria maritima*	4–7	6–12" high 18" wide	F-sun P-shade	Moist to dry

LANDSCAPE INFORMATION: Seathrift has a neat and tidy habit with grass-like foliage. It has pink, white, or red flowers depending on the variety and it blooms in spring and often again in fall. Remove faded flowers to extend flower production. Sea thrift grows in poor, dry, well-drained soils and requires more moisture and shade in hot, dry areas. Seathrift naturally occurs in coastal areas and it seems to grow best where summers are cool.

WILDLIFE INFORMATION: The flowers attract painted lady butterflies and the mat-like vegetation provides cool cover for a variety of invertebrates, including slug-eating ground beetles. Deer browse the foliage.

Goat's beard (N) *Aruncus sylvester*	4–7	12–36" high	F-sun P-shade	Moist

LANDSCAPE INFORMATION: An attractive, robust plant that spreads slowly from underground rhizomes. It has ferny foliage and large plumes of tiny white flowers that occur in summer. Male plants have larger flowers. It occurs naturally at low and mid elevations in wet roadsides, forest edges, and along creeks. A nice plant for any moist area in sun or considerable shade.

WILDLIFE INFORMATION: Its flowers are visited by hummingbirds, mourning cloak butterflies, and native bees and wasps. The foliage is browsed by deer and elk.

Aster (N) *Aster* spp.	All	Varies	F-sun	Moist to dry

LANDSCAPE INFORMATION: The many horticultural varieties of asters range from alpine types that form compact mounds 6 inches high, to open-branched types 6 feet tall. Flowers come in white or shades of blue, red, pink, lavender, or purple, mostly with yellow centers. They bloom in summer and fall; some start flowering in spring. Asters are valuable for abundant color in large flower borders, among shrubs, and in grasslands. Compact dwarf or cushion types make tidy edges, mounds of color in rock gardens, and good container plants. Two native asters that occur naturally in lowland meadows, clearings, and streambanks west of the Cascade mountains are **Pacific aster** (*Aster chilensis*) and **Douglas aster** (*Aster subspicatus*). There are many other native species and most are easy to grow from seed or divisions.

WILDLIFE INFORMATION: Asters are members of the sunflower family and their "flower" is actually made up of numerous individual flowers. These attract a wide variety of flying insects including painted lady, red admiral, spring azure, orange sulphur, and woodland skipper butterflies.

Common name *Botanical name*	Hardiness Zones	Plant Size	Light Needs	Soil Needs
Balsamroot (N) *Balsamorhiza* spp.	All	1-2' tall and wide	F-sun	Dry

LANDSCAPE INFORMATION: A perennial wildflower with large arrow-shaped leaves and sunflower-like yellow flower heads in late spring. Two common species that occur naturally with native grasses, lupines, and ponderosa pine on foothill and mountain slopes east of the Cascade mountains are **deltoid balsamroot** (*Balsamorhiza deltoidea*) and **arrow-leaf balsamroot** (*Balsamorhiza sagittata*). **Deltoid balsamroot** also naturally occurs in a few dry, open, grassy areas on the west side and is the preferred choice for dryland landscapes west of the mountains. Balsamroot has a significant taproot that makes transplanting nearly impossible.

WILDLIFE INFORMATION: The flowers are popular with pollinating insects; deer and elk browse the leaves.

Common name *Botanical name*	Hardiness Zones	Plant Size	Light Needs	Soil Needs
Campanula (N) *Campanula* spp.	All	Varies	F-sun P-shade	Moist to dry

LANDSCAPE INFORMATION: The common garden campanulas are a varied group including trailers, creeping or tufted miniatures, and erect species that reach 5 feet tall. The flowers are generally bell shaped, but some are round and flat. Flowers are blue, lavender, purple, or white and the bloom period is from spring to fall. All grow best in rich, well-drained soil. One of the most common natives is **harebell** (*Campanula rotundifolia*). It grows between 6 and 20 inches tall and has attractive blue or lavender flowers that are 1 inch across. It occurs naturally along grassy roadsides, sunny streamsides, and in rock outcrops from sea level to timberline. It is easy to grow and a good addition to any drought-tolerant landscape.

WILDLIFE INFORMATION: Campanula flowers are visited by swallowtail butterflies, bumblebees, and hummingbirds.

Common name *Botanical name*	Hardiness Zones	Plant Size	Light Needs	Soil Needs
Bleeding heart (N) *Dicentra* spp.	All	8-18" high	P-shade	Moist

LANDSCAPE INFORMATION: Bleeding heart has graceful, fernlike foliage that looks nice with ferns and other woodland plants. The dainty pink or red heart-shaped flowers appear on leafless stems from April to June. Bleeding heart needs rich, light, moist, porous soil. Never let water stand around roots. Since foliage dies down in winter, mark clumps to avoid digging into roots in the dormant season. The westside native is *Dicentra formosa*. It grows in moist woodlands, ravines, and streamsides west of the Cascade mountains. **Steer's head** (*Dicentra uniflora*) occurs in the mountains on the eastside. Species with brighter, larger flowers are also available in nurseries.

WILDLIFE INFORMATION: The flowers attract hummingbirds and the leaves are eaten by clodius parnassians butterfly larvae. There is a little glob of fat on each seed which attracts ants and ensures that the seed is moved around and new patches are established.

Common name *Botanical name*	Hardiness Zones	Plant Size	Light Needs	Soil Needs
Purple coneflower *Echinacea purpurea*	All	2-3' high	F-sun	Moist, dry

LANDSCAPE INFORMATION: A rather coarse, stiff plant that forms clumps of erect stems. The leaves are oblong and 3 to 8 inches long. The showy flower heads have drooping purple rays, dark purple centers, and bloom over a long period in late summer. Use among shrubs or in wide borders with other robust perennials.

Common name *Botanical name*	Hardiness Zones	Plant Size	Light Needs	Soil Needs

WILDLIFE INFORMATION: The flowers attract painted lady, common wood nymph, and silvery blue butterflies.

Globe-thistle *Echinops exaltatus*	All	3' high	F–sun	Moist to dry, good drainage

LANDSCAPE INFORMATION: A rugged-looking, erect, rigidly branched plant with coarse, prickly, deeply cut, gray-green leaves. The small steel-blue flowers are produced in round, 2-inch heads. It flowers from midsummer to late fall.

WILDLIFE INFORMATION: The flowers attract many butterfly species that appear, or reappear, in late summer.

Fireweed (N) *Epilobium angustifolium*	All	3–4' high	F–sun	Moist to dry

LANDSCAPE INFORMATION: A vigorous perennial that spreads from seeds and rhizomes. It has lance-shaped leaves and clusters of rose-to-purple flowers on top of each flower stem. The seeds are fluffy and blow in the wind. Fireweed is commonly seen in dry disturbed areas, including clearings, roadsides, and recently burned sites. In landscapes it can compete with aggressive grasses in fields and similar sites. It spreads quickly and should never be expected to remain restricted to a small area.

WILDLIFE INFORMATION: The flowers are visited by hummingbirds, butterflies, and bees.

Buckwheat (N) *Eriogonum* spp.	Varies	3' high	F–sun	Dry, good drainage

LANDSCAPE INFORMATION: Wild-buckwheats are spreading ground covers or small shrubs with attractive compact clusters of white, pink, or yellow flowers in summer. They grow best in full sun in well-drained, loose, gravelly soil. They are very drought-tolerant once established, needing very little water and no fertilizer. There are several native buckwheats and most are found east of the Cascade mountains in dryland settings. **Sulfur flower** (*Eriogonum umbellatum*) is one species that occurs on both sides of the mountains at mid to high elevations. When choosing native wild-buckwheats for the landscapes, you'll have the best luck growing species that occur naturally in your area.

WILDLIFE INFORMATION: The flowers are visited by several flying pollinators including the following butterfly species: bat blue, blue copper, brown elfin, and painted lady.

Sea-holly *Eryngium amethystinum*	All	2–3' tall	F–sun	Moist to dry

LANDSCAPE INFORMATION: An erect, stiff-branched, thistle-like plant that blooms from July to September. It has striking oval, steel-blue or amethyst, ½-inch long flower heads that are surrounded by spiny blue bracts. The leaves are dark green and deeply cut. An excellent plant for coastal gardens but it can also be grown inland.

WILDLIFE INFORMATION: The flowers attract butterfly species that appear, or reappear, in late summer.

Common name *Botanical name*	Hardiness Zones	Plant Size	Light Needs	Soil Needs
Blanket-flower (N) *Gaillardia* spp.	All	6-24" tall	F-sun	Dry, good drainage

LANDSCAPE INFORMATION: An easy-to-grow perennial or wildflower that is very hardy and thrives in heat. The plants form clumps and have daisy-like flowers in warm colors—yellow, bronze, scarlet. The most common garden perennial is *Gaillardia grandiflora* with roughish gray-green foliage and flower heads that are 3 inches across. The native wildflower is *Gaillardia aristata*. It can grow on dryish sites almost anywhere and is easy to establish from seed or small plugs. With water, both bloom from June until frost. The plants live about three years and the native may reseed itself.

WILDLIFE INFORMATION: The flowers attract a variety of flying pollinators, including painted lady and woodland skipper butterflies.

Common name *Botanical name*	Hardiness Zones	Plant Size	Light Needs	Soil Needs
Coral bells (N) *Heuchera* spp.	All	1-3' high	F-sun P-shade	Moist to dry, good drainage

LANDSCAPE INFORMATION: Coral bells form compact, evergreen clumps of roundish leaves with scalloped edges. The nodding, bell-shaped flowers occur from April to August on slender, wiry stems that are 15 to 30 inches high. The dainty flowers come in different colors according to species and variety and include red, reddish pink, greenish, and white. The addition of organic matter to the soil is extremely beneficial to growing high quality specimens. Use as edging in rock gardens and in front of shrubs, as ground cover, and mass in borders. The closely related natives include two **alumroots**, of which *Heuchera micrantha* is the most available. **Alumroot** is an attractive plant with many small white flowers and foliage that is well-suited for a wild woodland edge, stream bank, or moist rock crevice.

WILDLIFE INFORMATION: Flowers attract hummingbirds.

Common name *Botanical name*	Hardiness Zones	Plant Size	Light Needs	Soil Needs
Lupine (N) *Lupinus* spp.	Varies	1-3' high	F-sun	Moist to dry, good drainage

LANDSCAPE INFORMATION: The leaves are divided up into many leaflets and the flowers are generally sweet-pea shaped, in dense spikes at the end of stems. **Russell lupines** are popular non-native garden perennials. They come in many sizes and flower colors and do well in moist perennial borders where summers aren't too hot and dry. Many beautiful native species occur throughout the Pacific Northwest. **Bigleaf lupine** (*Lupinus polyphyllus*) is a common coastal species and one of the native parents of the Russell lupine. It can grow over 3 feet tall and has dense flower clusters that are blue, purple, or reddish. **Silky Lupine** (*Lupinus sericeus*) is the most common lupine in the inland Pacific Northwest. It is found in bunchgrass areas and well-drained slopes with balsamroot and blanket-flower. The silky leaves are silvery green and the flowers are blue, lavender, rarely yellow or white, and silky.

WILDLIFE INFORMATION: The flowers attract hummingbirds. Silvery blue and other butterfly species use the flowers and the vegetation. Grouse, songbirds, and small mammals eat the seeds.

Common name *Botanical name*	Hardiness Zones	Plant Size	Light Needs	Soil Needs
Skunk cabbage (N) *Lysichiton americanus*	All	2-3' high	P-shade F-shade	Wet

LANDSCAPE INFORMATION: This bold freshwater wetland perennial has large fleshy leaves that can grow up to 3 feet long and 1 foot wide under shady conditions. Plants are smaller when grown in sun. The flower is a large yellow spathe (or sheath) that appears early in the spring before leaves come out. When the plants are fresh their scent is sweet but the wilting flowers and crushed leaves give good reason for the common name. Plant in wooded wetlands, the sides of creeks and ponds, marshes, and other areas that collect water and hold water.

WILDLIFE INFORMATION: Flowering heads and leaves are extensively browsed by elk; fruits are eaten by bear; roots are eaten by muskrats. The pungent odor of the flowers attracts pollinators, including carrion beetles and blowflies.

Bee balm *Monarda didyma*	All	2-3' high	F-sun	Moist

LANDSCAPE INFORMATION: The oval, 6-inch long, dark-green leaves of bee balm have a strong, pleasant odor, like a blend of mint and basil. In summer, the stems are topped by tight clusters of long-tubed scarlet flowers. The plant forms leafy clumps that spread rapidly at the edges but are not really invasive.

WILDLIFE INFORMATION: The flowers are visited by hummingbirds and painted lady butterflies.

Penstemon (N) *Penstemon* spp.	Varies	Varies	F-sun	Moist to dry, good drainage

LANDSCAPE INFORMATION: Penstemons vary from perennials to ground covers to small shrubs. All have colorful, tubular flowers. Bright reds and blues are the most common flower colors, but there are penstemons in soft pink, lilac, deep purple, or white. The garden hybrid varieties grow quickly, flower over a long period, tend to be easier to grow alongside other garden perennials than the natives, and should be grown as annuals in cold winter areas. There are many native penstemon species that can look wonderful and grow well in landscape where there is sun and good drainage. **Cardwell's penstemon** (*Penstemon cardwellii*) is a low shrub found in forested edges, cutbanks, and rocky slopes. It has bright-purple to deep-blue-violet flowers. The mat-forming **Davidson's penstemon** (*Penstemon davidsonii*) occurs naturally on talus or rocky outcrops in the mountains at mid to high elevations. It has lovely violet blue flowers in late spring or early summer. One of the better native species for mild, damp landscapes west of the mountains is the **coast penstemon** (*Penstemon serrulatus*).

WILDLIFE INFORMATION: The flowers attract hummingbirds, bumblebees, night-flying moths, and butterflies including swallowtails, common wood nymphs, and Lorquin's admirals.

Common name *Botanical name*	Hardiness Zones	Plant Size	Light Needs	Soil Needs
Cape-fuchsia *Phygelius capensis*	4–7	3–4' high and wide	F–sun	Moist

LANDSCAPE INFORMATION: Cape fuchsia is related to snapdragon and penstemon, but the attractive drooping flowers also suggest fuchsia. The plant is shrubby and it spreads by underground stems or rooting prostrate branches. The flowers are tubular, curved, and occur in loosely branched clusters at branch ends over a long period beginning in early summer. There are red and yellow forms. It is striking in a flower border when given plenty of space.

WILDLIFE INFORMATION: An early and long bloomer that attracts hummingbirds.

Common name *Botanical name*	Hardiness Zones	Plant Size	Light Needs	Soil Needs
Scabiosa (Pincushion flower) *Scabiosa* spp.	All	30" high	F–sun	Moist

LANDSCAPE INFORMATION: The stamens of scabiosa protrude beyond the curved surface of the flower cluster, giving an illusion of pins stuck into a cushion. They are easy to grow and blooms begin in early summer and continue until winter if flowers are kept from seeding. *Scabiosa caucasica* is very popular species with finely cut leaves and 2- to 3-inch flower clusters that are blue to bluish lavender, or white, depending on the variety. Scabiosas look nice in mixed flower borders with other perennials grown to attract butterflies.

WILDLIFE INFORMATION: The long blooming period and attractive flowers make scabiosa a great butterfly plant. Flowers are visited by painted lady, red admiral, and woodland skippers.

Common name *Botanical name*	Hardiness Zones	Plant Size	Light Needs	Soil Needs
Sedum (N) *Sedum* spp.	All	Varies	F–sun	Dry

LANDSCAPE INFORMATION: Native Pacific Northwest sedums occur from sea level to interior sagebrush areas. A choice sedum for interior landscapes is the **lance-leaf stonecrop** (*Sedum lanceolatum*). It has lance-shaped leaves, erect flowering stems to about six inches, and bright yellow flowers in summer. **Broad-leaved stonecrop** (*Sedum spathulifolium*) occurs on our coastal bluffs, rock outcrops, and in forest openings from sea level to mid elevations. It's a tight ground cover with plump, blue and red foliage that is held tight to the stems. The flowers occur in clusters of bright yellow from May to July. Both species are a delight in rock gardens, among stepping stones, and in coarse dry areas. The non-native **fall sedum** (*Sedum spectabile*) is a unique looking upright plant that grows to about 24 inches when in flower. The popular variety 'Autumn Joy' has succulent stems and blue green, roundish, fleshy, 3-inch leaves. The pink flowers occur in broad, dense clusters atop stems in late summer. Plants die down in winter but the flower stocks persist.

WILDLIFE INFORMATION: Several butterfly and bee species glean nectar from sedums. Fall sedum is particularly popular with honeybees. Phoebus parnassian butterfly larvae use lance-leaf stonecrop as a host plant. Brown elfin and moss elfin butterfly larvae use broad-leaved stonecrop as a host plant.

Common name *Botanical name*	Hardiness Zones	Plant Size	Light Needs	Soil Needs
Solomon's seal (N) *Smilacina* spp.	All	Varies	F-sun P-shade	Moist

LANDSCAPE INFORMATION: **Large Solomon's seal** (*Smilacina racemosa*) is an elegant looking woodland plant with single arching stalks, each with several 3- to 10-inch long leaves. Plants can reach 3 feet tall and the stalks are topped by fluffy, conical clusters of small, fragrant, creamy whiter flowers in March-May. The flowers are followed by red, purple-spotted berries. Plants are easy to grow and not overly aggressive. **Star-flowered Solomon's seal** (*Smilacina stellata*) reaches only about 10 inches high and has lance-shaped leaves and creamy star-like flowers. The fruit is dark blue or reddish-black when mature. It is also easy to grow and more of a spreader. Both species are at home in moist woods and woodland gardens, wet meadows, moist perennial borders, and around the edges of ponds. Plants will require some shade in hot inland areas.

WILDLIFE INFORMATION: Fruit may be eaten by grouse, band-tailed pigeons, thrushes, and probably some small mammals.

Common name *Botanical name*	Hardiness Zones	Plant Size	Light Needs	Soil Needs
Goldenrod (N) *Solidago* spp.	All	24-36"	F-sun	Moist to dry

LANDSCAPE INFORMATION: Goldenrod is a good plant for late-summer color when a plume of yellow flowers tops clumps of leafy stems. The garden varieties differ chiefly in size and in depth of yellow shading. Goldenrod grows best in soil that isn't too rich. The common native **Canada goldenrod** (*Solidago canadensis*) is tough enough to compete with aggressive grasses along roadsides, at the edges of thickets and forests, and in meadow plantings. Be warned that this species has creeping underground stems that can spread extremely fast in soil that is watered and amended with rich organic matter.

WILDLIFE INFORMATION: The flowers attract bumblebees and pine white, red admiral, and mylitta crescent butterflies, also small pollinators including syrphid flies and small wasps. The seeds are eaten by birds.

Common name *Botanical name*	Hardiness Zones	Plant Size	Light Needs	Soil Needs
Thyme *Thymus* spp.	All	Varies	F-sun P-shade	Moist to dry

LANDSCAPE INFORMATION: Thymes are low-growers and most are ground covers that form a thick mat of tiny evergreen leaves. They are the perfect plant to place in and around stepping stones. Most have purple flowers in late May. All are very drought-tolerant once established.

WILDLIFE INFORMATION: Flowers attract various species of bees and butterflies.

For other perennials and wildflowers, see the tables in Appendix B.

GRASSES (AND SOME GRASS-LIKE PLANTS)

Plants are listed alphabetically by their botanical name.
The botanical name first includes the *genus* and then the *species*.
When referencing more than one species in a Genus, the abbreviation (spp.) is used.
(N) = Plants that are native to Oregon, Washington, or southern British Columbia.
For definitions of headings, see the "Key" at the beginning of Appendix C.
For information on where native plants occur naturally, see the "Native Plant Occurrence Chart" in Appendix B.

Common name *Botanical name*	Hardiness Zones	Plant Size	Light Needs	Soil Needs
Wheatgrass (N) *Agropyron* spp.	Varies	12-36" high	F-sun	Moist to dry

LANDSCAPE INFORMATION: There are about a dozen species of wheatgrass native to the Pacific Northwest. **Thickspike wheatgrass** (*Agropyron dasystachyum*) is a sod-forming grass with good erosion-control qualities that is adapted to drier areas. **Blue-bunch wheatgrass** (*Agropyron spicatum*) is a long-lived, steely blue bunchgrass that is widespread east of the Cascade mountains where it grows with ponderosa pine, Douglas-fir, and in open prairies with Idaho fescue. It also is found with sagebrush and rabbitbrush in drier areas. **Slender wheatgrass** (*Agropyron trachycaulum*) is a loosely tufted bunchgrass that occurs in meadows, rocky slopes and sunny forest edges.

WILDLIFE INFORMATION: Wheatgrasses are eaten by deer, elk, and rabbits, and are used as nesting cover by ground-nesting birds. In many areas, wheatgrass is the only nesting cover for birds. **Blue-bunch wheatgrass** is used extensively by elk, whitetail deer, and mule deer throughout the winter and spring months. It is also an excellent range grass for sheep, cattle, and horses.

Sloughgrass (N) *Beckmania syzigachne*	All	36" high	F-sun	Wet to moist

LANDSCAPE INFORMATION: A cool-season annual or short-lived perennial grass found in marshes, pond edges, ditches, and other wet sites, primarily in Oregon. This distinctive grass has light green leaves up to ½-inch wide and dense seed clusters that are produced on cylindrical flowering stems in July and August. Sloughgrass is sometimes included in pasture mixes for wet sites and occasionally used to provide temporary erosion control. Plants tend to decline after four to five years and can quickly succumb to more aggressive grass species.

WILDLIFE INFORMATION: Plants produce copious amounts of seed that are eaten by ducks and geese, songbirds, and small mammals.

Sedge (N) *Carex* spp.	All	1-3' high	F-sun P- shade	Wet to moist

LANDSCAPE INFORMATION: Sedges are perennial, grass-like plants with strap-shaped leaves and inconspicuous flowers formed on a spike. More than 100 species are native to the Pacific Northwest. Sedges

Common name *Botanical name*	Hardiness Zones	Plant Size	Light Needs	Soil Needs

generally occur in standing water or saturated soils in marshes, pond edges, lake shores and stream-banks; they are also found in wet meadows and moist forest openings. Two species that can eventually form dense stands include **Lyngby's sedge** (*Carex lyngbyei*) found in coastal areas west of the Cascade mountains, and **inflated sedge** (*Carex vesicaria* var. *major*) found on both sides of the mountains. Two less aggressive sedges that form attractive clumps and occur on both sides of mountains are **Grey sedge** (*Carex canescens*) and **Kellog's sedge** (*Carex lenticularis*). Two sedges that form clumps and can grow in a moist woodland garden are **Dewey's sedge** (*Carex deweyana*) and **Henderson sedge** (*Carex hendersonii*).

WILDLIFE INFORMATION: Sedges produce a large crop of water-dispersed seeds that are eaten by mallards, pintails, widgeons, teal, and wood ducks; also grouse, pheasants, finches, towhees, juncos, and sparrows. Swans, geese, and ducks eat the new growth. Both old and fresh leaves are used as nesting material and some mat-forming species provide shelter and nesting sites. The stems are used as egg-case attachment sites by pond-breeding amphibians, including red-legged frogs and Northwestern salamanders.

Common name *Botanical name*	Hardiness Zones	Plant Size	Light Needs	Soil Needs
Tufted hairgrass (N) *Deschampsia cespitosa*	All	40" high	F-sun P-shade	Moist

LANDSCAPE INFORMATION: A tussock-forming bunchgrass with bright-green foliage and an attractive silky seed culm that occurs on tall erect stocks in early spring. Plant in moist meadows, wet fields, ditches, lake and pond margins, and areas in the landscape that are moist. Tufted hairgrass can also grow in dryer soil, but will look better with some summer water. Plants don't tolerate year-round flooding. Tufted hairgrass is quite ornamental but beware—it can get surprisingly tall and wide when grown in highly amended soil.

WILDLIFE INFORMATION: Plants form mounding clumps that provide dense nesting foliage over a long period of time. The stems and flower stalks tend to remain upright all winter providing perching spots for songbirds.

Common name *Botanical name*	Hardiness Zones	Plant Size	Light Needs	Soil Needs
Creeping spike-rush (N) *Eleocharis palustris*	All	3' high	F-sun	Moist to wet

LANDSCAPE INFORMATION: A perennial wetland plant with round stems that occur from underground runners that can spread aggressively and form dense colonies. Plant in moist, marshy fields and pastures, ditches, pond, and lake sides. It prefers semi-permanently saturated or flooded conditions. Creeping spike-rush is a good alternative to cattail, a larger, more aggressive plant, and can be kept confined by growing in the appropriate container. In a garden pond setting it makes an attractive wetland ornamental.

WILDLIFE INFORMATION: Plants produce large clusters of water-dispersed seeds that are eaten by swans, geese, mallards, pintails, gadwalls, widgeons, teal, rails, and coots. Underground tubers may also be eaten. The stems serve as perch sites for small birds, dragonflies, and other insects. The stems are used as egg-case attachment sites by pond-breeding amphibians, including Pacific treefrogs and long-toed salamanders.

Common name *Botanical name*	Hardiness Zones	Plant Size	Light Needs	Soil Needs
Wildrye (N) *Elymus* spp.	Varies	Varies	F-sun	Moist to dry

LANDSCAPE INFORMATION: **Basin wildrye** (*Elymus cinereus*) (Zones 1–3) is a striking, long-lived, clump-forming perennial grass reaching 6 feet. It is not tolerant of shallow soils but will grow well in high water table areas and it responds well to summer water. Even plants in wild areas stay green well into summer

Common name *Botanical name*	Hardiness Zones	Plant Size	Light Needs	Soil Needs

in inland areas where it occurs with rabbitbrush and sagebrush along streams and roads, in meadows, and at the edge of woods. In gardens, **Basin wildrye** makes a good mini-windbreak. **Blue wildrye** (*Elymus glaucus*) (Zones All) is an attractive, fast-growing bunchgrass that grows to about 3 feet. It has dark-green or blue-grey leaf color and is a plant of prairies, open woods, thickets, and moist to dry hillsides. Blue wildrye is some-what tolerant of shade. *Leymus arenarius* 'Glaucus', an attractive though very aggressive European beach grass, is sometimes mistakenly sold as *Elymus glaucus*. **Dune wildrye** (*Elymus mollis*) = (*Leymus mollis*) (Zones 4-7) is a fast-growing spreader that forms large clumps in sandy and gravelly coastal areas. This is a fine ornamental but will need to be confined where growing space is limited.

WILDLIFE INFORMATION: Basin wildrye provides excellent cover for many animals of the sagebrush country. Songbirds and small rodents eat the seeds of all wildrye species.

Common name *Botanical name*	Hardiness Zones	Plant Size	Light Needs	Soil Needs
Fescue (N) *Festuca* spp.	Varies	Varies	F-sun P-shade	Moist to dry

LANDSCAPE INFORMATION: **California fescue** (*Festuca californica*) (Zones 1-4) is a drought-tolerant bunchgrass with striking ornamental features. The plants form large clumps of wiry leaves which are evergreen even through the dry season, and tall, elegant flowering culms. The stalks are commonly 4 or 5 feet tall, and the tops droop gracefully. **Idaho fescue** (*Festuca idahoensis*) (Zones All) is a densely tufted bunchgrass that is very cold- and drought-tolerant. It is slow to establish from seed, but eventually makes an excellent blue-green ground cover with a flower head that reaches 10 to 30 inches tall. Two main subspecies occur in the Pacific Northwest: *Festuca idahoensis* var. *idahoensis* primarily occurs east of the Cascade mountains in dry grasslands and in open, south-facing slopes in the mountains. *Festuca idahoensis* var. *roemeri* occurs on the west side in open prairies, south-facing grassy outcrops, and in open, often rocky outcrop areas in the Douglas-fir and Garry oak forests. Sheep fescue, also know as blue fescue (*Festuca ovina*), is a common ornamental grass that resembles Idaho fescue. **Red fescue** (*Festuca rubra*) (Zones All) is a loosely tufted perennial with a flower height of about 20 inches. It prefers moist, cool areas and acid soil and naturally occurs along stream banks, meadows, river flats, clearings, and roadsides from sea level to high elevations. While normally a bunching grass, in conditions of more moisture and fertility it will behave much like a turf grass and can take periodic mowing. Like Idaho fescue, red fescue contains subspecies and in areas where the integrity of the local gene pool is of concern, attention needs to be placed on species selection. Many introduced cultivars of red fescue are used for turf, pasture, and erosion control.

WILDLIFE INFORMATION: Plants are palatable to elk and deer. The small seeds are eaten by songbirds and small mammals. Several species of skipper butterflies use fescue as their host plant.

Common name *Botanical name*	Hardiness Zones	Plant Size	Light Needs	Soil Needs
Tall mannagrass (N) *Glyceria elata*	4-7	5' tall	F-sun P-shade	Moist to wet

LANDSCAPE INFORMATION: A loosely tufted perennial with erect, rather succulent stems and narrow, ribbon-like leaves. It grows singularly or in clumps from rhizomes and occurs in wet meadows, moist woods, in shallow streams, ponds and lakes, ditches, and around springs

WILDLIFE INFORMATION: It provides food and cover for muskrats, and waterfowl including gadwall and mallards.

Common name *Botanical name*	Hardiness Zones	Plant Size	Light Needs	Soil Needs
Rush (N) *Juncus* spp.	Varies	8-36" high	F-sun P-shade	Wet soils or shallow water

LANDSCAPE INFORMATION: There are dozens of rush species native to the Pacific Northwest. **Soft rush** (*Juncus effusus*) grows in all zones and is easily recognized. It's a tufted perennial plant with round stems that generally grows in large clumps. It is commonly found in wet fields, pastures, wet roadsides, and along the edges of ponds and lakes. It even grows well where ground traffic has compacted the soil and it can be difficult to remove.

WILDLIFE INFORMATION: Seeds are eaten by mallards, pintails, widgeons, teal, wood ducks, grouse, pheasants, house finches, towhees, sparrows, and juncos. Muskrats feed on rootstalks.

Bulrush (N) *Scirpus* spp.	All	1-5' tall	F-sun P-shade	Wet

LANDSCAPE INFORMATION: There are many species of bullrush throughout the Pacific Northwest. **Hardstem bulrush** (*Scirpus acutus*) is very similar to **soft-stem bulrush** (*Scirpus validus*). Both grow to about 4 feet and have stout, round stems topped by flower clusters. **Small-fruited bulrush** (*Scirpus microcarpus*) is a robust, coarse plant that grows singly or in loose clumps from rhizomes. The flowers are compact grayish-brown spikelets clustering at the ends of short stems. Plant bulrush in marshes and the mucky margins of lakes and ponds; also in wet ditches and clearings. Plant with rushes and sedges.

WILDLIFE INFORMATION: Birds which eat the hard seeds include swans, geese, pintail, widgeon, teal, mallard, common goldeneye, pheasants and song sparrows. Stems and underground parts are eaten by muskrats and geese. Larger species furnish nesting cover for ducks, geese, marsh wrens and red-winged blackbirds. Mature plants provide concealment for raccoons, otter, muskrats, and other mammals.

Cattail (N) *Typha latifolia*	All	3-9' high	F-sun P-shade	Wet to moist

LANDSCAPE INFORMATION: The long, strap-like leaves of cattail grow from fibrous rhizomes. The erect stems reach 8 feet and support brown, cigar-shaped female flowers which persist for months (the male flowers are above the female flowers and quickly disintegrate). Cattail grows in a wide range of habitats, from seasonally saturated soil to areas with standing water up to 2 feet deep. It is often found in marshes and around lake margins. It is likely to form large monocultures in disturbed areas, to the detriment of other desirable native plants including sedges and bulrushes.

WILDLIFE INFORMATION: Geese, beaver, and muskrats eat the rootstocks. Muskrats use the leaves and stems to construct "houses," which are used as nesting sites by ducks, geese, and terns. Cattail stands provide nesting sites for diving ducks, coots, grebes, rails, red-winged blackbirds, yellow-headed blackbirds, and marsh wrens.

For other grasses and grass-like plants, see the Tables in Appendix B.

CONSTRUCTION PLANS FOR NEST BOXES, ROOST BOXES, AND BIRD FEEDERS

ROBIN, BARN SWALLOW, AND PHOEBE NEST PLATFORM

Lumber Detail

9½"

13" — Back

8"

7" — Floor

8½" — Roof

Side

Extra Side

Side

Materials
1 – 1 x 10 x 4' rough cedar board
12 – 1¼" outdoor wood screws

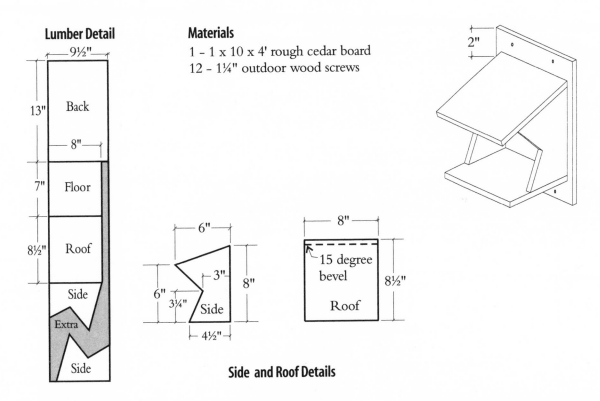

2"

Side and Roof Details

6"

3"

6"   3¼"   Side   8"

4½"

8"

15 degree bevel

Roof

8½"

Washington Department of Fish and Wildlife

BASIC SONGBIRD NEST BOX

Materials

1 - 1 x 6 x 6' rough cedar board
18 - 1¼" outdoor wood screws or #7 galvanized nails
Wire to keep side door shut

Lumber Detail

5½"	⅜"
4"	Floor
10"	Side
12"	Side
8½"	Roof
2"	
10"	Front
17"	Back
	Extra

Pivot screw works as a hinge. To allow the side to open easily, the pivot screw on the opposite side needs to be at the same level as the one in front.

Two screws and wire to keep side door shut.

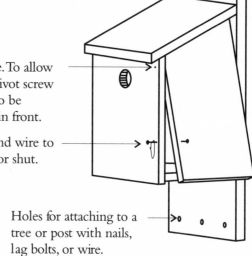

Holes for attaching to a tree or post with nails, lag bolts, or wire.

Exact Round Entry Hole Diameters

Chickadees	1 to 1⅛" ★
House wrens	1⅛" ★
Violet-green swallows	1¼" ★
Tree swallows	1¼"
Nuthatches	1¼" ★
Bluebirds	1½"

★ These species can also use the Optional Entry Hole (see detail).

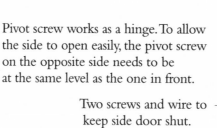

¼"

Detail of Two Slant Cuts

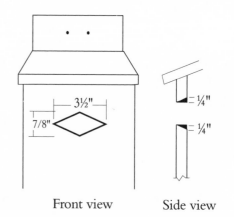

3½"

7/8"

¼"

¼"

Front view   Side view

Optional Entry Hole

NOTE: This diamond-shaped entry hole is designed to prevent access by house sparrows. To work properly, it is extremely important that the final entry hole be made to the above dimensions. To accommodate violet-green swallows, file down the area inside of the entry hole, as shown in the side view.

Assembly Sequence

Washington Department of Fish and Wildlife

BARN OWL NEST BOX

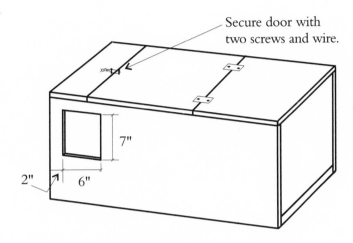

Secure door with two screws and wire.

7"

2"   6"

Plywood Detail

36"

		15¾"
Back	
Entry hole →	Front	15¾"
Lid	Roof	12½"

Extra

5" 13" 18"

Board Detail

20 – ½" diameter drain holes in bottom

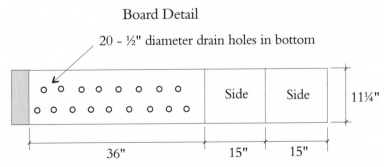

| | Side | Side | 11¼" |

36"   15"   15"

Mounting Options

1. Securely knot one end of a 5' long piece of nylon rope. Thread it through a small hole made in the top of one side so that the knot is inside the box. Then thread the rope through a small hole at the other side and knot the rope inside. The box can be hung inside a barn or other building about 20' above the ground. If placed in a tree, first wrap the rope around a sturdy limb before threading through the second hole.

2. Place box on a cross-beam with the front facing the inside of the building. Nail the box through the bottom to the cross-beam.

3. To permit direct access to the box from the outside, place the box against the inside of a wall after cutting a 6" x 6" entrance in the barn wall at the level of the entry on the box.

Materials

1 - 4 x 8 x ⅝" exterior plywood
1 - 1 x 12 x 6' rough cedar board
2 galvanized hinges with galvanized screws
5' of nylon mounting rope (optional)
35 - 1⅝" outdoor wood screws
or 35 #7 galvanised nails
Two coats of dark latex exterior paint
(if mounted outside)

FLICKER NEST BOX

Lumber Detail

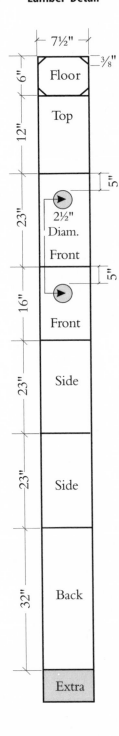

Materials

1 - 1 x 8 x 12' rough cedar board
20 - 1¼" outdoor wood screws
Wire to keep box shut
2 galvanized hinges

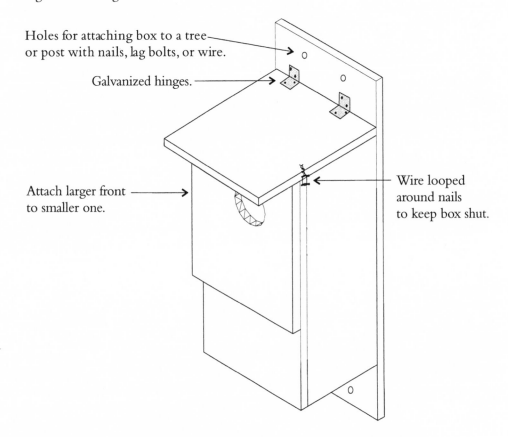

Holes for attaching box to a tree or post with nails, lag bolts, or wire.

Galvanized hinges.

Attach larger front to smaller one.

Wire looped around nails to keep box shut.

Tips on Mounting and Care of Flicker Nest Box

1. When possible, face box to the south. Mount box 6' high or more and keep the flyway in front of the box open.
2. Tilt top of box forward about 15 degrees to make it easier for the adults to feed the young, and babies to climb to the hole.
3. Fill box to the top with wood shavings (animal bedding can be purchased at a pet store). Fine sawdust is not recomended because it remains damp and is difficult for birds to work with. Pack-in the shavings to discourage starlings and sparrows from nesting in the box. Excavating a cavity is an essential part of the Flicker's mating behavior.

DOWNY AND HAIRY WOODPECKER NEST BOX

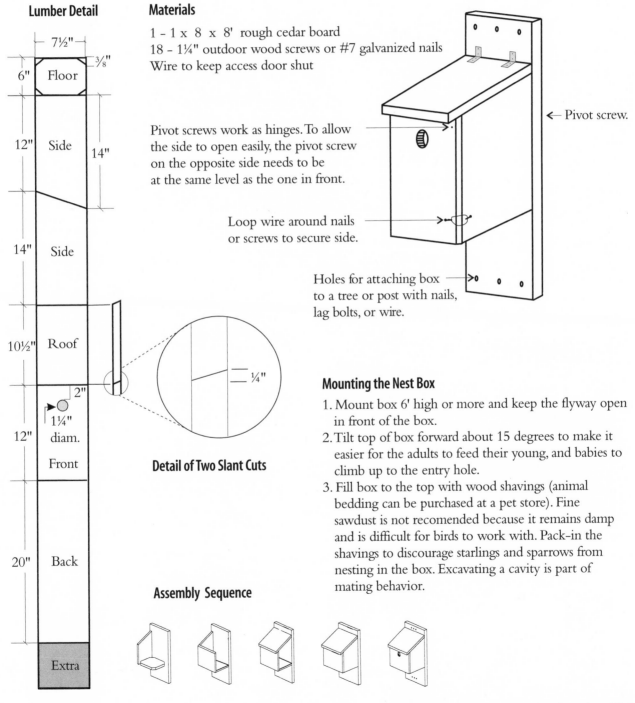

Lumber Detail

7½"
⅜"
6" Floor
14"
12" Side
14" Side
10½" Roof
2"
1¼" diam.
12" Front
20" Back
Extra

Detail of Two Slant Cuts

¼"

Materials

1 - 1 x 8 x 8' rough cedar board
18 - 1¼" outdoor wood screws or #7 galvanized nails
Wire to keep access door shut

Pivot screws work as hinges. To allow the side to open easily, the pivot screw on the opposite side needs to be at the same level as the one in front.

Loop wire around nails or screws to secure side.

← Pivot screw.

Holes for attaching box to a tree or post with nails, lag bolts, or wire.

Mounting the Nest Box

1. Mount box 6' high or more and keep the flyway open in front of the box.
2. Tilt top of box forward about 15 degrees to make it easier for the adults to feed their young, and babies to climb up to the entry hole.
3. Fill box to the top with wood shavings (animal bedding can be purchased at a pet store). Fine sawdust is not recomended because it remains damp and is difficult for birds to work with. Pack-in the shavings to discourage starlings and sparrows from nesting in the box. Excavating a cavity is part of mating behavior.

Assembly Sequence

Washington Department of Fish and Wildlife

KESTREL, SAW-WHET OWL, AND SCREECH-OWL NEST BOX

Lumber Detail

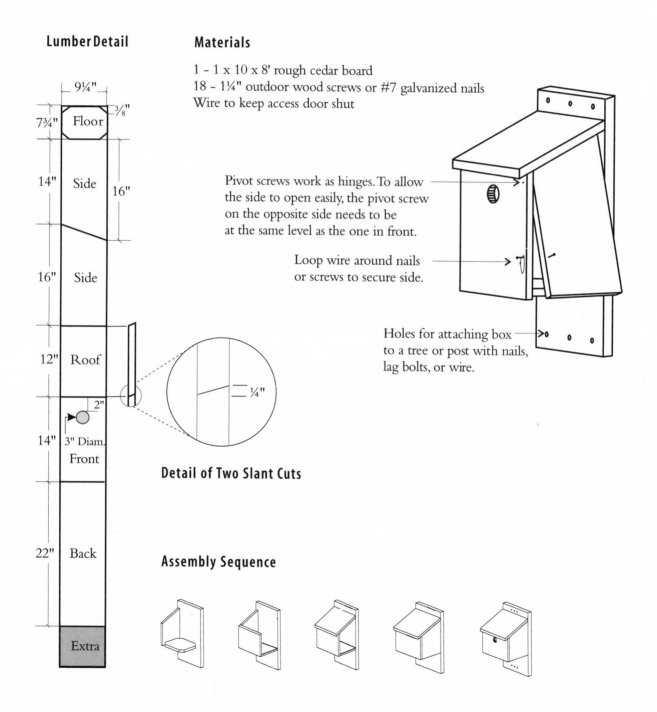

9¼"

7¾"   Floor   ⅜"

14"   Side   16"

16"   Side

12"   Roof

14"   3" Diam.   2"
      Front

22"   Back

Extra

Materials

1 – 1 x 10 x 8' rough cedar board
18 – 1¼" outdoor wood screws or #7 galvanized nails
Wire to keep access door shut

Pivot screws work as hinges. To allow the side to open easily, the pivot screw on the opposite side needs to be at the same level as the one in front.

Loop wire around nails or screws to secure side.

Holes for attaching box to a tree or post with nails, lag bolts, or wire.

¼"

Detail of Two Slant Cuts

Assembly Sequence

DOUGLAS SQUIRREL (CHICKAREE) AND FLYING SQUIRREL NEST BOX

Lumber Detail

Floor	5¾"
Baffle	5½" (3")
Top	10"
Front	9" (1½", 2¼")
Side	9"
Side	9"
Back	14"
Support	2½"
Ledge	4½"

7¼"

Materials

1 – 1 x 8 x 6' rough cedar board
20 – 1¼" outdoor wood screws

The pivot screws work as hinges. To allow the side to open easily, the pivot screws need to be at the same level as the one in front.

The baffle

Two screws and wire to keep side door shut.

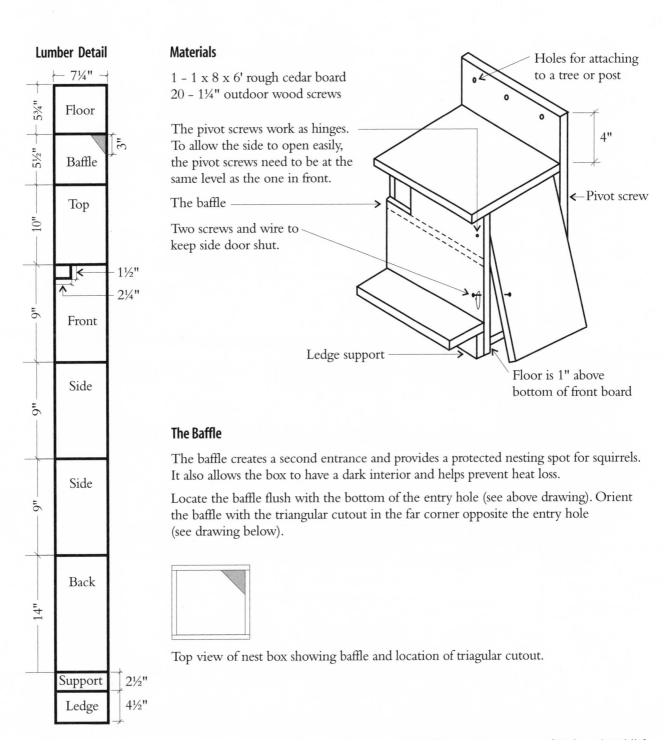

Holes for attaching to a tree or post

4"

Pivot screw

Ledge support

Floor is 1" above bottom of front board

The Baffle

The baffle creates a second entrance and provides a protected nesting spot for squirrels. It also allows the box to have a dark interior and helps prevent heat loss.

Locate the baffle flush with the bottom of the entry hole (see above drawing). Orient the baffle with the triangular cutout in the far corner opposite the entry hole (see drawing below).

Top view of nest box showing baffle and location of triagular cutout.

ROOST BOX FOR SONGBIRDS

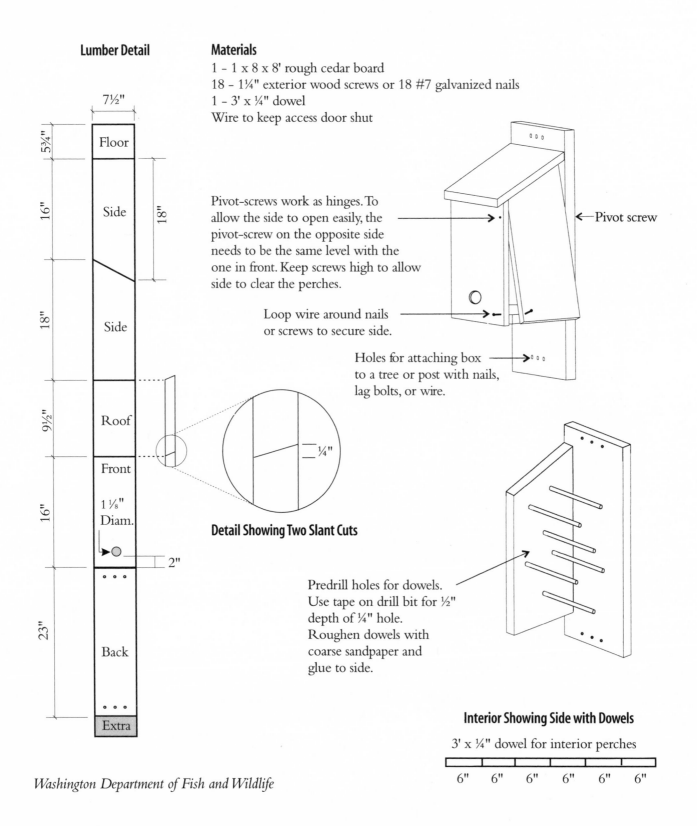

Lumber Detail

7½"

5¾"

Floor

16"

Side

18"

18"

Side

9½"

Roof

Front

16"

1⅛" Diam.

2"

23"

Back

Extra

Materials

1 - 1 x 8 x 8' rough cedar board
18 - 1¼" exterior wood screws or 18 #7 galvanized nails
1 - 3' x ¼" dowel
Wire to keep access door shut

Pivot-screws work as hinges. To allow the side to open easily, the pivot-screw on the opposite side needs to be the same level with the one in front. Keep screws high to allow side to clear the perches.

Loop wire around nails or screws to secure side.

Holes for attaching box to a tree or post with nails, lag bolts, or wire.

← Pivot screw

¼"

Detail Showing Two Slant Cuts

Predrill holes for dowels. Use tape on drill bit for ½" depth of ¼" hole. Roughen dowels with coarse sandpaper and glue to side.

Interior Showing Side with Dowels

3' x ¼" dowel for interior perches

6"   6"   6"   6"   6"   6"

Washington Department of Fish and Wildlife

WOOD DUCK NEST BOX

Materials

1 – 1 x 12 x 12' rough cedar board
20 – 1½" outdoor wood screws or 20 #7 galvanized nails
Wire to keep side door closed

Lumber Detail

Pivot screw works as a hinge. To allow the side to open easily, the pivot screw on the opposite side needs to be level with the one in front.

Two screws and wire to keep side door shut.

Holes for attaching box to a tree or post with nails, lag bolts, or wire.

Detail of Two Slant Cuts

Assembly Sequence

Entrance Detail

Roughen interior area with the back of a claw hammer.

Front interior

Detail of Interior Ladder

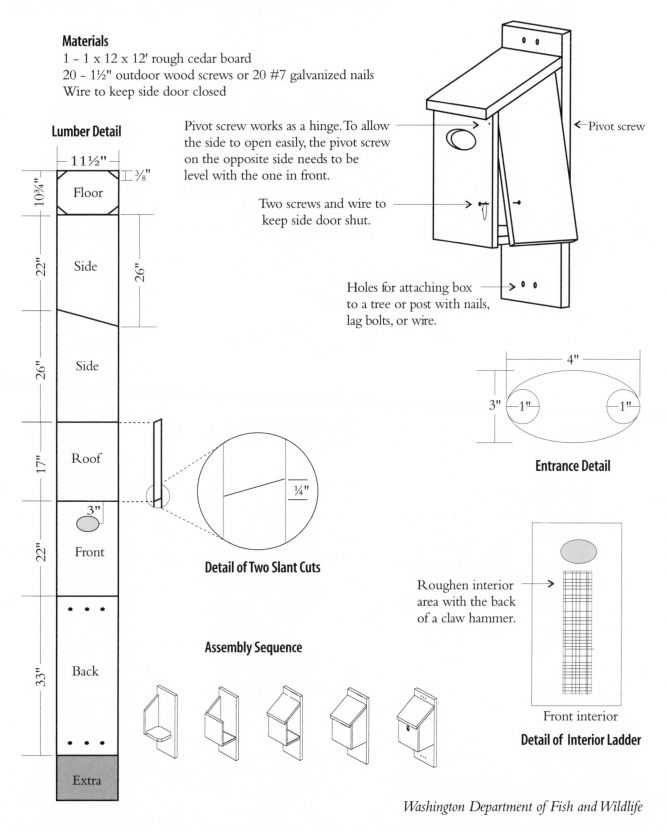

Washington Department of Fish and Wildlife

ECONOMY BAT HOUSES

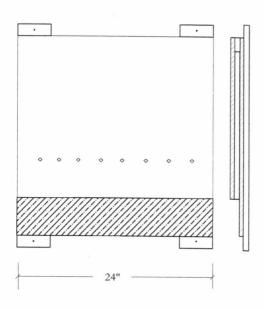

Small Economy Bat Box
(front and side views)

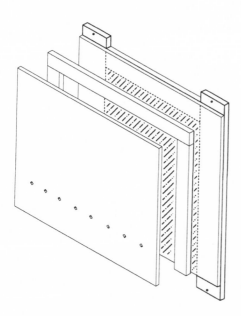

Small Economy Bat Box
(interior view)

Washington Department of Fish and Wildlife

Adapted from Bat Conservation International,
PO Box 162603, Austin, TX 78716
www.batcon.org (512) 327-9721

Small Economy Bat House

Materials (makes one)
¼ sheet of ½" exterior plywood
1 - 1 x 2 x 8' fir board (furring strip)
1 - 1 x 4 x 8' cedar board (mounting board)
25 - 1¼" outdoor wood screws
1 pint of latex paint

Construction Procedure
1. Cut plywood into two pieces: 26½" x 24" and 21½" x 24".
2. Cut furring strip into one 24" and two 20¼" pieces.
3. Cut cedar board into one 17" and two 30" pieces.
4. Screw back to furring. Start with the 24" piece on top.
5. Roughen all sides of plywood, including the back but not the front exterior, with a claw hammer or other tool. Remove any splintered wood.
6. Screw front to furring, top piece first.
7. Attach cedar mounting boards to back with screws entering through plywood and furring.
8. Paint with dark exterior latex paint at least twice. If necessary, roughen landing area below front sheet again.
9. Caulk all seams that aren't tight with paintable silicone caulk.

Large Economy Bat House

Materials (makes one)
½ sheet (2' x 8') of ½" exterior plywood
2 - 1 x 2 x 8' fir board (furring strip)
2 - 1 x 4 x 8' cedar board
40 - 1¼" outdoor wood screws
1 pint of latex paint

Construction Procedure
1. Cut plywood into two pieces: 51" x 24" and 45" x 24".
2. Cut furring strip into one 17" and two 43¾" pieces.
3. Cut cedar board into one 17" and two 58" pieces.
4. Follow steps 4 thru 9 for Small Economy Bat House.

Optional Modifications to Economy Bat Houses
• In hot climates, drill 8 to 10, ½" ventilation holes in the front of the box approximately 5" up from the bottom. Vent holes may not be necessary in cooler climates.
• Attach a 1 x 4 board to the top as a roof (recommended).
• Attach a 1 x 4 board in back of the box at the top and in between the two furring strips to create a chamber.
• Two bat boxes can be placed back-to-back and mounted on a pole. Build two houses the same size. Drill 4 - ¾" holes in the back of each to permit movement of bats between the two houses. The holes should be about 10" from the bottom edge of the back piece.

POST BAT HOUSE

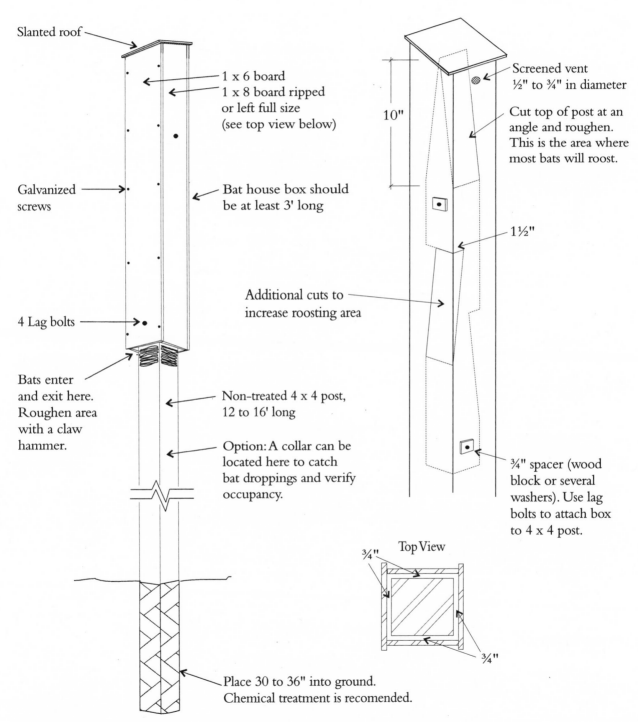

Slanted roof

1 x 6 board
1 x 8 board ripped
or left full size
(see top view below)

Galvanized
screws

Bat house box should
be at least 3' long

4 Lag bolts

Bats enter
and exit here.
Roughen area
with a claw
hammer.

Non-treated 4 x 4 post,
12 to 16' long

Option: A collar can be
located here to catch
bat droppings and verify
occupancy.

Additional cuts to
increase roosting area

Place 30 to 36" into ground.
Chemical treatment is recomended.

Screened vent
½" to ¾" in diameter

Cut top of post at an
angle and roughen.
This is the area where
most bats will roost.

10"

1½"

¾" spacer (wood
block or several
washers). Use lag
bolts to attach box
to 4 x 4 post.

Top View

¾"

¾"

Adapted from Daniel Boone National Forest
1700 Bypass Road, Winchester, KY 40391

Washington Department of Fish and Wildlife

SUET FEEDERS FOR BIRDS

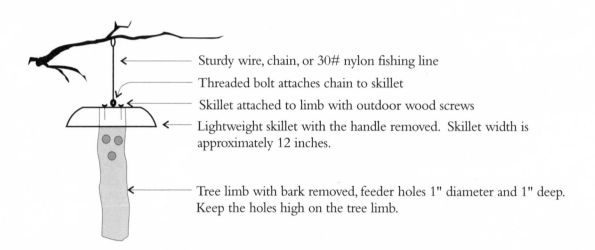

Sturdy wire, chain, or 30# nylon fishing line

Threaded bolt attaches chain to skillet

Skillet attached to limb with outdoor wood screws

Lightweight skillet with the handle removed. Skillet width is approximately 12 inches.

Tree limb with bark removed, feeder holes 1" diameter and 1" deep. Keep the holes high on the tree limb.

Woodpecker Feeder #1

Wooden spring-type clothespin

Wood screw

2"

Cedar tube with 3/16" wall. Drill hole approximately 1" deep

Clip-On Feeder

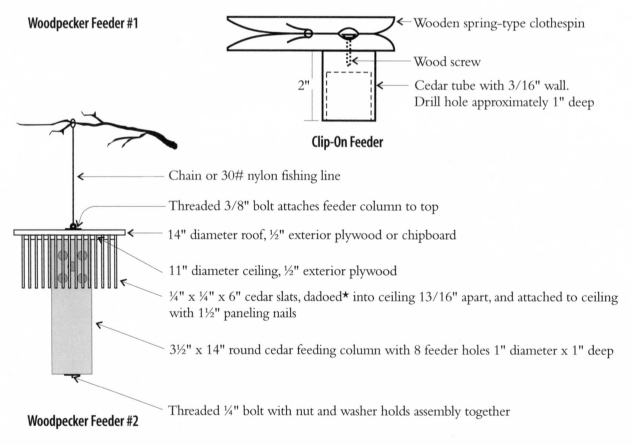

Chain or 30# nylon fishing line

Threaded 3/8" bolt attaches feeder column to top

14" diameter roof, ½" exterior plywood or chipboard

11" diameter ceiling, ½" exterior plywood

¼" x ¼" x 6" cedar slats, dadoed★ into ceiling 13/16" apart, and attached to ceiling with 1½" paneling nails

3½" x 14" round cedar feeding column with 8 feeder holes 1" diameter x 1" deep

Threaded ¼" bolt with nut and washer holds assembly together

Woodpecker Feeder #2

★ Dado cut ¼" x ¼" slots into the ceiling perimeter. Before attaching the ceiling to the roof, nail the slats into the dadoes, using 1½" annular ring paneling nails.

Designs by Ken Short, Bainbridge Island, WA

Washington Department of Fish and Wildlife

RESOURCES

BOOKS, MAGAZINES, NEWSLETTERS, VIDEOS, ORGANIZATIONS, AND INTERNET RESOURCES

Wildlife Gardening and Garden Design
Books

Baines, Chris. *How to Make a Wildlife Garden*. Elm Tree Books, London, 1985.

Cox, Jeff. *Landscaping with Nature*. Rodale Press, Emmaus, PA, 1991.

Dennis, John. *The Wildlife Gardener*. Knopf, New York, 1985.

Kress, Stephen W. *National Audubon Society: The Bird Garden*. D.K., New York, 1995.

Merilees, Bill. *Attracting Backyard Wildlife*. Voyager Press, Stillwater, MN, 1989.

Munro, Mike. *Northwest Landscaping: A Practical Guide to Creating the Garden You've Always Wanted*. Alaska Northwest, Anchorage and Seattle, 1992.

Schneck, Marcus. *Your Backyard Wildlife Garden*. Rodale Press, Emmaus, PA, 1992.

Smyser, Carol A. *Nature's Design:*

A Practical Guide to Natural Landscaping. Rodale Press, Emmaus, PA, 1982.

Stein, Sara. *Noah's Garden*. Houghton Mifflin, Boston, 1993.

Stein, Sara. *Planting Noah's Garden*. Houghton Mifflin, Boston, 1997.

Tufts, Craig, and Peter Loewer. *Gardening for Wildlife*. Rodale Press, Emmaus, PA, 1995.

Magazines/Newsletters

Wildlife Gardening News, 12536 NW Oakridge Road, Yamhill, OR 97148-8115.

Organizations

Backyard Wildlife Sanctuary Program, Dept. of Fish and Wildlife, 16018 Mill Creek Blvd. Mill Creek, WA 98012.★

British Columbia Wildlife Federation, 303-19292 60th Ave., Surrey, BC V3S 8E5.

National Wildlife Federation, Backyard Wildlife Habitat Program, 8925 Leesburg Pike, Vienna, VA 22184-0001.★

Naturescape British Columbia,

PO Box 9354 Stn. Prov. Govt. Victoria, BC, V8W 9M1.★

Wildlife Resources, 5130 W. Running Brook Rd., Columbia, MD 21044.

★ Organizations with Wildlife Habitat Certification Programs

Internet Resources

National Wildlife Federation's Backyard Wildlife Habitat Program, www.nwf.org

Naturescape British Columbia, www.env.gov.bc.ca/hctf/nature.htm

United States Department of Agriculture Backyard Conservation, www.nrcs.usda.gov

Washington Department of Fish and Wildlife's Backyard Wildlife Sanctuary Program, www.wa.gov/wdfw

Wildlife Resources, www.erols.com/urbanwildlife

Wildlife and Wildlife Gardening, www.wildwords.com/

WildOne's Handbook, www.epa.gov/greenacres/wildones/

Landscape Maintenance
Books

Ellis, Barbara, and F. Marshal Bradley. *The Organic Gardener's Handbook of Natural Insect and Disease Control*. Rodale Garden Books, Emmaus, PA, 1992.

Olkowski, William, et al. *Common Sense Pest Control*. Taunton Press, Newtown, CT, 1991.

Turnball, Cass. *Landscape Design, Renovation and Maintenance*. Betterway Publications, Inc., Box 219, Crozet, VA 22932.

Organizations and Agencies

IPM (integrated pest management) Practitioners Association, IPM Associates, PO Box 21108, Eugene, OR 97402.

State University Cooperative Extension Service (See "Public Agencies" in Appendix E for addresses and phone numbers).

Washington Toxics Coalition, 4516 University Way NE, Seattle, WA 98105. Fact Sheets available by mail: Aphids, Tent Caterpillars, Garden Insects, Slugs, Carpenter Ants.

Internet Resources

Information on both cultural and chemical remedies for the most common yard and garden plant problems:

British Columbia. http://www.env.gov.bc.ca/

Oregon: http://www.oda.state.or.us/Plant/noxiousweeds.html

Pacific Northwest. http://pep.wsu.edu/hortsense/

Washington and Oregon State Noxious Weeds Lists:

Washington: http://www.midpacific.wsu.edu/weedlist.html

Wildlife Plants

Books

Anderson, Charles B. *Native Plant Alliance: A Manual of Native Plant Communities for Urban Areas of the Pacific Northwest.* 2nd printing, 1996. Cascade Biomes, PO Box 22419, Seattle 98122-0419, (206) 322-0258.

Alexander Martin C., and H. Zim and A. Nelson. *American Wildlife Plants: A Guide to Wildlife Food Habitats.* Dover, New York, 1951.

Benyus, Janine M. *The Field Guide to Wildlife Habitats of the Western*

United States. Fireside Books, New York, 1989.

Grant, John A., and C. L. Grant. *Trees and Shrubs for Pacific Northwest Gardens.* 2nd ed. Timber Press, Portland, OR, 1990.

Kruckeberg, Arthur. *Gardening with Native Plants of the Pacific Northwest.* 2nd ed. University of Washington Press, 1996.

Parish, Roberta, and Ray Coupe and Dennis Lloyd. *Plants of Southern Interior British Columbia.* Lone Pine Publishing, 1996.

Pettinger, April. *Native Plants in the Coastal Garden: A Guide for Gardeners in British Columbia and the Pacific Northwest.* Whitecap Books, Vancouver/Toronto, 1996.

Pojar, Jim, and Andy MacKinnon. *Plants of the Pacific Northwest Coast.* BC Ministry of Forests and Lone Pine, Redmond, WA, Vancouver and Edmonton, BC, 1994.

Sunset Western Garden Book. 6th ed. Sunset Publishing, Menlo Park, CA, 1995.

Journals and Videos

Cordillera: The Journal of British Columbia Natural History. Subscription Dept., Box 473, Vernon, BC V1T 6M4.

Hortus First: A North America Native Plant Directory and Journal. Hortus First, PO Box 2870, Wilsonville, OR 97070-2870, (800) 704-7927, www.hortuswest.com.

Native Plants of Washington Videotape: North Cascades Institute, 2105 State Route 20, Sedro-Woolley, WA 98284-9394, (360) 856-5700, ext. 209.

Organizations

British Columbia Native Plant Society, Diane Gertzen, Nursery Extension Services, Ministry

of Forests, 14275 96th Ave. Surrey, BC, Canada V3V 7Z2 email: Diane.Gertzen@gems3.gov.bc.ca

Idaho Native Plant Society, PO Box 9451, Boise, ID 83707

Oregon Native Plant Society, 2584 NW Savier St., Portland, OR 97210-2412

Washington Native Plant Society, PO Box 28690, Seattle, WA 98118-8690, (206) 760-8022.

Internet Resources

Pacific Northwest Native Wildlife Gardening, http://www.teleport.com/~allyn/natives/

Identifying, Propagating and Landscaping with Native Plants, http://gardening.wsu.edu/text/nwnative.htm

Landscaping with Native Plants, http://www.epa.gov/greenacres

Washington Native Plant Society, http://www.wnps.org.

Pacific Northwest Natural History (General)

Books

Kozloff, Eugene N. *Plants and Animals of the Pacific Northwest.* University of Washington Press, Seattle, 1976.

Kruckeberg, Arthur R. *The Natural History of Puget Sound Country.* University of Washington Press, Seattle, 1991.

Mathews, Daniel. *Cascade/Olympic Natural History.* Raven Editions & Portland Audubon Society, Portland, OR, 1988.

Schwartz, S. *Nature in the Northwest: An Introduction to the Natural History and Ecology of the Northwest United States from the Rockies to the Pacific.* Prentice Hall, NJ, 1983.

Mammals

Books

Earle, Ann. *Zipping, Zapping, Zooming Bats.* New York: HarperCollins Publishers, 1995.

Ingles, L. G. *Mammals of the Pacific States.* Stanford University Press, Stanford, CA, 1965.

Larrison, Earl J. *Mammals of the Northwest.* Seattle Audubon Society, Seattle, 1976.

Master, C., et al. *Natural History of Oregon Coast Mammals.* USDA Forest Service General Technical Report, PNW 133, Corvallis, OR, 1981.

Nagorsen, David W., and Mark R. Brigham. *Bats of British Columbia.* UBC Press, Vancouver, BC, 1993.

Also see "Bat Houses."

Birds (General)

Books

Campbell, R. Wayne, et al. *Wildlife Habitat Handbooks for the Southern Interior Ecoprovince. Vol 2: Species Notes for Selected Birds.* Wildlife Report No. R16, Ministry of the Environment, Victoria, BC, 1988.

Ehrlich, Paul R., et al. *The Birder's Handbook: A Field Guide to the Natural History of North American Birds.* Simon and Schuster, New York, 1988.

Gaussoin, Bret, and Janice Lapsansky. *The Barn Owl and Pellet Book.* Pellets, 3004 Pinewood, Bellingham, WA, 1988.

Nehls, Harry B. *Familiar Birds of the Northwest.* Audubon Society of Portland, Portland, OR, 1989.

Pederson, Richard J., and Ron Shay. *Hawk, Eagle and Osprey Management on Small Woodlands.* World Forestry Center, 4033 SW Canyon Road, Portland, OR 97221.

Peterson, Roger Tory. *A Field Guide to Western Birds.* Houghton Mifflin, Boston, 1990.

Scott, Shirley L., ed. *National Geographic Society Field Guide to the Birds of North America.* National Geographic Society, Washington, DC, 1987.

Udvardy, Miklos D. F. *Audubon Society Field Guide to North American Birds* (Western Region) Knopf, New York, 1977.

Magazines

Birder's World, Subscription Dept., 434 W Downer Pl., Aurora, IL 60506-9919.

Bird Watcher's Digest, PO Box 110, Marietta, OH 45750, (800) 879-2473.

Washington Birder, PO Box 191, Moxee City, WA 98936.

Organizations

Audubon Society of Portland: 5151 NW Cornell Road, Portland, OR 97210, (503) 292-6855.

National Audubon Society chapters: For local chapter information look in your phone directory under "Audubon Society."

Washington Audubon Society Regional Office: Box 462, Olympia WA 98507, (360) 786-8020.

Washington Ornithological Society, P O Box 85786, Seattle, WA 98145.

Internet Resources

Birding in the Pacific Northwest: http://www.scn.org/earth/tweeters

Introduction to the Birds: www.ucmp.berkeley.edu/diapsids/birds/birdintro.html

National Audubon Society and the Cornell Lab of Ornithology: http://birdsource.cornell.edu/

Hummingbirds and Hummingbird Gardening

Books

Grant, V., and K. A. *Grant. Hummingbirds and their Flowers.* Columbia University Press, New York, 1986.

Skelly, Flora Johnson, and Brett Johnson. *Gardening for Hummingbirds in Western Washington.* Wild Words, Redmond, WA, 1998. Available from Wild Words, Box 464, 23316 NE Redmond-Fall City Road, Redmond, WA 98053.

Stokes, Lillian, and Donald Strokes. *Attracting, Identifying and Enjoying Hummingbirds.* Little, Brown, and Co., 1989.

Tekulsky, Mathew. *The Hummingbird Garden.* Crown, New York, 1990.

Thies, Elena. *Hummingbirds in Your Garden.* Elena Thies, 12536 NW Oakridge Road, Yamhill, OR 97148-8115.

Toops, Connie. *Hummingbirds: Jewels in Flight.* Voyageur Press, 1992.

Reptiles and Amphibians

Books

Corkran, Charlotte C., and Chris Thoms. *Amphibians of Oregon, Washington, and British Columbia.* Lone Pine, Redmond, WA, 1996.

Dove, Louise E. *Reptiles and Amphibians.* Urban Wildlife Manager's Notebook No. 6. Urban Wildlife Resources, 5130 W Running Brook Rd., Columbia, MD 21044.

Leonard, William P., et al. *Amphibians of Washington and Oregon.* Seattle Audubon Society, Seattle, 1993.

Nussbaum, Ronald A., et al. *Amphibians and Reptiles of the Pacific Northwest*. University of Idaho Press, Moscow, 1993.

Storm, R. M., and W. P. Leonard (coordinating editors). *Reptiles of Washington and Oregon*. Seattle Audubon Society, Seattle, 1995.

Fish

Books

Griggs, Jim. *Trout in Small Woodland Areas*. Woodland Fish and Wildlife Project. World Forestry Center, 4033 SW Canyon Road, Portland, OR 97221, (503) 228-1367.

Smith, C. Lavett. *Fish Watching: An Outdoor Guide to Freshwater Fishes*. Cornell University, 1994.

Yates, Steven A. *Adopting a Stream: A Northwest Handout*. The Adopt-A-Stream Foundation, 1988.

Wydoski, Richard S., and Richard Whitney. *Inland Fishes of Washington*. University of Washington Press, Seattle, 1979.

Insects (General)

Books

Akre, Roger D., et al. *Insects Did It First*. Insects First Assoc., NW 1005 Fisk, Pullman, WA 99163.

Buchmann, Stephen L., and Gary Paul Nabhan. *The Forgotten Pollinators*. Island Press, 1996.

Dove, Louise E. *Urban Insects I: Dragonflies and Damselflies in Your Backyard Pond*. 1986. Urban Wildlife Resources, 5130 W Running Brook Rd., Columbia, MD 21044.

Dove, Louise E. *Urban Insects III: Stinging Insects and You*. Urban Wildlife Manager's Notebook No. 13, Urban Wildlife Resources, 5130 W Running Brook Rd., Columbia, MD 21044.

Griffin, Brian L. *Humblebee Bumblebee*. Knox Cellars, Bellingham, WA, 1997. (Available from Knox Cellars, see Supplies.)

Griffin, Brian L. *The Orchard Mason Bee*. Knox Cellars, Bellingham, WA, 1993. (Available from Knox Cellars, see Supplies.)

Kirby, Peter. *Habitat Management for Invertebrates: A Practical Handbook*. The Royal Society for the Protection of Birds, The Lodge, Sandy, Bedfordshire SG19 2DL, England, 1992.

Leahy, Christopher. *Insects: A Simplified Field Guide to the Common Insects of North America*. Peterson First Guides, Houghton Mifflin, Boston, 1987.

Stokes, Donald. *A Guide to Observing Insect Lives*. Little, Brown & Co., Boston, 1983.

Pamphlets

State University Cooperative Extension Service has many inexpensive pamphlets for sale. (See "Public Agencies" in Appendix E for addresses and phone numbers.)

Supplies

Ben Meadows Co. 3589 Broad St., Atlanta, GA 30341, (800) 241-6401, mail@benmeadows.com

BioQuip Products, Inc. 17803 La Salle Ave., Gardena, CA 90248, (310) 324-0620. Butterfly nets and other supplies.

Entomo-Logic, 9807 NE 140th St., Bothell, WA 98011-5132, (206) 820-8037, easugden@msn.com

Insect Lore. PO Box 1535, Shafter, CA 93263, (800) LIVE BUG. A catalog offering science and nature materials, including painted lady caterpillars that children can raise, and lots of other butterfly and insect goodies.

Knox Cellars, 25724 NE 10th Street, Redmond, WA 98053 (360) 733-3283, knoxclr@accessone.com web site http://www.accessone.com/~knoxclr

Organizations

Scarabs Society. Editor: Rod Crawford, Burke Museum, University of Washington, P O Box 353010, Seattle, WA 98195. puffinus@u.washington.edu. Includes subscription to *The Scarabogram*.

The Xerces Society. 4828 SE Hawthorne Blvd., Portland, OR 97215, (503) 232-6639.

Internet Resources

http://gardening.wsu.edu/library/inse006/inse006.htm

http://www.accessone.com/~knoxclr/omb.htm

http://www.seattleu.edu/~ciscoe/archive/b0298.html

Butterflies, Moths, and Caterpillars

Books

Brewer, J., and D. Winter. *Butterflies and Moths: A Companion to Your Field Guide*. Prentice Hall, New York, 1986.

Brooklyn Botanical Garden. *Butterfly Gardens*. Brooklyn Botanical Garden, Handbook #143, 1995.

Emmel, Thomas C. *Butterfly Gardening*. Friedman/Fairfax, 1997.

Layberry, Ross A., and P. Hall, J. Lafontaine. *The Butterflies of Canada*. University of Toronto Press, 1998.

Mikula, Rick. *Garden Butterflies of North America*. Willow Creek Press, 1997.

Pyle, Robert Michael. *Handbook for Butterfly Watchers*. Houghton Mifflin, Boston, 1992.

Sedenko, Jerry. *The Butterfly Garden*. Villard, New York, 1991.

Skelly, Flora Johnson, and Brett Johnson. *Butterfly Gardening in Western Washington*. Wild Words, Redmond, WA, 1998. Available from Wild Words, Box 464, 23316 NE Redmond-Fall City Road, Redmond, WA 98053.

Xerces Society/Smithsonian Institution. *Butterfly Gardening: Creating Summer Magic in Your Garden*. 2nd ed. Sierra Club Books, San Francisco, CA, 1999.

Field Guides

Mitchell, R.T., and H. S. Zim. Golden Nature Guide Series. *Butterflies and Moths: A Guide to the More Common American Species*. Golden Press, Western Press Company, Racine, WI, 1991.

Opler, Paul, and Amy Bartlett Wright. *Peterson First Guide to Butterflies and Moths*. Houghton Mifflin, Boston, 1994.

Pyle, Robert Michael. *Field Guide to the Butterflies of Cascadia*. Seattle Audubon, Seattle, WA, 1999.

Wright, Amy. *Peterson First Guide to Caterpillars*. Houghton Mifflin, Boston, 1993.

Publications

Butterfly Gardeners' Quarterly. PO Box 30931, Seattle, WA 98103. http:/ButterflyGardeners.com.

Supplies

See "Insects (General)"

Organizations

North American Butterfly Association, Inc. (NABA). 4 Delaware Road, Morristown, NJ 07960, www.naba.org

Wildlife Watching/ Tracking

Books

Brown, Tom Jr. *Tom Brown's Field Guide to Nature Observation and Tracking*. Berkeley Books, New York, 1983.

Halfpenny, James. A *Field Guide to Mammal Tracking in North America*. Johnson Books, 1986.

Hudson, Wendy. *Naturewatch— A Resource for Enhancing Wildlife Viewing Areas*. Falcon Press, 1992.

La Tourrette, Joe. *Wildlife Watchers Handbook*. Henry Holt and Co. Inc., 1997.

Pandell, Karen, and C. Stall. *Animal Tracks of the Pacific Northwest*. Mountaineers Press, Seattle, 1992.

Rezendes, Paul. *Tracking and the Art of Seeing: How to Read Animal Tracks and Signs*. Camden House, 1992.

Stokes, Donald W., and Lillian Q. Stokes. *Guide to Animal Tracking and Behavior*. Little Brown & Co, Boston, 1987.

Organizations

Wilderness Awareness School, www.natureoutlet.com

Ponds

Books

Allison, James. *Water in the Garden: A Complete Guide to the Design and Installation of Ponds, Fountains, Streams, and Waterfalls*. Little, Brown, Boston, 1991.

Clafin, Edward B. *Garden Pools and Fountains*. Ortho Books, San Ramon, CA, 1988.

Grinstein, Dawn Tucker. *For Your Garden: Pools, Ponds, and Waterways*. Grove Weidenfeld, New York, 1991.

Matson, Tim. *Earth Ponds: The Country Pond Maker's Guide to Building, Maintenance, and Restoration*. Countryman Press, Woodstock, VT, 1991.

Nash, Helen. *The Pond Doctor: Planning and Maintaining a Healthy Water Garden*. Sterling, New York, 1994.

Natural Resource Conservation Service (NRCS) *Ponds: Planning, Design, Construction*. Ag. Handbook #590, USDA.

Prescott, G. W. *How to Know the Aquatic Plants*. 2nd ed. Brown, Dubuque, IA, 1980.

Plants and Equipment

Lilypons Water Gardens, PO Box 10, Dept. 4692, Buckeystown, MD 21717-0010, (800) 999-5459.

Moorehaven Water Gardens, 3006 York Road, Everett, WA 98204, (425) 743-6888.

Oasis Water Gardens, 404 South Brandon, Seattle, WA 98108, (206) 767-9776.

Rainbow's End, 39 Second Creek, Superior, MT 59872, (800) 759-5475.

Nest Structures for Birds

Books

Campbell, Scott. *The Complete Book of Birdhouse Construction for Woodworkers*. Dover, New York, 1984.

Harrison, Hal H. *Western Bird Nests*. A Peterson Field Guide. Houghton Mifflin, Boston, 1979.

Henderson, Carrol. *Woodworking for Wildlife: Homes for Birds and Mammals*. Available from Minnesota's Bookstore, 117 University Ave., St. Paul, MN 55155, (800) 657-375.

Martin, Chester O., and Wilma A. Mitchell. *Osprey Nest Platforms*. US Army Corps of Engineers Wildlife Resources Management Manual, Technical Report EL-86-21, July 1986.

McNeil, Don. *The Birdhouse Book*. Pacific Search Press, Seattle, 1979.

Organizations

Prescott Bluebird Recovery Project. Pat Johnston, President, 7717 SW 50th, Portland, OR, 97219, (503) 246-1337

Bats and Bat Houses

Books

Tuttle, Merlin D., and Donna L. Hensley. *The Bat House Builders Book*. Bat Conservation International, Austin, TX, 1996.
Also see "Mammals."

Organizations

Bat Conservation International, PO Box 162603, Austin, TX 78716-2603, (800) 538-BATS, http://www.batcon.org
Bats Northwest, PO Box 19558, Seattle, WA 98109, (206) 256-0406.

Internet Resources

All about bats: www.nyx.net/~jbuzbee/bat_house.html
Bat house enthusiasts: www.batcon.org/bhra/bhratop.html

Bird Feeders

Books

Alder, Bill Jr. *Impeccable Birdfeeding*. Chicago Review Press, Chicago, 1992.

Burton, Robert. *National Audubon Society North American Birdfeeder Handbook*. Dorling Kindersley, New York, 1992.

Campbell, Scott D. *Easy-to-Make Bird Feeders for Woodworkers*. Dover, New York, 1989.

Henderson, Carrol L. *Wild About Birds: The DNR Bird Feeding Guide*. State of Minnesota, Department of Natural Resources, 1995. To order call (800) 657-3757.

Kress, Stephen W. *The Audubon Society Guide to Attracting Birds*. Scribner's, New York, 1985.

Waldon, Bob. *Feeding Winter Birds in the Pacific Northwest*. Mountaineers, Seattle, 1994.

Catalogs

Droll Yankees, 27 Mill Road, Foster, RI 02825.
Duncraft, 102 Fisherville Road, Penacook, NH 03303-9020.
Hyde Bird Feeder Company, PO Box 168, Waltham, MA 02254.

Magazines/Newsletters

Around the Bird Feeder, PO Box 225, Mystic, CT 06355. The quarterly magazine of the Bird Feeder's Association.
The Dick E. Bird News, PO Box 377, Acme, MI 49610. A newspaper with lots of practical information filled with humor.
Wild Bird, Subscription Dept., PO Box 52898, Boulder, CO 80323-2898.

Organizations

Project Feeder Watch, Cornell Laboratory of Ornithology, 159 Sapsucker Woods Road, Ithaca, NY 14850, (607) 254-2414.

Canada: Project Feeder Watch, Long Point Bird Observatory, PO Box 160, Port Rowan, ON N0E 1M0.

Internet Resources

Wildbirds Unlimited Internet Store: www.wbu.com

Dead Trees (Snags) and Down Wood

Books

Bull, Evelyn, C. Parks, and T. Torgerson. *Trees and Logs Important to Wildlife in the Interior Columbia River Basin*. USDA Forest Service, General Report PNW-GTR-391. May, 1997.

Brown, E. R., ed. *Management of Wildlife and Fish Habitats in Forests of Western Oregon and Washington, Vols. 1 and 2*. Pub. No. R6F&WL1921985, USDA Forest Service, Pacific Northwest Region, Portland, OR, 1985.

Guideline for Selecting Live or Dead Standing Tree Wildlife Habitat. Pub. No. R6F&WL2191986. USDA Forest Service, Pacific Northwest Region, Portland, OR, 1986.

Maser, Chris, and James M. Trappe. *The Seen and Unseen World of the Fallen Tree*. Pacific Northwest Forest and Range Experimental Station, USDA Forest Service, General Technical Report PNW164, 1984.

Maser, Chris, et al. *From the Forest to the Sea: The Story of Fallen Trees*. PNWGTR229, Pacific Northwest Research Station, US Dept. of Agriculture, Portland, OR, 1988.

Pederson, Richard J. *Managing Small Woodlands for Cavity-Nesting Birds*. USDA Forest Service, Pacific Northwest Region, Fish and Wildlife

Division, PO Box 3623, Portland, OR 97208.

Thomas, Jack Ward. *Wildlife Habitats in Managed Forests: The Blue Mountains of Oregon and Washington.* Agricultural Handbook No. 553, USDA Forest Service, Washington, DC, 1979.

Washington Department of Fish and Wildlife. *Priority Habitat Management Recommendations: Snags.* Draft, 1997. Washington Department of Fish and Wildlife, Olympia, WA.

Tapes

Creating Cavities in Trees...One Facet of Ecosystem Management. One tape cassette. USDA Forest Service, Pacific Northwest Research Station, Portland, OR, 1993.

Organizations

International Society of Arboriculture, Pacific Northwest Chapter. (800) 335-4391, http://www.teleport.com/~pnwisa/

Wildlife-Related Problems

Books and Pamphlets

Bird, D. M. *City Critters: How to Live with Urban Wildlife.* Eden Press, Montreal, Quebec, Canada, 1986.

Coleman, John S., S. Temple, and S. Craven. *Cats and Wildlife: A Conservation Dilemma.* Publications Room 170, 630 W. Miffin St., Madison, WI 53703, (608) 262-3346.

Hodge, G. R., ed. *Wild Neighbors: The Humane Approach to Living with Wildlife.* The Humane Society of the United States, 1997. Available from: Urban Wildlife Resources, 5130 W.

Running Brook Rd., Columbia, MD 21044.

Progressive Animal Welfare Society (PAWS). *Coexisting with Wildlife Series.* PO Box 1037, Lynnwood, WA 98046, (425) 787-2500.

Public Agencies

(See "Public Agencies" in Appendix E for contact information.)

British Columbia Ministry of Environment, Lands and Parks

State Department of Fish and Wildlife

State University Cooperative Extension Services

Internet Resources

American Bird Conservancy, http://abc@abcbirds.org

Centers for Disease Control: http://www.cdc.gov./

HOWL Wildlife Center: www.paws.org/wildlife

Urban wildlife resources: http://www.erols.com/urbanwildlife

Wetlands and Wetland Gardens

Books

Cooke, Sara Spear. *A Field Guide to the Common Wetland Plants of Western Washington and Northwest Oregon.* Seattle Audubon Society, 1997.

Environmental Protection Agency. *A Citzen's Guide to Wetland Restoration.* EPA Publication No. 910/R-94-006, Washington, DC, 1994.

Guard, Jennifer. *Wetland Plants of Washington and Oregon.* Lone Pine, Redmond, WA, 1995.

Maynard, Chris, and Shawn Ultican. *Wetland Walk: A Guide to Washington's Wetlands.* Pub. #89-30. Department of Ecology, Seattle, 1991.

Stevens, Michelle L., and Ron Vanbianchi. *Restoring Wetlands in Washington: A Guidebook for Wetland Restoration, Planning, and Implementation.* Publication No. 93-17. Department of Ecology, Olympia, WA, 1993.

Washington Department of Natural Resources. *Recognizing Wetlands and Wetland Indicator Plants on Forested Lands.* Publication No. 500, Olympia, WA, 1993.

Yates, S. *Adopting a Wetland: A Northwest Guide.* Snohomish County Planning and Community Development, Everett, WA, 1989.

Public Agencies

(See "Public Agencies" in Appendix E for contact information.)

Oregon Division of State Lands

State Department of Fish and Wildlife

State University Cooperative Extension Service

U.S. Army Corp of Engineers

U.S. Environmental Protection Agency, wetland hotline: 1-800-832-7828

U.S. Natural Resources Conservation Service

Washington Department of Ecology

Private Organizations

British Columbia Wetlands Network (BC WETNET), c/o Friends of Boundary Bay, PO Box 1441, Station A, Delta, BC V4M 3Y8, (604) 940-1540 or (800) 4WETNET.

Ducks Unlimited Canada, South Coastal Office, W.R.P.S. Box 39530, White Rock, BC, V4A 9P3, (604) 591-1104.

Society for Ecological Restoration, 3644 Albion Place North, Seattle, WA 98103, www.halcyon.com/sernw/

Washington Wetlands Network (WETNET) 8028 – 35th Ave. NE, Seattle, WA 98115, (206) 524-4570.

Woodlands and Woodland Landscapes

Books

Fazio, James. R. *The Woodland Steward: A Practical Guide to Small Private Forests.* The Woodland Press, Moscow, ID.

Hebda. Richard J., and Fran Aitkens. *Garry Oak–Meadows Colloquium.* Available from: Fran Aitkens, Apt. 4, 921 Foul Bay Rd., Victoria, BC V8S 4H9.

Morgan, Robin. *A Technical Guide to Urban and Community Forestry.* World Forestry Center, Portland, OR, 1993.

Oregon State University, *The Woodland Workbook.* Publications Orders, Agricultural Communications, Oregon State University, Administrative Services A422, Corvallis, OR 97331-2119.

Reid, Collins & Assoc. *Managing Your Woodland: A Non-forester's Guide to Small-scale Forestry in British Columbia.* BC Ministry of Forests, Victoria, 1992.

Washington Department of Natural Resources. Backyard Forest Stewardship/Wildfire Safety Kit, free on request (888) STEWKIT; Forestry Practices Illustrated, free on request (800) 527-3305, press "0."

Washington State University. *Consulting Foresters Directory for Washington Landowners.* Bulletin No. EB1303. Available from Washington State University Cooperative Extension (see "Public Agencies" in Appendix E).

Woodland Fish and Wildlife Project. Several publications on managing small woodlands for wildlife are available from Washington State University Cooperative Extension (see "Public Agencies" in Appendix E).

Also see "Dead Trees and Down Wood."

Public Agencies

(See "Public Agencies" in Appendix E for contact information.)

British Columbia Ministry of Forests

Oregon Department of Forestry

State University Cooperative Extension

Washington Department of Natural Resources

Private Organizations

Forestry Canada, Information Services, 506 West Burnside Road, Victoria, BC V8Z 1M53.

Oregon Small Woodlands Association, PO Box 3079, Salem, OR 97302, (503) 588-1813.

Oregon Urban and Community Forest Council, PO Box 13074, Salem, OR 97309-1074.

Washington Farm Forestry Association, PO Box 7663, Olympia, WA 98507, (360) 459-0984.

Internet Resources

Department of Natural Resource Forest Stewardship: www.wa.gov/dnr/base/assistance.html

TreeLink, a community forestry resource: www.treelink.org/

Washington State University Cooperative Extension: http://caheinfo.wsu.edu/

Washington State University Natural Resources Extension: http://ext.nrs.wsu.edu

Woodland Fish and Wildlife: www.dfw.state.or.us/odfwhtml/woodland/woodland.html

Creeks, Streams, and Other Waterways

Books and Pamphlets

Brown, E. R., ed. *Management of Wildlife and Fish Habitats in Forests of Western Oregon and Washington*, Vols. 1 and 2. Pub. No. R6-F&WL-192-1985, USDA Forest Service, Pacific Northwest Region, Portland OR, 1985.

Department of Fisheries and Oceans. *The Streamkeepers Handbook: A Practical Guide to Stream and Wetland Care.* Vancouver, BC, 1995.

Hunt, Robert L. *Trout Stream Therapy.* University of Wisconsin Press, Madison, 1993.

Hunter, Christopher J. *Better Trout Habitat: A Guide to Stream Restoration and Management.* Montana Land Reliance.

Kahan, J. *Streamside Planting Guide for Western Washington.* Harza Northwest, Bellevue, WA, 1995.

Yates, S. *Adopting a Stream: A Northwest Handbook.* University of Washington Press, Seattle, 1988.

Public Agencies

(See "Public Agencies" in Appendix E for contact information.)

BC Ministry of Environment, Lands and Parks

State Department of Fish and Wildlife

U.S. Farm Service Agency

U.S. Natural Resources Conservation Service

Washington State Department of Natural Resources

Private Organizations

Adopt-A-Stream Foundation, 600-128th Street SE, Everett, WA 98208-6353.

British Columbia Conservation Foundation, 12411-60th Avenue, Surrey, BC, (604) 594-6752.

Northwest Steelhead and Salmon Council of Trout Unlimited, PO Box 2137, Olympia, WA 98507.

Grasslands and Grassland Landscapes

Books

Bureau of Land Management. *Watchable Wildflowers: a Columbia Basin Guide*. BLM, 1997.

Borman, Herbert, et al. *Redesigning the American Lawn*. Yale University Press, New Haven, CT, 1993.

Clark, Lewis J. *Wild Flowers of the Pacific Northwest*. Gray's, Sidney, BC, 1976.

Daubenmire, Rexford F. *Steppe Vegetation of Washington*. Technical Bulletin No. 62, Washington Agricultural Experiment Station, Pullman, 1970.

Daniels, Stevie. *The Wild Lawn Handbook: Alternatives to the Traditional Front Lawn*. Macmillan, New York, 1995.

Larrison, Earl J., et al. *Washington Wildflowers*. Seattle Audubon Society, Seattle, 1974.

Taylor, Ronald J. *Sagebrush Country: A Wildflower Sanctuary*. Mountain Press, Missoula, MT, 1992.

Martin, L. C. *The Wildflower Meadow Book*. East Woods Press, Charlotte, NC, 1986.

Washington State University Cooperative Extension Service. *Lawns: Sound Gardening—Gardening with an Eye on Water Quality*. No. KCSG10, Pullman, WA.

PACIFIC NORTHWEST NATIVE PLANT NURSERIES

A dapted from *Hortus West: A Western North America Native Plant Directory and Journal* (Vol. 9: Issue 2, 1998), a publication listing native plant sources in the western states with particular emphasis on the Pacific Northwest. For current information, contact: Hortus West, PO Box 2870, Wilson-ville, OR 97070-2870, 1-800 704-7927, www.hortuswest.com.

British Columbia

Alpenflora Gardens, **Surrey**
(604) 576-2464
Birch Creek Nursery, **Prince George**
(250) 964-1684
C. E. Jones & Associates, Ltd., **Victoria**
(604) 383-8375
Edgemont Farms, **Salmon Arm**,
(250) 833-4679
Gabriola Growing Co., **Gabriola Island**
(604) 247-8204
Ground Effects, **Langley**
(604) 530-7710
Kimoff Wholesale Nursery, **Victoria**
(604) 544-2297
Linnaea Nurseries, Ltd., **Langley**
(604) 857-2139
Meadow Sweet, **Langley**
(604) 530-2611
Mosterman Plant Propagators,
Chilliwack (640) 823-4749
Natural Legacy Seeds, **Armstrong**
(604) 546-9799
Pacific Rim Native Plants, Ltd., **Sardis**
(604) 792-1891
Peel's Nurseries, Ltd., **Pitt Meadows**
(604) 465-7627
Reid, Collins Nurseries Ltd., **Mission**
(800) 820-7381
Sagebrush Native Plant Nursery, **Oliver**
(604) 498-8898
Slug Hollow Nursery, **Crofton**,
(250) 246-2520
Streamside Native Plants, **Courtenay**
(250) 338-7509
The Greenhouse, **Sidney**
(604) 655-4391
Thimble Farms, **Ganges**
(604) 537-5788
VanDusen Botanical Gardens,
Vancouver (604) 878-9274

Oregon

Althouse Nursery, **Cave Junction**
(541) 592-2395
American Ornamental Perennials, **Eagle Creek** (541) 637-3096
Balance Restoration Nursery, **Lorane**
(541) 942-5530
Blooming Nursery, Inc., **Cornelius**
(800) 257-0719
Bosky Dell Natives, **West Linn**
(503) 638-5945
Callahan Seed, **Central Point**
(541) 855-1164
Cascadian Nurseries, Inc., **Portland**
(503) 645-3350
Chehalem Mountain Nursery, Inc.,
Hillsboro (503) 628-0376
Clackamas Community College,
Oregon City (503) 657-8400
Curry Native Plants, **Port Orford**
(541) 332-5635
Doak Creek Native Plant Nursery,
Eugene (541) 484-9206
Down to Earth, **Eugene**
(541) 342-6820
Drakes Crossing Nursery, **Silverton**
(503) 873-4932
D. Wells Farms, **Hubbard**
(503) 982-1012
Emerald Seed and Supply, **Portland**
(800) 826-8873
Ferris Nursery, **South Beach**
(541) 867-4806
Flora Lan Nursery, **Forest Grove**
(503) 357-8386
Forestfarm, **Williams**
(541) 846-7269
Fruit of the Bloom, **Springfield**
(541) 726-8997
Garden Gate Nursery, **Colton**
(503) 824-2532
Gossler Farms Nursery, **Springfield**
(541) 746-3922
Goodwin Creek Gardens, **Williams**
(541) 846-7357
Green Hills Nursery, **Beaver**
(503) 398-5965
Greer Gardens, **Eugene**
(800) 548-0111

Hansen Nursery, **Donald**
(503) 678-5409
Harold Miller, Landscape Nursery,
Jefferson (503) 399-1599
Hells Canyon Plant Company,
Sherwood (503) 538-2133
Heritage Seedlings, Inc., **Salem**
(503) 585-9835
Hobbs and Hopkins Ltd., **Portland**
(800) 345-3295
Holden Wholesale Growers, **Silverton**
(503) 873-5940
Huckleberry Lane Nursery, **North Bend** (541) 756-7328
Hughes Water Gardens, **Tualatin**
(503) 638-1709
Janzen Farms, **Dayton**
(503) 868-7353
Joy Creek Nursery, **Scappoose**
(503) 543-7474
Krueger's Tree Farms, **North Plains**
(503) 647-1000
Log House Plants, **Cottage Grove**
(541) 942-2288
Mahonia Vineyards & Nursery, **Salem**
(503) 363-9654
Mar-Lyn Farms, **Canby**
(503) 266-2112
Mary's Peak Nursery, **Philomath**
(541) 929-3448
Meadow Lake Nursery Co.,
McMinnville (503) 435-2000
Mineral Springs, **Carlton**
(503) 852-6129
Monument County, **Monument**
(503) 623-5534
Mo's Nursery, **Mulino**
(503) 829-7643
Mt. Angel Nursery, **Mt. Angel**
(503) 845-6570
Nature's Garden, **Scio**
(503) 394-3217
Nichols Garden Nursery, **Albany**
(541) 928-9280
Oakhill Farms, **Roseburg**
(541) 459-1361
Oakhill Farms Native Plant Nursery,
Oakland (541) 459-1361
Oisinn, Ltd., **Hillsboro**
(503) 647-0252
Oregon Department of Forestry, **Elkton**
(541) 584-2214
Pacific Natives Wholesale, **Lincoln City**
(541) 994-6767
Pacific Northwest Natives, **Albany**
(541) 928-8239
Pleasant Hill Nursery, **Pleasant Hill**
(541) 746-7178
Portland Nursery, **Portland**
(503) 231-7123
Quail Ridge Nursery, **Molalla**
(503) 829-3105
Rare Plant Research, **Portland**
(503) 762-0288

Red's Rhodies, **Sherwood**
(503) 625-6331

Richard Bush's Nursery, **Canby**
(503) 266-9251

Rogue House Seed, **Central Point**
(541) 664-1775

Round Butte Seed Growers, Inc.,
Culver (541) 546-5222

R & S Nursery Inc., **Hillsboro**
(800) 628-8804

Russell Graham Plants, **Salem**
(503) 362-1135

Sage Creek Gardens, **Bend**
(541) 385-3336

Samuel J. Rich Nursery, **Aurora**
(503) 678-2828

Serendipity Nursery, **Canby**
(503) 651-2122

Sevenoaks Native Nursery, **Corvallis**
(541) 745-5540

Siskiyou Rare Plant Nursery, **Medford**
(541) 772-6846

Squaw Mountain Gardens, **Estacada**
(503) 630-5458

Stonecrop Gardens, **Albany**
(541) 928-8652

Teufel Nursery, **Portland**
(503) 646-1111

Trillium Gardens, **Pleasant Hill**
(541) 937-3073

Tree of Life Nursery, **Lostine**
(503) 546-5222

Valley Growers, **Canby**
(503) 651-3535

Wallace W. Hansen, **Salem**
(503) 581-2638

Westlake Nursery, **Westlake**
(541) 997-3383

Whitman Farms, **Salem**
(503) 585-8728

Wichita Nursery, Inc., **Canby**
(503) 651-2279

Wild Garden Seed, **Philomath**
(541) 929-4068

Willowell Nursery, **Tigard**
(503) 731-1308

Wood's Native Plants, **Parkdale**
(503) 352-7497

Washington

Abundant Life Seed Foundation,
Port Townsend (360) 385-5660

Aldrich Berry Farm and Nursery,
Mossyrock (360) 983-3138

Bear Creek Nursery, **Northport**
(509) 732-4417

Black Lake Organic Nursery, **Olympia**
(360) 786-0537

Blossom & Bloomers, **Spokane**
(509) 922-1344

Briggs Nurseries, **Olympia**
(800) 999-9972

Burnt Ridge Nursery, **Onalaska**
(360) 985-2873

Classic Nursery, **Redmond**
(425) 885-5678

Cloud Mountain Farm, **Everson**
(360) 966-0921

Collector's Nursery, **Battle Ground**
(360) 574-3832

Colvos Creek Nursery, **Vashon Island**
(206) 749-9508

Dutch Tuch Acres, **LaCenter**
(360) 263-1505

Fancy Fronds, **Gold Bar**
(360) 793-1472

Fir Run Nursery, **Puyallup**
(206) 648-4731

Firetrail Nursery, **Marysville**
(360) 652-9021

Firstline Seeds, Inc., **Moses Lake**
(509) 765-1772

Foliage Gardens, **Bellevue**
(425) 747-2998

Forest Floor Recovery, **Lummi Island**
(360) 758-2778

Fourth Corner Nurseries, **Bellingham**
(360) 734-0079

Frosty Hollow Eco. Restoration,
Langley (360) 579-2332

Furney's Nursery, Inc., **Des Moines**
(206) 624-0634

Grassland West Co., **Clarkston**
(509) 758-9100

Green Man Gardens, **Mercer Island**
(206) 232-5734

Heaths and Heathers, **Shelton**
(360) 427-5318

Heronswood Nursery Ltd., **Kingston**
(360) 297-4172

IFA Nurseries, Inc., **Toledo**
(360) 864-2803

Inland NW Native Plants, **Spokane**
(509) 448-7992

Inside Passage, **Port Townsend**
(800) 361-9657

J & J Landscape Co., **Bothell**
(425) 486-3677

Judd Creek Nursery, **Vashon**
(206) 463-9641

Kinder Gardens Nursery, **Othello**
(509) 488-5017

Madrona Nursery, **Seattle**
(206) 323-8325

Madronamai Nursery Co., **Everson**
(360) 592-2200

Mary's Country Garden, **Kent**
(253) 639-1243

Maxwelton Valley Gardens, **Clinton**
(360) 579-1770

Milestone Nursery, **Lyle**
(509) 365-5222

Moses Lake Conservation District,
Moses Lake (509) 765-5333

Mt. Tahoma Nursery, **Graham**
(253) 847-9827

Native Origins Nursery, **Raymond**
(360) 942-0027

Natives Northwest Co., **Mossyrock**
(360) 983-3138

Newell Wholesale Nursery, **Ethel**
(360) 985-2460

Northwest Native Plants, **Kapowsin**
(253) 846-1137

Nothing but NW Natives, **Battle
Ground** (360) 666-3023

Oyster Bay Nursery, **Olympia**
(360) 866-0809

Pacific Natives & Ornamentals, **Bothell**
(425) 483-8108

Pacific Plant Company, **Port Angeles**
(360) 457-1536

Peninsula Gardens, **Gig Harbor**
(253) 851-8115

Plantas Nativa, **Bellingham**
(360) 715-9655

Plants of the Wild, **Tekoa**
(509) 284-2848

Quartzite Mountain Nursery,
Chewelah (509) 935-6880

Rainier Seeds Inc., **Davenport**
(800) 828-8873

Raintree Nursery, **Morton**
(360) 496-6400

Shore Road Nursery, **Port Angeles**
(360) 457-1536

Silvaseed Company, **Roy**
(253) 843-2246

Soos Creek Gardens, **Renton**
(425) 226-9308

Sound Native Plants, **Olympia**
(360) 352-4122

Spring Creek Nursery, Inc., **Deer Park**
(509) 276-8278

Storm Lake Growers, **Snohomish**
(360) 794-4842

Sunbreak Nursery, **Bellingham**
(360) 384-3763

Swanson's Nursery, **Seattle**
(206) 782-2543

Sweetbriar Nursery, **Woodinville**
(425) 821-2222

Syverson Seed, Inc., **Ridgefield**
(360) 887-4094

Tissues & Liners, Inc., **Woodinville**
(425) 885-5050

Wabash Farms, **Enumclaw**
(360) 825-7051

Warm Beach Groundcovers, **Stanwood**
(360) 652-5833

Watershed Garden Works, **Longview**
(360) 423-6456

Weyerhaeuser Company, **Rochester**
(800) 732-4769

Wildside Growers, **Lynden**
(360) 398-7158

Wilkins Nursery, **Vashon**
(206) 463-3050

Woodbrook Nursery, **Gig Harbor**
(253) 265-6471

PUBLIC AGENCIES

British Columbia

British Columbia Ministry of Environment, Lands and Parks (250) 387-1161, http://www.env.gov.bc.ca/

Manages natural habitats, wildlife, and water resources for ecological diversity and the economic and recreational opportunities they provide. Refer to the Phone Directory for your nearest Regional Office.

British Columbia Ministry of Fisheries (250) 387-4573, http://www.elp.gov.bc.ca/fsh/

Manages fish and fish habitat (including urban salmon projects). Refer to the Phone Directory for your nearest Regional Office.

British Columbia Ministry of Forests (250) 387-5255, http://www.for.gov.bc.ca/

Offers technical assistance plus numerous forest practices guidebooks on a wide variety of topics for private landowners.

Canadian Wildlife Service (604) 940-4700, http://www.pyr.ec.gc.ca

Handles wildlife issues, migratory birds, significant habitats and endangered species.

Oregon

Oregon Department of Environmental Quality (503) 229-5696, (800) 452-4011, http://www.deq.state.or.us

The mission of the ODEQ is to be an active force to restore, enhance, and maintain the quality of Oregon's air, water, and land.

Oregon Department of Fish and Wildlife (503) 872-5255, http://www.dfw.state.or.us/

Responsible for managing of fish and wildlife resources, and regulating commercial and recreational harvest.

Refer to the Phone Directory for your nearest Regional Office.

Oregon Department of Forestry (503) 945-7200, http://www.odf.state.or.us

Responsible for fire protection and insect and disease management on state and private forests; provides forestry assistance to private forest owners; enforces Oregon forest laws, and has forestry information for the public.

Oregon Division of State Lands (503) 378-3805, (541) 388-6112

Administers State Removal-Fill Law; source for National Wetlands Inventory maps; offers wetland identification assistance.

Oregon State University Cooperative Extension (541) 737-2513.

Many publications are available from Peavy Hall 119, Forestry Extension, Oregon State University, Corvallis, OR 97331-2119. For local offices look under "Government" in your phone directory.

Washington

County Conservation Districts

Non-regulatory districts found in almost every county in Oregon and Washington. The board of supervisors is comprised of local landowners and employees who often specialize in conservation practices. Look under "Government" in your phone directory.

Washington Department of Ecology (360) 407-6000, http://www.wa.gov/ecology/ecyhome.html

Has considerable information on water quality enhancement, permit assistance, and wetland restoration.

Washington Department of Fish and Wildlife (360) 902-2200, http://www.wa.gov/wdfw

Maintains current list of threatened and endangered animal species in Washington State and provides technical assistance and publications on fish and wildlife enhancement projects. Refer to the Phone Directory for your nearest Regional Office.

Washington Department of Natural Resources (WDNR) (360) 902-1000, http://www.wa.gov/dnr

Extensive experience with riparian area management on forest lands; provides assistance, advice, and financial aid to small forest owners for habitat enhancement projects.

Natural Heritage Program (360) 902-1340

Maintains current list of state-listed rare plants—endangered, threatened, and sensitive.

Washington State University Co-operative Extension (800) 723-1763, (509) 335-2857, http://caheinfo.wsu.edu

Many publications are available from: Cooperative Extension Publications Building, Bulletin Department, Washington State University, Pullman, WA 99164-5912. For local offices, look under "Government" in your phone directory.

United States

United States Army Corps of Engineers
Oregon:
Portland (503) 808-5150
Washington:
Seattle (206) 764-3495
Walla Walla (509) 522-6427

United States Department of Agriculture
USDA strives to enhance the quality of life for the American people by caring for agricultural, forest, and range lands. http://www.usda.gov/
Oregon: (503) 986-4551
Washington: (360) 902-1800

United States Environmental Protection Agency
Offers information on a variety of subjects, including streams, wetlands, and beneficial landscaping.
Washington: (206) 553-1200 or (800) 424-4EPA. National wetland hotline: (800) 832-7828.
Oregon: (503) 326-3250
http://www.epa.gov/

United States Farm Service Agency
Local offices provide limited cost-share assistance to rural landowners for projects such as streambank protection and riparian enhancement; most share space with the local Natural Resources Conservation Service office.
Oregon: (503) 692-6830
Washington: (509) 353-2307

United States Fish and Wildlife Service
Specializes in fish and wildlife habitat enhancement; is very familiar with the permit requirements for restoring riparian areas; also has permitting responsibilities. Can assist with habitat improvements and wetland restoration. http://www.fw.gov
Oregon: (503) 231-6158
Washington: (360) 753-9440

United States Natural Resources Conservation Service
Non-regulatory agency with more than 50 years' experience in soil and water conservation and streambank stabilization projects, plus expertise in conservation plantings and erosion control. http://www.nrcs.usda.gov/
Oregon: (503) 414-3003
Washington: (509) 353-2337
Refer to the Phone Directory for your nearest Regional Office.